Free! Online PDF version
BNi HOME BUILDER ⌐
2020 COSTBOOK

Check out the demo by going to
www.ConstructionWorkZone.com/costbooks

Want a customized version of this costbook in PDF format that's based on your own actual labor and markup rates? You've got it!

Just call us at **1.888.264.2665** to confirm your purchase* and we will activate your digital costbook for you. You can download it, print it and share it with the other members of your project team. You can also use it to search for individual cost items.

New! Simpl•Est Unit Price

Simpl•Est Unit Price is a new online resource that truly makes estimating <u>simple</u>!

With **Simpl•Est Unit Price** you can create a complete estimate just by clicking on individual items in the *BNi General Construction Costbook.* These items are then saved for you in the Cloud and you can easily download them to a special Excel template that lets you adjust quantities, add comments, modify labor rates and update your entire estimate at the click of a mouse.

You can start your 30-day free trial of **Simpl•Est Unit Price** by going to www.dcd.com/simpl_est

What's more if you call us to confirm your purchase of this book, we will extend your access to **Simpl•Est Unit Price** for a full year...a $154.95 value! Call us* at **1.888.264.2665**

*Call M-F between 9 am and 7 pm Eastern time

2020

28TH EDITION

A **DESIGN COST DATA** COMPANY

DATA YOU CAN TRUST

HOME
BUILDER'S
COSTBOOK

DATA YOU CAN TRUST

EDITOR-IN-CHIEF

William D. Mahoney, P.E.

TECHNICAL SERVICES

Tony De Augustine
Joan Hamilton
Anthony Jackson
Eric Mahoney, AIA
Ana Varela

GRAPHIC DESIGN

Robert O. Wright Jr.

BNi Publications, Inc.

VISTA
990 PARK CENTER DRIVE, SUITE E
VISTA, CA 92081

1-888-BNI-BOOK (1-888-264-2665)
www.bnibooks.com

ISBN 978-1-55701-985-1

Table of Contents

Preface

For over 73 years, BNi Building News has been dedicated to providing construction professionals with timely and reliable information. Based on this experience, our staff has researched and compiled thousands of up-to-the-minute costs for the **BNi Costbooks**. This book is an essential reference for contractors, engineers, architects, facility managers — any construction professional who must provide an estimate for any type of building project.

Whether working up a preliminary estimate or submitting a formal bid, the costs listed here can be quickly and easily tailored to your needs. All costs are based on prevailing labor rates. Overhead and profit should be included in all costs. Man-hours are also provided.

All data is categorized according to the 16 division format. This industry standard provides an all-inclusive checklist to ensure that no element of a project is overlooked. In addition, to make specific items even easier to locate, there is a complete alphabetical index.

The "Features of this Book" section presents a clear overview of the many features of this book. Included is an explanation of the data, sample page layout and discussion of how to best use the information in the book.

Of course, all buildings and construction projects are unique. The information provided in this book is based on averages from well-managed projects with good labor productivity under normal working conditions (eight hours a day). Other circumstances affecting costs such as overtime, unusual working conditions, savings from buying bulk quantities for large projects, and unusual or hidden costs must be factored in as they arise.

The data provided in this book is for estimating purposes only. Check all applicable federal, state and local codes and regulations for local requirements.

Format

All data is categorized according to the 16 division format. This industry standard provides an all-inclusive checklist to ensure that no element of a project is overlooked.

Format *(Continued)*

Features of this Book

The construction estimating information in this book is divided into two main sections: Costbook Pages and Man-Hour Tables. Each section is organized according to the 16 division format. In addition there are extensive Supporting References.

Sample pages with graphic explanations are included before the Costbook pages. These explanations, along with the discussions below, will provide a good understanding of what is included in this book and how it can best be used in construction estimating.

Material Costs

The material costs used in this book represent national averages for prices that a contractor would expect to pay plus an allowance for freight (if applicable) and handling and storage. These costs reflect neither the lowest or highest prices, but rather a typical average cost over time. Periodic fluctuations in availability and in certain commodities (e.g. copper, conduit) can significantly affect local material pricing. In the final estimating and bidding stages of a project when the highest degree of accuracy is required, it is best to check local, current prices.

Labor Costs

Labor costs include the basic wage, plus commonly applicable taxes, insurance and markups for overhead and profit. The labor rates used here to develop the costs are typical average prevailing wage rates. Rates for different trades are used where appropriate for each type of work.

Fixed government rates and average allowances for taxes and insurance are included in the labor costs. These include employer-paid Social Security/Medicare taxes (FICA), Worker's Compensation insurance, state and federal unemployment taxes, and business insurance.

Please note, however, most of these items vary significantly from state to state and within states. For more specific data, local agencies and sources should be consulted.

Equipment Costs

Costs for various types and pieces of equipment are included in Division 1 - General Requirements and can be included in an estimate when required either as a total "Equipment" category or with specific appropriate trades. Costs for equipment are included when appropriate in the installation costs in the Costbook pages.

Overhead and Profit

Included in the labor costs are allowances for overhead and profit for the contractor/employer whose workers are performing the specific tasks. No cost allowances or fees are included for management of subcontractors by the general contractor or construction manager. These costs, where appropriate, must be added to the costs as listed in the book.

The allowance for overhead is included to account for office overhead, the contractors' typical costs of doing business. These costs normally include in-house office staff salaries and benefits, office rent and operating expenses, professional fees, vehicle costs and other operating costs which are not directly applicable to specific jobs. It should be noted for this book that office overhead as included should be distinguished from project overhead, the General Requirements (Division 1) which are specific to particular projects. Project overhead should be included on an item by item basis for each job.

Depending on the trade, an allowance of 10-15 percent is incorporated into the labor/installation costs to account for typical profit of the installing contractor. See Division 1, General Requirements, for a more detailed review of typical profit allowances.

Features of this Book *(Continued)*

Adjustments to Costs

The costs as presented in this book attempt to represent national averages. Costs, however, vary among regions, states and even between adjacent localities.

In order to more closely approximate the probable costs for specific locations throughout the U.S., a table of Geographic Multipliers is provided. These adjustment factors are used to modify costs obtained from this book to help account for regional variations of construction costs. Whenever local current costs are known, whether material or equipment prices or labor rates, they should be used if more accuracy is required.

Hours (Man-Hours)

These productivities represent typical installation labor for thousands of construction items. The data takes into account all activities involved in normal construction under commonly experienced working conditions such as site movement, material handling, start-up, etc.

Editor's Note: This **Costbook** is intended to provide accurate, reliable, average costs and typical productivities for thousands of common construction components. The data is developed and compiled from various industry sources, including government, manufacturers, suppliers and working professionals. The intent of the information is to provide assistance and guidelines to construction professionals in estimating. The user should be aware that local conditions, material and labor availability and cost variations, economic considerations, weather, local codes and regulations, etc., all affect the actual cost of construction. These and other such factors must be considered and incorporated into any and all construction estimates.

Sample Costbook Page

In order to best use the information in this book, please review this sample page and read the "Features In This Book" section.

Division

Broadscope Category

Material Cost
Material cost represents average contractor prices plus an allowance for freight, handling and storage.

Mediumscope Category (First 5 Digits)

Installation Cost
Installation cost includes basic wage rates, markups for taxes, insurance overhead and profit and also includes equipment costs where appropriate.

Detailed Descriptions
Complete descriptions of items may include information listed above a particular line. Review of the whole category is recommended for a complete description.

Total Cost
The total cost is the sum of material and installation costs. This total represents typical contractors' costs including overhead and profit, but does not include markups for the general contractor or construction management fees.

Unit of Measurement
Each item (and cost) is defined in terms of the common estimating unit. All costs are listed in dollars per unit.

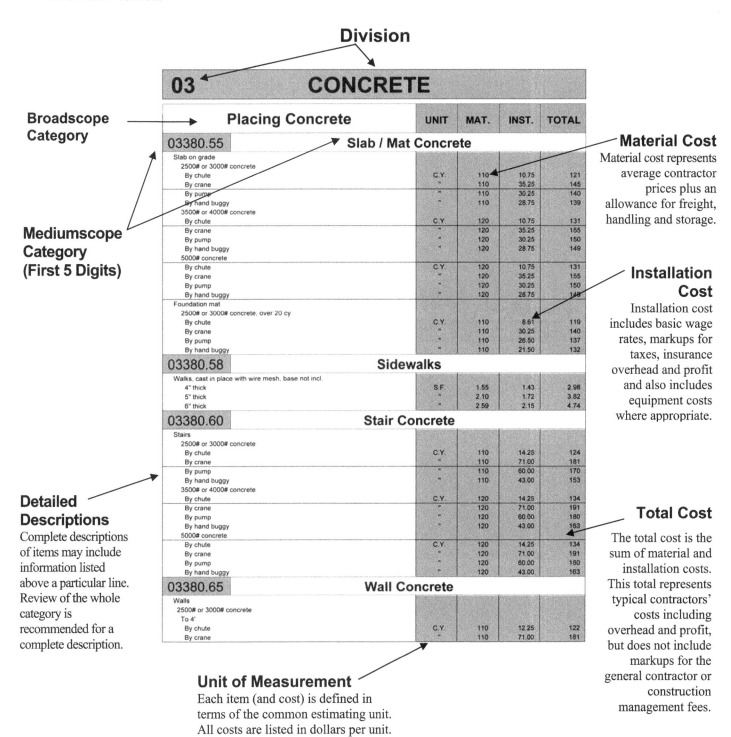

03 CONCRETE

Placing Concrete

	UNIT	MAT.	INST.	TOTAL
03380.55 Slab / Mat Concrete				
Slab on grade				
2500# or 3000# concrete				
By chute	C.Y.	110	10.75	121
By crane		110	35.25	145
By pump	"	110	30.25	140
By hand buggy	"	110	28.75	139
3500# or 4000# concrete				
By chute	C.Y.	120	10.75	131
By crane	"	120	35.25	155
By pump	"	120	30.25	150
By hand buggy	"	120	28.75	149
5000# concrete				
By chute	C.Y.	120	10.75	131
By crane	"	120	35.25	155
By pump	"	120	30.25	150
By hand buggy	"	120	28.75	149
Foundation mat				
2500# or 3000# concrete, over 20 cy				
By chute	C.Y.	110	8.61	119
By crane	"	110	30.25	140
By pump	"	110	26.50	137
By hand buggy	"	110	21.50	132
03380.58 Sidewalks				
Walks, cast in place with wire mesh, base not incl.				
4" thick	S.F.	1.55	1.43	2.98
5" thick	"	2.10	1.72	3.82
6" thick	"	2.59	2.15	4.74
03380.60 Stair Concrete				
Stairs				
2500# or 3000# concrete				
By chute	C.Y.	110	14.25	124
By crane	"	110	71.00	181
By pump	"	110	60.00	170
By hand buggy	"	110	43.00	153
3500# or 4000# concrete				
By chute	C.Y.	120	14.25	134
By crane	"	120	71.00	191
By pump	"	120	60.00	180
By hand buggy	"	120	43.00	163
5000# concrete				
By chute	C.Y.	120	14.25	134
By crane	"	120	71.00	191
By pump	"	120	60.00	180
By hand buggy	"	120	43.00	163
03380.65 Wall Concrete				
Walls				
2500# or 3000# concrete				
To 4'				
By chute	C.Y.	110	12.25	122
By crane	"	110	71.00	181

Requirements	UNIT	MAT.	INST.	TOTAL
01020.10		Allowances		
Overhead				
$20,000 project				
Minimum	PCT			15.00
Average	"			20.00
Maximum	"			40.00
$100,000 project				
Minimum	PCT			12.00
Average	"			15.00
Maximum	"			25.00
$500,000 project				
Minimum	PCT			10.00
Average	"			12.00
Maximum	"			20.00
Profit				
$20,000 project				
Minimum	PCT			10.00
Average	"			15.00
Maximum	"			25.00
$100,000 project				
Minimum	PCT			10.00
Average	"			12.00
Maximum	"			20.00
$500,000 project				
Minimum	PCT			5.00
Average	"			10.00
Maximum	"			15.00
Professional fees				
Architectural				
$100,000 project				
Minimum	PCT			5.00
Average	"			10.00
Maximum	"			20.00
$500,000 project				
Minimum	PCT			5.00
Average	"			8.00
Maximum	"			12.00
Structural engineering				
Minimum	PCT			2.00
Average	"			3.00
Maximum	"			5.00
Mechanical engineering				
Minimum	PCT			4.00
Average	"			5.00
Maximum	"			15.00
Taxes				
Sales tax				
Minimum	PCT			4.00
Average	"			5.00
Maximum	"			10.00
Unemployment				
Minimum	PCT			3.00
Average	"			6.50
Maximum	"			8.00

Requirements	UNIT	MAT.	INST.	TOTAL
01020.10 **Allowances** *(Cont.)*				
Social security (FICA)	PCT			7.85
01050.10 **Field Staff**				
Superintendent				
Minimum	YEAR			99,141
Average	"			123,943
Maximum	"			148,881
Foreman				
Minimum	YEAR			65,895
Average	"			105,383
Maximum	"			123,383
Bookkeeper/timekeeper				
Minimum	YEAR			38,117
Average	"			49,777
Maximum	"			64,401
Watchman				
Minimum	YEAR			28,401
Average	"			37,993
Maximum	"			47,958
01330.10 **Surveying**				
Surveying				
Small crew	DAY			1,010
Average crew	"			1,520
Large crew	"			2,000
Lot lines and boundaries				
Minimum	ACRE			730
Average	"			1,520
Maximum	"			2,480
01380.10 **Job Requirements**				
Job photographs, small jobs				
Minimum	EA			130
Average	"			200
Maximum	"			460
Large projects				
Minimum	EA			660
Average	"			990
Maximum	"			3,310
01410.10 **Testing**				
Testing concrete, per test				
Minimum	EA			23.50
Average	"			39.25
Maximum	"			78.00
01500.10 **Temporary Facilities**				
Barricades, temporary				
Highway				
Concrete	LF	14.75	4.51	19.26
Wood	"	5.10	1.80	6.90
Steel	"	5.29	1.50	6.79

01 GENERAL

Requirements	UNIT	MAT.	INST.	TOTAL
01500.10 **Temporary Facilities** *(Cont.)*				
Pedestrian barricades				
Plywood	SF	4.54	1.50	6.04
Chain link fence	"	3.85	1.50	5.35
Trailers, general office type, per month				
Minimum	EA			250
Average	"			410
Maximum	"			820
Crew change trailers, per month				
Minimum	EA			150
Average	"			160
Maximum	"			250
01505.10 **Mobilization**				
Equipment mobilization				
Bulldozer				
Minimum	EA			220
Average	"			460
Maximum	"			780
Backhoe/front-end loader				
Minimum	EA			130
Average	"			230
Maximum	"			510
Truck crane				
Minimum	EA			560
Average	"			860
Maximum	"			1,490
01525.10 **Construction Aids**				
Scaffolding/staging, rent per month				
Measured by lineal feet of base				
10' high	LF			14.50
20' high	"			26.50
30' high	"			37.00
Measured by square foot of surface				
Minimum	SF			0.64
Average	"			1.11
Maximum	"			1.99
Tarpaulins, fabric, per job				
Minimum	SF			0.30
Average	"			0.51
Maximum	"			1.32
01570.10 **Signs**				
Construction signs, temporary				
Signs, 2' x 4'				
Minimum	EA			41.75
Average	"			100
Maximum	"			350
Signs, 4' x 8'				
Minimum	EA			88.00
Average	"			230
Maximum	"			980
Signs, 8' x 8'				

Requirements	UNIT	MAT.	INST.	TOTAL
01570.10 Signs *(Cont.)*				
Minimum	EA			110
Average	"			350
Maximum	"			3,540
01600.10 Equipment				
Air compressor				
60 cfm				
By day	EA			110
By week	"			320
By month	"			980
300 cfm				
By day	EA			230
By week	"			700
By month	"			2,140
Air tools, per compressor, per day				
Minimum	EA			44.50
Average	"			56.00
Maximum	"			78.00
Generators, 5 kw				
By day	EA			110
By week	"			330
By month	"			1,020
Heaters, salamander type, per week				
Minimum	EA			130
Average	"			190
Maximum	"			400
Pumps, submersible				
50 gpm				
By day	EA			89.00
By week	"			270
By month	"			800
Pickup truck				
By day	EA			170
By week	"			490
By month	"			1,510
Dump truck				
6 cy truck				
By day	EA			440
By week	"			1,330
By month	"			4,010
10 cy truck				
By day	EA			550
By week	"			1,670
By month	"			5,010
16 cy truck				
By day	EA			890
By week	"			2,670
By month	"			8,020
Backhoe, track mounted				
1/2 cy capacity				
By day	EA			910
By week	"			2,780
By month	"			8,240

Requirements		UNIT	MAT.	INST.	TOTAL
01600.10	**Equipment** *(Cont.)*				
Backhoe/loader, rubber tired					
1/2 cy capacity					
By day		EA			550
By week		"			1,670
By month		"			5,010
3/4 cy capacity					
By day		EA			670
By week		"			2,000
By month		"			6,010
Bulldozer					
75 hp					
By day		EA			780
By week		"			2,340
By month		"			7,020
Cranes, crawler type					
15 ton capacity					
By day		EA			1,000
By week		"			3,010
By month		"			9,020
Truck mounted, hydraulic					
15 ton capacity					
By day		EA			940
By week		"			2,840
By month		"			8,190
Loader, rubber tired					
1 cy capacity					
By day		EA			670
By week		"			2,000
By month		"			6,020

Site Remediation	UNIT	MAT.	INST.	TOTAL
02115.66 **Septic Tank Removal**				
Remove septic tank				
1000 gals	EA		220	220
2000 gals	"		270	270

Site Preparation	UNIT	MAT.	INST.	TOTAL
02210.10 **Soil Boring**				
Borings, uncased, stable earth				
2-1/2" dia.	LF		33.50	33.50
4" dia.	"		38.25	38.25
Cased, including samples				
2-1/2" dia.	LF		44.50	44.50
4" dia.	"		76.00	76.00
Drilling in rock				
No sampling	LF		70.00	70.00
With casing and sampling	"		89.00	89.00
Test pits				
Light soil	EA		450	450
Heavy soil	"		670	670
02210.20 **Rough Grading**				
Site grading, cut & fill, sandy clay, 200' haul, 75 hp dozer	CY		4.18	4.18
Spread topsoil by equipment on site	"		4.65	4.65
Site grading (cut and fill to 6") less than 1 acre				
75 hp dozer	CY		6.98	6.98
1.5 cy backhoe/loader	"		10.50	10.50

Demolition	UNIT	MAT.	INST.	TOTAL
02220.10 **Complete Building Demolition**				
Wood frame	CF		0.39	0.39
02220.15 **Selective Building Demolition**				
Partition removal				
Concrete block partitions				
8" thick	SF		3.00	3.00
Brick masonry partitions				
4" thick	SF		2.25	2.25
8" thick	"		2.82	2.82

Demolition	UNIT	MAT.	INST.	TOTAL
02220.15 **Selective Building Demolition** *(Cont.)*				
Stud partitions				
Metal or wood, with drywall both sides	SF		2.25	2.25
Door and frame removal				
Wood in framed wall				
2'6"x6'8"	EA		32.25	32.25
3'x6'8"	"		37.50	37.50
Ceiling removal				
Acoustical tile ceiling				
Adhesive fastened	SF		0.45	0.45
Furred and glued	"		0.37	0.37
Suspended grid	"		0.28	0.28
Drywall ceiling				
Furred and nailed	SF		0.50	0.50
Nailed to framing	"		0.45	0.45
Window removal				
Metal windows, trim included				
2'x3'	EA		45.00	45.00
3'x4'	"		56.00	56.00
4'x8'	"		110	110
Wood windows, trim included				
2'x3'	EA		25.00	25.00
3'x4'	"		30.00	30.00
6'x8'	"		45.00	45.00
Concrete block walls, not including toothing				
4" thick	SF		2.50	2.50
6" thick	"		2.65	2.65
8" thick	"		2.82	2.82
Rubbish handling				
Load in dumpster or truck				
Minimum	CF		1.00	1.00
Maximum	"		1.50	1.50
Rubbish hauling				
Hand loaded on trucks, 2 mile trip	CY		42.00	42.00
Machine loaded on trucks, 2 mile trip	"		26.75	26.75

Selective Site Demolition	UNIT	MAT.	INST.	TOTAL
02225.13 **Core Drilling**				
Concrete				
6" thick				
3" dia.	EA		44.25	44.25
8" thick				
3" dia.	EA		62.00	62.00

02 SITE CONSTRUCTION

Selective Site Demolition	UNIT	MAT.	INST.	TOTAL
02225.15 — **Curb & Gutter Demolition**				
Removal, plain concrete curb	LF		6.67	6.67
Plain concrete curb and 2' gutter	"		9.21	9.21
02225.20 — **Fence Demolition**				
Remove fencing				
Chain link, 8' high				
For disposal	LF		2.25	2.25
For reuse	"		5.64	5.64
Wood				
4' high	SF		1.50	1.50
Masonry				
8" thick				
4' high	SF		4.51	4.51
6' high	"		5.64	5.64
02225.25 — **Guardrail Demolition**				
Remove standard guardrail				
Steel	LF		8.90	8.90
Wood	"		6.85	6.85
02225.30 — **Hydrant Demolition**				
Remove and reset fire hydrant	EA		1,340	1,340
02225.40 — **Pavement And Sidewalk Demolition**				
Concrete pavement, 6" thick				
No reinforcement	SY		17.75	17.75
With wire mesh	"		26.75	26.75
With rebars	"		33.50	33.50
Sidewalk, 4" thick, with disposal	"		8.90	8.90
02225.42 — **Drainage Piping Demolition**				
Remove drainage pipe, not including excavation				
12" dia.	LF		11.25	11.25
18" dia.	"		14.00	14.00
02225.43 — **Gas Piping Demolition**				
Remove welded steel pipe, not including excavation				
4" dia.	LF		16.75	16.75
5" dia.	"		26.75	26.75
02225.45 — **Sanitary Piping Demolition**				
Remove sewer pipe, not including excavation				
4" dia.	LF		10.75	10.75
02225.48 — **Water Piping Demolition**				
Remove water pipe, not including excavation				
4" dia.	LF		12.25	12.25

02 SITE CONSTRUCTION

Selective Site Demolition	UNIT	MAT.	INST.	TOTAL
02225.50	**Saw Cutting Pavement**			
Pavement, bituminous				
2" thick	LF		2.09	2.09
3" thick	"		2.61	2.61
Concrete pavement, with wire mesh				
4" thick	LF		4.02	4.02
5" thick	"		4.36	4.36
Plain concrete, unreinforced				
4" thick	LF		3.49	3.49
5" thick	"		4.02	4.02

Site Clearing	UNIT	MAT.	INST.	TOTAL
02230.10	**Clear Wooded Areas**			
Clear wooded area				
Light density	ACRE		6,680	6,680
Medium density	"		8,910	8,910
Heavy density	"		10,690	10,690
02230.50	**Tree Cutting & Clearing**			
Cut trees and clear out stumps				
9" to 12" dia.	EA		530	530
To 24" dia.	"		670	670
24" dia. and up	"		890	890
Loading and trucking				
For machine load, per load, round trip				
1 mile	EA		110	110
3 mile	"		120	120
5 mile	"		130	130
10 mile	"		180	180
20 mile	"		270	270
Hand loaded, round trip				
1 mile	EA		260	260
3 mile	"		300	300
5 mile	"		350	350
10 mile	"		420	420
20 mile	"		520	520

02 SITE CONSTRUCTION

Earthwork, Excavation & Fill	UNIT	MAT.	INST.	TOTAL
02315.10		**Base Course**		
Base course, crushed stone				
3" thick	SY	3.19	0.73	3.92
4" thick	"	4.29	0.79	5.08
6" thick	"	6.43	0.86	7.29
Base course, bank run gravel				
4" deep	SY	3.02	0.77	3.79
6" deep	"	4.62	0.84	5.46
Prepare and roll sub base				
Minimum	SY		0.73	0.73
Average	"		0.92	0.92
Maximum	"		1.23	1.23
02315.20		**Borrow**		
Borrow fill, F.O.B. at pit				
Sand, haul to site, round trip				
10 mile	CY	22.75	14.75	37.50
20 mile	"	22.75	24.50	47.25
30 mile	"	22.75	37.00	59.75
Place borrow fill and compact				
Less than 1 in 4 slope	CY	22.75	7.38	30.13
Greater than 1 in 4 slope	"	22.75	9.84	32.59
02315.30		**Bulk Excavation**		
Excavation, by small dozer				
Large areas	CY		2.09	2.09
Small areas	"		3.49	3.49
Trim banks	"		5.23	5.23
Hydraulic excavator				
1 cy capacity				
Light material	CY		4.45	4.45
Medium material	"		5.34	5.34
Wet material	"		6.67	6.67
Blasted rock	"		7.63	7.63
1-1/2 cy capacity				
Light material	CY		1.84	1.84
Medium material	"		2.46	2.46
Wet material	"		2.95	2.95
Wheel mounted front-end loader				
7/8 cy capacity				
Light material	CY		3.69	3.69
Medium material	"		4.21	4.21
Wet material	"		4.92	4.92
Blasted rock	"		5.90	5.90
1-1/2 cy capacity				
Light material	CY		2.10	2.10
Medium material	"		2.27	2.27
Wet material	"		2.46	2.46
Blasted rock	"		2.68	2.68
2-1/2 cy capacity				
Light material	CY		1.73	1.73
Medium material	"		1.84	1.84
Wet material	"		1.96	1.96
Blasted rock	"		2.10	2.10

Earthwork, Excavation & Fill	UNIT	MAT.	INST.	TOTAL
02315.30 **Bulk Excavation** *(Cont.)*				
Track mounted front-end loader				
1-1/2 cy capacity				
Light material	CY		2.46	2.46
Medium material	"		2.68	2.68
Wet material	"		2.95	2.95
Blasted rock	"		3.28	3.28
2-3/4 cy capacity				
Light material	CY		1.47	1.47
Medium material	"		1.64	1.64
Wet material	"		1.84	1.84
Blasted rock	"		2.10	2.10
02315.40 **Building Excavation**				
Structural excavation, unclassified earth				
3/8 cy backhoe	CY		19.75	19.75
3/4 cy backhoe	"		14.75	14.75
1 cy backhoe	"		12.25	12.25
Foundation backfill and compaction by machine	"		29.50	29.50
02315.45 **Hand Excavation**				
Excavation				
To 2' deep				
Normal soil	CY		50.00	50.00
Sand and gravel	"		45.00	45.00
Medium clay	"		56.00	56.00
Heavy clay	"		64.00	64.00
Loose rock	"		75.00	75.00
To 6' deep				
Normal soil	CY		64.00	64.00
Sand and gravel	"		56.00	56.00
Medium clay	"		75.00	75.00
Heavy clay	"		90.00	90.00
Loose rock	"		110	110
Backfilling foundation without compaction, 6" lifts	"		28.25	28.25
Compaction of backfill around structures or in trench				
By hand with air tamper	CY		32.25	32.25
By hand with vibrating plate tamper	"		30.00	30.00
1 ton roller	"		52.00	52.00
Miscellaneous hand labor				
Trim slopes, sides of excavation	SF		0.07	0.07
Trim bottom of excavation	"		0.09	0.09
Excavation around obstructions and services	CY		150	150
02315.50 **Roadway Excavation**				
Roadway excavation				
1/4 mile haul	CY		2.95	2.95
2 mile haul	"		4.92	4.92
5 mile haul	"		7.38	7.38
Spread base course	"		3.69	3.69
Roll and compact	"		4.92	4.92

Earthwork, Excavation & Fill	UNIT	MAT.	INST.	TOTAL
02315.60			**Trenching**	
Trenching and continuous footing excavation				
By gradall				
1 cy capacity				
Light soil	CY		4.21	4.21
Medium soil	"		4.54	4.54
Heavy/wet soil	"		4.92	4.92
Loose rock	"		5.36	5.36
Blasted rock	"		5.67	5.67
By hydraulic excavator				
1/2 cy capacity				
Light soil	CY		4.92	4.92
Medium soil	"		5.36	5.36
Heavy/wet soil	"		5.90	5.90
Loose rock	"		6.56	6.56
Blasted rock	"		7.38	7.38
1 cy capacity				
Light soil	CY		3.47	3.47
Medium soil	"		3.69	3.69
Heavy/wet soil	"		3.93	3.93
Loose rock	"		4.21	4.21
Blasted rock	"		4.54	4.54
1-1/2 cy capacity				
Light soil	CY		3.10	3.10
Medium soil	"		3.28	3.28
Heavy/wet soil	"		3.47	3.47
Loose rock	"		3.69	3.69
Blasted rock	"		3.93	3.93
2 cy capacity				
Light soil	CY		2.95	2.95
Medium soil	"		3.10	3.10
Heavy/wet soil	"		3.28	3.28
Loose rock	"		3.47	3.47
Blasted rock	"		3.69	3.69
Hand excavation				
Bulk, wheeled 100'				
Normal soil	CY		50.00	50.00
Sand or gravel	"		45.00	45.00
Medium clay	"		64.00	64.00
Heavy clay	"		90.00	90.00
Loose rock	"		110	110
Trenches, up to 2' deep				
Normal soil	CY		56.00	56.00
Sand or gravel	"		50.00	50.00
Medium clay	"		75.00	75.00
Heavy clay	"		110	110
Loose rock	"		150	150
Trenches, to 6' deep				
Normal soil	CY		64.00	64.00
Sand or gravel	"		56.00	56.00
Medium clay	"		90.00	90.00
Heavy clay	"		150	150
Loose rock	"		230	230
Backfill trenches				

Earthwork, Excavation & Fill	UNIT	MAT.	INST.	TOTAL
02315.60 **Trenching** *(Cont.)*				
With compaction				
By hand	CY		37.50	37.50
By 60 hp tracked dozer	"		2.61	2.61
02315.70 **Utility Excavation**				
Trencher, sandy clay, 8" wide trench				
18" deep	LF		2.32	2.32
24" deep	"		2.61	2.61
36" deep	"		2.99	2.99
Trench backfill, 95% compaction				
Tamp by hand	CY		28.25	28.25
Vibratory compaction	"		22.50	22.50
Trench backfilling, with borrow sand, place & compact	"	22.75	22.50	45.25
02315.75 **Gravel And Stone**				
F.O.B. PLANT, material only				
No. 21 crusher run stone	CY			33.00
No. 26 crusher run stone	"			33.00
No. 57 stone	"			33.00
No. 67 gravel	"			33.00
No. 68 stone	"			33.00
No. 78 stone	"			33.00
No. 78 gravel, (pea gravel)	"			33.00
No. 357 or B-3 stone	"			33.00
Structural & foundation backfill				
No. 21 crusher run stone	TON			24.50
No. 26 crusher run stone	"			24.50
No. 57 stone	"			24.50
No. 67 gravel	"			24.50
No. 68 stone	"			24.50
No. 78 stone	"			24.50
No. 78 gravel, (pea gravel)	"			24.50
No. 357 or B-3 stone	"			24.50
02315.80 **Hauling Material**				
Haul material by 10 cy dump truck, round trip distance				
1 mile	CY		5.81	5.81
2 mile	"		6.98	6.98
5 mile	"		9.52	9.52
10 mile	"		10.50	10.50
20 mile	"		11.75	11.75
30 mile	"		14.00	14.00

Soil Stabilization & Treatment	UNIT	MAT.	INST.	TOTAL
02340.05 **Soil Stabilization**				
Straw bale secured with rebar	LF	7.54	1.50	9.04
Filter barrier, 18" high filter fabric	"	1.82	4.51	6.33
Sediment fence, 36" fabric with 6" mesh	"	4.32	5.64	9.96
Soil stabilization with tar paper, burlap, straw and stakes	SF	0.36	0.06	0.42
02360.20 **Soil Treatment**				
Soil treatment, termite control pretreatment				
Under slabs	SF	0.38	0.25	0.63
By walls	"	0.38	0.30	0.68
02370.40 **Riprap**				
Riprap				
Crushed stone blanket, max size 2-1/2"	TON	35.25	78.00	113
Stone, quarry run, 300 lb. stones	"	44.25	72.00	116
400 lb. stones	"	46.00	67.00	113
500 lb. stones	"	48.00	63.00	111
750 lb. stones	"	49.75	59.00	109
Dry concrete riprap in bags 3" thick, 80 lb. per bag	BAG	5.96	3.92	9.88

Piles And Caissons	UNIT	MAT.	INST.	TOTAL
02455.60 **Steel Piles**				
H-section piles				
8x8				
36 lb/ft				
30' long	LF	21.75	13.50	35.25
40' long	"	21.75	10.75	32.50
Tapered friction piles, fluted casing, up to 50'				
With 4000 psi concrete no reinforcing				
12" dia.	LF	21.50	8.05	29.55
14" dia.	"	24.75	8.26	33.01
02455.65 **Steel Pipe Piles**				
Concrete filled, 3000# concrete, up to 40'				
8" dia.	LF	24.00	11.50	35.50
10" dia.	"	31.00	12.00	43.00
12" dia.	"	36.00	12.50	48.50
Pipe piles, non-filled				
8" dia.	LF	21.75	8.95	30.70
10" dia.	"	27.25	9.20	36.45
12" dia.	"	33.25	9.47	42.72
Splice				
8" dia.	EA	97.00	90.00	187
10" dia.	"	110	90.00	200
12" dia.	"	120	110	230
Standard point				

Piles And Caissons	UNIT	MAT.	INST.	TOTAL
02455.65 **Steel Pipe Piles** *(Cont.)*				
8" dia.	EA	130	90.00	220
10" dia.	"	170	90.00	260
12" dia.	"	180	110	290
Heavy duty point				
8" dia.	EA	230	110	340
10" dia.	"	320	110	430
12" dia.	"	340	150	490
02455.80 **Wood And Timber Piles**				
Treated wood piles, 12" butt, 8" tip				
25' long	LF	18.00	16.00	34.00
30' long	"	19.25	13.50	32.75
35' long	"	19.25	11.50	30.75
40' long	"	19.25	10.00	29.25
02465.50 **Prestressed Piling**				
Prestressed concrete piling, less than 60' long				
10" sq.	LF	20.75	6.71	27.46
12" sq.	"	29.00	7.00	36.00
Straight cylinder, less than 60' long				
12" dia.	LF	27.00	7.32	34.32
14" dia.	"	36.50	7.49	43.99

Utility Services	UNIT	MAT.	INST.	TOTAL
02510.10 **Wells**				
Domestic water, drilled and cased				
4" dia.	LF	30.50	81.00	112
6" dia.	"	33.50	90.00	124
02510.40 **Ductile Iron Pipe**				
Ductile iron pipe, cement lined, slip-on joints				
4"	LF	18.25	7.42	25.67
6"	"	21.25	7.85	29.10
8"	"	27.75	8.34	36.09
Mechanical joint pipe				
4"	LF	19.75	10.25	30.00
6"	"	23.50	11.25	34.75
8"	"	31.00	12.25	43.25
Fittings, mechanical joint				
90 degree elbow				
4"	EA	230	30.00	260
6"	"	300	34.75	335
8"	"	430	45.00	475
45 degree elbow				
4"	EA	200	30.00	230

Utility Services	UNIT	MAT.	INST.	TOTAL
02510.40 **Ductile Iron Pipe** *(Cont.)*				
6"	EA	270	34.75	305
8"	"	380	45.00	425
02510.60 **Plastic Pipe**				
PVC, class 150 pipe				
4" dia.	LF	5.31	6.67	11.98
6" dia.	"	10.00	7.22	17.22
8" dia.	"	16.00	7.63	23.63
Schedule 40 pipe				
1-1/2" dia.	LF	1.34	2.65	3.99
2" dia.	"	1.99	2.82	4.81
2-1/2" dia.	"	3.01	3.00	6.01
3" dia.	"	4.09	3.22	7.31
4" dia.	"	5.78	3.76	9.54
6" dia.	"	11.00	4.51	15.51
90 degree elbows				
1"	EA	1.12	7.52	8.64
1-1/2"	"	2.14	7.52	9.66
2"	"	3.35	8.20	11.55
2-1/2"	"	10.25	9.02	19.27
3"	"	12.25	10.00	22.25
4"	"	19.75	11.25	31.00
6"	"	62.00	15.00	77.00
45 degree elbows				
1"	EA	1.72	7.52	9.24
1-1/2"	"	3.01	7.52	10.53
2"	"	3.91	8.20	12.11
2-1/2"	"	10.25	9.02	19.27
3"	"	15.75	10.00	25.75
4"	"	25.50	11.25	36.75
6"	"	63.00	15.00	78.00
Tees				
1"	EA	1.48	9.02	10.50
1-1/2"	"	2.86	9.02	11.88
2"	"	4.12	10.00	14.12
2-1/2"	"	13.50	11.25	24.75
3"	"	18.00	13.00	31.00
4"	"	29.25	15.00	44.25
6"	"	98.00	18.00	116
Couplings				
1"	EA	0.91	7.52	8.43
1-1/2"	"	1.30	7.52	8.82
2"	"	2.01	8.20	10.21
2-1/2"	"	4.42	9.02	13.44
3"	"	6.91	10.00	16.91
4"	"	9.02	11.25	20.27
6"	"	28.50	15.00	43.50
Drainage pipe				
PVC schedule 80				
1" dia.	LF	2.03	2.65	4.68
1-1/2" dia.	"	2.46	2.65	5.11
ABS, 2" dia.	"	3.14	2.82	5.96
2-1/2" dia.	"	4.47	3.00	7.47

Utility Services	UNIT	MAT.	INST.	TOTAL
02510.60 — **Plastic Pipe** *(Cont.)*				
3" dia.	LF	5.26	3.22	8.48
4" dia.	"	7.17	3.76	10.93
6" dia.	"	12.00	4.51	16.51
8" dia.	"	16.00	7.03	23.03
10" dia.	"	21.25	8.34	29.59
12" dia.	"	34.75	8.90	43.65
90 degree elbows				
1"	EA	3.38	7.52	10.90
1-1/2"	"	4.22	7.52	11.74
2"	"	5.09	8.20	13.29
2-1/2"	"	12.25	9.02	21.27
3"	"	12.50	10.00	22.50
4"	"	22.25	11.25	33.50
6"	"	48.75	15.00	63.75
45 degree elbows				
1"	EA	5.50	7.52	13.02
1-1/2"	"	6.99	7.52	14.51
2"	"	8.66	8.20	16.86
2-1/2"	"	16.25	9.02	25.27
3"	"	17.25	10.00	27.25
4"	"	32.75	11.25	44.00
6"	"	76.00	15.00	91.00

Sanitary Sewer	UNIT	MAT.	INST.	TOTAL
02530.20 — **Vitrified Clay Pipe**				
Vitrified clay pipe, extra strength				
6" dia.	LF	5.73	12.25	17.98
8" dia.	"	6.87	12.75	19.62
10" dia.	"	10.50	13.25	23.75
02530.30 — **Manholes**				
Precast sections, 48" dia.				
Base section	EA	360	220	580
1'0" riser	"	100	180	280
1'4" riser	"	120	190	310
2'8" riser	"	180	210	390
4'0" riser	"	340	220	560
2'8" cone top	"	220	270	490
Precast manholes, 48" dia.				
4' deep	EA	700	530	1,230
6' deep	"	1,070	670	1,740
7' deep	"	1,220	760	1,980
8' deep	"	1,380	890	2,270
10' deep	"	1,540	1,070	2,610
Cast-in-place, 48" dia., with frame and cover				

02 SITE CONSTRUCTION

Sanitary Sewer	UNIT	MAT.	INST.	TOTAL
02530.30 — **Manholes** *(Cont.)*				
5' deep	EA	630	1,340	1,970
6' deep	"	830	1,530	2,360
8' deep	"	1,210	1,780	2,990
10' deep	"	1,410	2,140	3,550
Brick manholes, 48" dia. with cover, 8" thick				
4' deep	EA	670	550	1,220
6' deep	"	840	610	1,450
8' deep	"	1,080	690	1,770
10' deep	"	1,340	780	2,120
Frames and covers, 24" diameter				
300 lb	EA	410	45.00	455
400 lb	"	430	50.00	480
Steps for manholes				
7" x 9"	EA	17.25	9.02	26.27
8" x 9"	"	22.00	10.00	32.00
02530.40 — **Sanitary Sewers**				
Clay				
6" pipe	LF	9.41	8.90	18.31
PVC				
4" pipe	LF	4.05	6.67	10.72
6" pipe	"	8.11	7.03	15.14
02540.10 — **Drainage Fields**				
Perforated PVC pipe, for drain field				
4" pipe	LF	2.71	5.93	8.64
6" pipe	"	5.08	6.36	11.44
02540.50 — **Septic Tanks**				
Septic tank, precast concrete				
1000 gals	EA	1,020	450	1,470
2000 gals	"	2,740	670	3,410
Leaching pit, precast concrete, 72" diameter				
3' deep	EA	780	330	1,110
6' deep	"	1,370	380	1,750
8' deep	"	1,740	450	2,190
02630.70 — **Underdrain**				
Drain tile, clay				
6" pipe	LF	4.52	5.93	10.45
8" pipe	"	7.21	6.21	13.42
Porous concrete, standard strength				
6" pipe	LF	5.14	5.93	11.07
8" pipe	"	5.56	6.21	11.77
Corrugated metal pipe, perforated type				
6" pipe	LF	7.13	6.67	13.80
8" pipe	"	8.43	7.03	15.46
Perforated clay pipe				
6" pipe	LF	5.98	7.63	13.61
8" pipe	"	8.02	7.85	15.87
Drain tile, concrete				
6" pipe	LF	4.08	5.93	10.01
8" pipe	"	6.35	6.21	12.56

02 SITE CONSTRUCTION

Sanitary Sewer	UNIT	MAT.	INST.	TOTAL
02630.70 Underdrain *(Cont.)*				
Perforated rigid PVC underdrain pipe				
4" pipe	LF	2.10	4.45	6.55
6" pipe	"	4.04	5.34	9.38
8" pipe	"	6.17	5.93	12.10
Underslab drainage, crushed stone				
3" thick	SF	0.33	0.89	1.22
4" thick	"	0.45	1.02	1.47
6" thick	"	0.68	1.11	1.79
Plastic filter fabric for drain lines	"	0.50	0.45	0.95

Paving	UNIT	MAT.	INST.	TOTAL
02740.20 Asphalt Surfaces				
Asphalt wearing surface, flexible pavement				
1" thick	SY	4.52	2.68	7.20
1-1/2" thick	"	6.82	3.22	10.04
Binder course				
1-1/2" thick	SY	6.45	2.98	9.43
2" thick	"	8.58	3.66	12.24
Bituminous sidewalk, no base				
2" thick	SY	9.84	3.14	12.98
3" thick	"	14.75	3.33	18.08

Rigid Pavement	UNIT	MAT.	INST.	TOTAL
02750.10 Concrete Paving				
Concrete paving, reinforced, 5000 psi concrete				
6" thick	SY	28.75	25.25	54.00
7" thick	"	33.50	26.75	60.25
8" thick	"	38.25	28.75	67.00

Site Improvements	UNIT	MAT.	INST.	TOTAL
02810.40	**Lawn Irrigation**			
Residential system, complete				
Minimum	ACRE			19,250
Maximum	"			36,630
02820.10	**Chain Link Fence**			
Chain link fence, 9 ga., galvanized, with posts 10' o.c.				
4' high	LF	7.38	3.22	10.60
5' high	"	9.87	4.10	13.97
6' high	"	11.25	5.64	16.89
Corner or gate post, 3" post				
4' high	EA	86.00	15.00	101
5' high	"	95.00	16.75	112
6' high	"	110	19.50	130
Gate with gate posts, galvanized, 3' wide				
4' high	EA	95.00	110	205
5' high	"	120	150	270
6' high	"	150	150	300
Fabric, galvanized chain link, 2" mesh, 9 ga.				
4' high	LF	4.01	1.50	5.51
5' high	"	4.91	1.80	6.71
6' high	"	6.87	2.25	9.12
Line post, no rail fitting, galvanized, 2-1/2" dia.				
4' high	EA	28.00	13.00	41.00
5' high	"	30.50	14.00	44.50
6' high	"	33.25	15.00	48.25
Vinyl coated, 9 ga., with posts 10' o.c.				
4' high	LF	7.98	3.22	11.20
5' high	"	9.50	4.10	13.60
6' high	"	11.25	5.64	16.89
Gate, with posts, 3' wide				
4' high	EA	110	110	220
5' high	"	130	150	280
6' high	"	150	150	300
Fabric, vinyl, chain link, 2" mesh, 9 ga.				
4' high	LF	3.76	1.50	5.26
5' high	"	4.59	1.80	6.39
6' high	"	6.43	2.25	8.68
Swing gates, galvanized, 4' high				
Single gate				
3' wide	EA	200	110	310
4' wide	"	220	110	330
6' high				
Single gate				
3' wide	EA	270	150	420
4' wide	"	290	150	440
02880.70	**Recreational Courts**			
Walls, galvanized steel				
8' high	LF	16.00	9.02	25.02
10' high	"	18.75	10.00	28.75
12' high	"	21.75	11.75	33.50
Vinyl coated				
8' high	LF	15.25	9.02	24.27

Site Improvements	UNIT	MAT.	INST.	TOTAL
02880.70 Recreational Courts *(Cont.)*				
10' high	LF	18.75	10.00	28.75
12' high	"	20.75	11.75	32.50
Gates, galvanized steel				
Single, 3' transom				
3'x7'	EA	370	230	600
4'x7'	"	390	260	650
5'x7'	"	540	300	840
6'x7'	"	580	360	940
Vinyl coated				
Single, 3' transom				
3'x7'	EA	730	230	960
4'x7'	"	790	260	1,050
5'x7'	"	790	300	1,090
6'x7'	"	820	360	1,180

Planting	UNIT	MAT.	INST.	TOTAL
02910.10 Topsoil				
Spread topsoil, with equipment				
Minimum	CY		14.75	14.75
Maximum	"		18.50	18.50
By hand				
Minimum	CY		45.00	45.00
Maximum	"		56.00	56.00
Area prep. seeding (grade, rake and clean)				
Square yard	SY		0.36	0.36
By acre	ACRE		1,800	1,800
Remove topsoil and stockpile on site				
4" deep	CY		12.25	12.25
6" deep	"		11.25	11.25
Spreading topsoil from stock pile				
By loader	CY		13.50	13.50
By hand	"		150	150
Top dress by hand	SY		1.47	1.47
Place imported top soil				
By loader				
4" deep	SY		1.47	1.47
6" deep	"		1.64	1.64
By hand				
4" deep	SY		5.01	5.01
6" deep	"		5.64	5.64
Plant bed preparation, 18" deep				
With backhoe/loader	SY		3.69	3.69
By hand	"		7.52	7.52

Planting	UNIT	MAT.	INST.	TOTAL
02920.10 **Fertilizing**				
Fertilizing (23#/1000 sf)				
By square yard	SY	0.03	0.15	0.18
By acre	ACRE	190	780	970
Liming (70#/1000 sf)				
By square yard	SY	0.03	0.20	0.23
By acre	ACRE	190	1,040	1,230
02920.30 **Seeding**				
Mechanical seeding, 175 lb/acre				
By square yard	SY	0.23	0.12	0.35
By acre	ACRE	930	620	1,550
450 lb/acre				
By square yard	SY	0.59	0.15	0.74
By acre	ACRE	2,310	780	3,090
Seeding by hand, 10 lb per 100 s.y.				
By square yard	SY	0.66	0.15	0.81
By acre	ACRE	2,580	750	3,330
Reseed disturbed areas	SF	0.06	0.22	0.28
02930.10 **Plants**				
Euonymus coloratus, 18" (Purple Wintercreeper)	EA	2.87	7.52	10.39
Hedera Helix, 2-1/4" pot (English ivy)	"	1.19	7.52	8.71
Liriope muscari, 2" clumps	"	5.01	4.51	9.52
Santolina, 12"	"	5.74	4.51	10.25
Vinca major or minor, 3" pot	"	0.93	4.51	5.44
Cortaderia argentia, 2 gallon (Pampas Grass)	"	18.25	4.51	22.76
Ophiopogan japonicus, 1 quart (4" pot)	"	5.01	4.51	9.52
Ajuga reptans, 2-3/4" pot (carpet bugle)	"	0.93	4.51	5.44
Pachysandra terminalis, 2-3/4" pot (Japanese Spurge)	"	1.27	4.51	5.78
02930.30 **Shrubs**				
Juniperus conferia litoralis, 18"-24" (Shore Juniper)	EA	42.00	18.00	60.00
Horizontalis plumosa, 18"-24" (Andorra Juniper)	"	44.50	18.00	62.50
Sabina tamar-iscfolia-tamarix juniper, 18"-24"	"	44.50	18.00	62.50
Chin San Jose, 18"-24" (San Jose Juniper)	"	44.50	18.00	62.50
Sargenti, 18"-24" (Sargent's Juniper)	"	42.00	18.00	60.00
Nandina domestica, 18"-24" (Heavenly Bamboo)	"	28.25	18.00	46.25
Raphiolepis Indica Springtime, 18"-24"	"	30.25	18.00	48.25
Osmanthus Heterophyllus Gulftide, 18"-24"	"	32.50	18.00	50.50
Ilex Cornuta Burfordi Nana, 18"-24"	"	37.00	18.00	55.00
Glabra, 18"-24" (Inkberry Holly)	"	34.75	18.00	52.75
Azalea, Indica types, 18"-24"	"	39.25	18.00	57.25
Kurume types, 18"-24"	"	44.00	18.00	62.00
Berberis Julianae, 18"-24" (Wintergreen Barberry)	"	25.75	18.00	43.75
Pieris Japonica Japanese, 18"-24"	"	25.75	18.00	43.75
Ilex Cornuta Rotunda, 18"-24"	"	30.50	18.00	48.50
Juniperus Horiz. Plumosa, 24"-30"	"	28.00	22.50	50.50
Rhodopendrow Hybrids, 24"-30"	"	75.00	22.50	97.50
Aucuba Japonica Varigata, 24"-30"	"	25.50	22.50	48.00
Ilex Crenata Willow Leaf, 24"-30"	"	28.00	22.50	50.50
Cleyera Japonica, 30"-36"	"	32.75	28.25	61.00
Pittosporum Tobira, 30"-36"	"	38.25	28.25	66.50
Prumus Laurocerasus, 30"-36"	"	71.00	28.25	99.25

Planting	UNIT	MAT.	INST.	TOTAL
02930.30 **Shrubs** *(Cont.)*				
Ilex Cornuta Burfordi, 30"-36" (Burford Holly)	EA	37.50	28.25	65.75
Abelia Grandiflora, 24"-36" (Yew Podocarpus)	"	25.75	22.50	48.25
Podocarpos Macrophylla, 24"-36"	"	42.00	22.50	64.50
Pyracantha Coccinea Lalandi, 3'-4' (Firethorn)	"	24.00	28.25	52.25
Photinia Frazieri, 3'-4' (Red Photinia)	"	38.00	28.25	66.25
Forsythia Suspensa, 3'-4' (Weeping Forsythia)	"	24.00	28.25	52.25
Camellia Japonica, 3'-4' (Common Camellia)	"	42.25	28.25	70.50
Juniperus Chin Torulosa, 3'-4' (Hollywood Juniper)	"	45.00	28.25	73.25
Cupressocyparis Leylandi, 3'-4'	"	37.75	28.25	66.00
Ilex Opaca Fosteri, 5'-6' (Foster's Holly)	"	150	37.50	188
Opaca, 5'-6' (American Holly)	"	220	37.50	258
Nyrica Cerifera, 4'-5' (Southern Wax Myrtles)	"	47.75	32.25	80.00
Ligustrum Japonicum, 4'-5' (Japanese Privet)	"	37.50	32.25	69.75
02930.60 **Trees**				
Cornus Florida, 5'-6' (White flowering Dogwood)	EA	110	37.50	148
Prunus Serrulata Kwanzan, 6'-8' (Kwanzan Cherry)	"	120	45.00	165
Caroliniana, 6'-8' (Carolina Cherry Laurel)	"	140	45.00	185
Cercis Canadensis, 6'-8' (Eastern Redbud)	"	100	45.00	145
Koelreuteria Paniculata, 8'-10' (Goldenrain Tree)	"	170	56.00	226
Acer Platanoides, 1-3/4"-2" (11'-13')	"	230	75.00	305
Rubrum, 1-3/4"-2" (11'-13') (Red Maple)	"	170	75.00	245
Saccharum, 1-3/4"-2" (Sugar Maple)	"	300	75.00	375
Fraxinus Pennsylvanica, 1-3/4"-2"	"	150	75.00	225
Celtis Occidentalis, 1-3/4"-2"	"	220	75.00	295
Glenditsia Triacantos Inermis, 2"	"	200	75.00	275
Prunus Cerasifera 'Thundercloud', 6'-8'	"	120	45.00	165
Yeodensis, 6'-8' (Yoshino Cherry)	"	120	45.00	165
Lagerstroemia Indica, 8'-10' (Crapemyrtle)	"	200	56.00	256
Crataegus Phaenopyrum, 8'-10'	"	310	56.00	366
Quercus Borealis, 1-3/4"-2" (Northern Red Oak)	"	180	75.00	255
Quercus Acutissima, 1-3/4"-2" (8'-10')	"	170	75.00	245
Saliz Babylonica, 1-3/4"-2" (Weeping Willow)	"	85.00	75.00	160
Tilia Cordata Greenspire, 1-3/4"-2" (10'-12')	"	380	75.00	455
Malus, 2"-2-1/2" (8'-10') (Flowering Crabapple)	"	180	75.00	255
Platanus Occidentalis, (12'-14')	"	280	90.00	370
Pyrus Calleryana Bradford, 2"-2-1/2"	"	220	75.00	295
Quercus Palustris, 2"-2-1/2" (12'-14') (Pin Oak)	"	240	75.00	315
Phellos, 2-1/2"-3" (Willow Oak)	"	260	90.00	350
Nigra, 2"-2-1/2" (Water Oak)	"	230	75.00	305
Magnolia Soulangeana, 4'-5' (Saucer Magnolia)	"	130	37.50	168
Grandiflora, 6'-8' (Southern Magnolia)	"	180	45.00	225
Cedrus Deodara, 10'-12' (Deodare Cedar)	"	300	75.00	375
Gingko Biloba, 10'-12' (2"-2-1/2")	"	280	75.00	355
Pinus Thunbergi, 5'-6' (Japanese Black Pine)	"	110	37.50	148
Strobus, 6'-8' (White Pine)	"	120	45.00	165
Taeda, 6'-8' (Loblolly Pine)	"	100	45.00	145
Quercus Virginiana, 2"-2-1/2" (Live Oak)	"	270	90.00	360

Planting	UNIT	MAT.	INST.	TOTAL
02935.10 **Shrub & Tree Maintenance**				
Moving shrubs on site				
3' high	EA		45.00	45.00
4' high	"		50.00	50.00
Moving trees on site				
6' high	EA		59.00	59.00
8' high	"		67.00	67.00
10' high	"		89.00	89.00
Palm trees				
10' high	EA		89.00	89.00
40' high	"		530	530
02935.30 **Weed Control**				
Weed control, bromicil, 15 lb./acre, wettable powder	ACRE	310	230	540
Vegetation control, by application of plant killer	SY	0.02	0.18	0.20
Weed killer, lawns and fields	"	0.26	0.09	0.35
02945.20 **Landscape Accessories**				
Steel edging, 3/16" x 4"	LF	1.29	0.56	1.85
Landscaping stepping stones, 15"x15", white	EA	5.83	2.25	8.08
Wood chip mulch	CY	40.75	30.00	70.75
2" thick	SY	2.49	0.90	3.39
4" thick	"	4.70	1.28	5.98
6" thick	"	7.04	1.64	8.68
Gravel mulch, 3/4" stone	CY	32.25	45.00	77.25
White marble chips, 1" deep	SF	0.63	0.45	1.08
Peat moss				
2" thick	SY	3.46	1.00	4.46
4" thick	"	6.66	1.50	8.16
6" thick	"	10.25	1.88	12.13
Landscaping timbers, treated lumber				
4" x 4"	LF	3.58	1.50	5.08
6" x 6"	"	8.32	1.61	9.93
8" x 8"	"	10.00	1.88	11.88

Formwork	UNIT	MAT.	INST.	TOTAL
03110.05 **Beam Formwork**				
Beam forms, job built				
Beam bottoms				
1 use	SF	5.15	9.60	14.75
4 uses	"	1.92	8.47	10.39
5 uses	"	1.75	8.22	9.97
Beam sides				
1 use	SF	3.68	6.40	10.08
5 uses	"	1.56	5.23	6.79
03110.15 **Column Formwork**				
Column, square forms, job built				
8" x 8" columns				
1 use	SF	4.33	11.50	15.83
5 uses	"	1.53	9.93	11.46
12" x 12" columns				
1 use	SF	3.95	10.50	14.45
5 uses	"	1.29	9.14	10.43
Round fiber forms, 1 use				
10" dia.	LF	5.65	11.50	17.15
12" dia.	"	6.95	11.75	18.70
03110.18 **Curb Formwork**				
Curb forms				
Straight, 6" high				
1 use	LF	2.58	5.76	8.34
5 uses	"	0.94	4.80	5.74
Curved, 6" high				
1 use	LF	2.79	7.20	9.99
5 uses	"	1.13	5.87	7.00
03110.25 **Equipment Pad Formwork**				
Equipment pad, job built				
1 use	SF	4.50	7.20	11.70
3 uses	"	2.16	6.40	8.56
5 uses	"	1.34	5.76	7.10
03110.35 **Footing Formwork**				
Wall footings, job built, continuous				
1 use	SF	2.09	5.76	7.85
3 uses	"	1.21	5.23	6.44
5 uses	"	0.93	4.80	5.73
03110.50 **Grade Beam Formwork**				
Grade beams, job built				
1 use	SF	3.29	5.76	9.05
3 uses	"	1.44	5.23	6.67
5 uses	"	1.00	4.80	5.80

Formwork	UNIT	MAT.	INST.	TOTAL
03110.53 **Pile Cap Formwork**				
Pile cap forms, job built				
Square				
1 use	SF	3.74	7.20	10.94
5 uses	"	1.25	5.76	7.01
03110.55 **Slab / Mat Formwork**				
Mat foundations, job built				
1 use	SF	3.27	7.20	10.47
3 uses	"	1.39	6.40	7.79
5 uses	"	0.94	5.76	6.70
Edge forms				
6" high				
1 use	LF	3.30	5.23	8.53
3 uses	"	1.39	4.80	6.19
5 uses	"	0.95	4.43	5.38
03110.65 **Wall Formwork**				
Wall forms, exterior, job built				
Up to 8' high wall				
1 use	SF	3.52	5.76	9.28
3 uses	"	1.71	5.23	6.94
5 uses	"	1.29	4.80	6.09
Retaining wall forms				
1 use	SF	3.27	6.40	9.67
3 uses	"	1.50	5.76	7.26
5 uses	"	1.11	5.23	6.34
Column pier and pilaster				
1 use	SF	3.87	11.50	15.37
5 uses	"	1.74	8.22	9.96
03110.90 **Miscellaneous Formwork**				
Keyway forms (5 uses)				
2 x 4	LF	0.30	2.88	3.18
2 x 6	"	0.43	3.20	3.63
Bulkheads				
Walls, with keyways				
3 piece	LF	6.28	5.76	12.04
Ground slab, with keyway				
2 piece	LF	5.88	4.11	9.99
3 piece	"	7.19	4.43	11.62
Chamfer strips				
Wood				
1/2" wide	LF	0.28	1.28	1.56
3/4" wide	"	0.36	1.28	1.64
1" wide	"	0.49	1.28	1.77
PVC				
1/2" wide	LF	1.26	1.28	2.54
3/4" wide	"	1.36	1.28	2.64
1" wide	"	1.98	1.28	3.26

03 CONCRETE

Reinforcement		UNIT	MAT.	INST.	TOTAL
03210.05	**Beam Reinforcing**				
Beam-girders					
#3 - #4		TON	1,980	1,470	3,450
#5 - #6		"	1,740	1,180	2,920
03210.15	**Column Reinforcing**				
Columns					
#3 - #4		TON	1,980	1,680	3,660
#5 - #6		"	1,740	1,310	3,050
03210.20	**Elevated Slab Reinforcing**				
Elevated slab					
#3 - #4		TON	1,980	730	2,710
#5 - #6		"	1,740	650	2,390
03210.25	**Equip. Pad Reinforcing**				
Equipment pad					
#3 - #4		TON	1,980	1,180	3,160
#5 - #6		"	1,740	1,070	2,810
03210.35	**Footing Reinforcing**				
Footings					
#3 - #4		TON	1,980	980	2,960
#5 - #6		"	1,740	840	2,580
#7 - #8		"	1,650	730	2,380
Straight dowels, 24" long					
3/4" dia. (#6)		EA	5.51	5.88	11.39
5/8" dia. (#5)		"	4.76	4.90	9.66
1/2" dia. (#4)		"	3.59	4.20	7.79
03210.45	**Foundation Reinforcing**				
Foundations					
#3 - #4		TON	1,980	980	2,960
#5 - #6		"	1,740	840	2,580
#7 - #8		"	1,650	730	2,380
03210.50	**Grade Beam Reinforcing**				
Grade beams					
#3 - #4		TON	1,980	900	2,880
#5 - #6		"	1,740	780	2,520
#7 - #8		"	1,650	690	2,340
03210.53	**Pile Cap Reinforcing**				
Pile caps					
#3 - #4		TON	1,980	1,470	3,450
#5 - #6		"	1,740	1,310	3,050
#7 - #8		"	1,650	1,180	2,830

Reinforcement	UNIT	MAT.	INST.	TOTAL
03210.55	**Slab / Mat Reinforcing**			
Bars, slabs				
#3 - #4	TON	1,980	980	2,960
#5 - #6	"	1,740	840	2,580
Wire mesh, slabs				
Galvanized				
4x4				
W1.4xW1.4	SF	0.49	0.39	0.88
W2.0xW2.0	"	0.63	0.42	1.05
W2.9xW2.9	"	0.89	0.45	1.34
W4.0xW4.0	"	1.32	0.49	1.81
6x6				
W1.4xW1.4	SF	0.45	0.29	0.74
W2.0xW2.0	"	0.63	0.32	0.95
W2.9xW2.9	"	0.86	0.34	1.20
W4.0xW4.0	"	0.93	0.39	1.32
03210.65	**Wall Reinforcing**			
Walls				
#3 - #4	TON	1,980	840	2,820
#5 - #6	"	1,740	730	2,470
Masonry wall (horizontal)				
#3 - #4	TON	1,980	2,350	4,330
#5 - #6	"	1,740	1,960	3,700
Galvanized				
#3 - #4	TON	3,370	2,350	5,720
#5 - #6	"	3,190	1,960	5,150
Masonry wall (vertical)				
#3 - #4	TON	1,980	2,940	4,920
#5 - #6	"	1,740	2,350	4,090
Galvanized				
#3 - #4	TON	3,370	2,940	6,310
#5 - #6	"	3,190	2,350	5,540

Accessories	UNIT	MAT.	INST.	TOTAL
03250.40	**Concrete Accessories**			
Expansion joint, poured				
Asphalt				
1/2" x 1"	LF	0.88	0.90	1.78
1" x 2"	"	2.76	0.98	3.74
Expansion joint, premolded, in slabs				
Asphalt				
1/2" x 6"	LF	0.98	1.12	2.10
1" x 12"	"	1.64	1.50	3.14
Cork				
1/2" x 6"	LF	1.95	1.12	3.07

Accessories	UNIT	MAT.	INST.	TOTAL
03250.40 **Concrete Accessories** *(Cont.)*				
1" x 12"	LF	7.41	1.50	8.91
Neoprene sponge				
1/2" x 6"	LF	2.87	1.12	3.99
1" x 12"	"	10.50	1.50	12.00
Polyethylene foam				
1/2" x 6"	LF	1.11	1.12	2.23
1" x 12"	"	5.11	1.50	6.61
Polyurethane foam				
1/2" x 6"	LF	1.46	1.12	2.58
1" x 12"	"	3.22	1.50	4.72
Polyvinyl chloride foam				
1/2" x 6"	LF	3.13	1.12	4.25
1" x 12"	"	6.75	1.50	8.25
Rubber, gray sponge				
1/2" x 6"	LF	4.87	1.12	5.99
1" x 12"	"	21.25	1.50	22.75
Asphalt felt control joints or bond breaker, screed joints				
4" slab	LF	1.32	0.90	2.22
6" slab	"	1.65	1.00	2.65
8" slab	"	2.15	1.12	3.27
Waterstops				
Polyvinyl chloride				
Ribbed				
3/16" thick x				
4" wide	LF	1.57	2.25	3.82
6" wide	"	2.38	2.50	4.88
1/2" thick x				
9" wide	LF	6.32	2.82	9.14
Ribbed with center bulb				
3/16" thick x 9" wide	LF	5.31	2.82	8.13
3/8" thick x 9" wide	"	6.24	2.82	9.06
Dumbbell type, 3/8" thick x 6" wide	"	6.32	2.50	8.82
Plain, 3/8" thick x 9" wide	"	8.39	2.82	11.21
Center bulb, 3/8" thick x 9" wide	"	10.00	2.82	12.82
Rubber				
Vapor barrier				
4 mil polyethylene	SF	0.05	0.15	0.20
6 mil polyethylene	"	0.08	0.15	0.23
Gravel porous fill, under floor slabs, 3/4" stone	CY	21.75	75.00	96.75
Reinforcing accessories				
Beam bolsters				
1-1/2" high, plain	LF	0.58	0.58	1.16
Galvanized	"	1.27	0.58	1.85
3" high				
Plain	LF	0.83	0.73	1.56
Galvanized	"	2.05	0.73	2.78
Slab bolsters				
1" high				
Plain	LF	0.62	0.29	0.91
Galvanized	"	1.25	0.29	1.54
2" high				
Plain	LF	0.69	0.32	1.01
Galvanized	"	1.46	0.32	1.78

Accessories	UNIT	MAT.	INST.	TOTAL

03250.40 Concrete Accessories (Cont.)

Chairs, high chairs				
3" high				
Plain	EA	1.67	1.47	3.14
Galvanized	"	1.84	1.47	3.31
8" high				
Plain	EA	2.82	1.68	4.50
Galvanized	"	4.79	1.68	6.47
Continuous, high chair				
3" high				
Plain	LF	2.32	0.39	2.71
Galvanized	"	2.87	0.39	3.26

Cast-in-place Concrete	UNIT	MAT.	INST.	TOTAL

03300.10 Concrete Admixtures

Concrete admixtures				
Water reducing admixture	GAL			12.00
Set retarder	"			25.75
Air entraining agent	"			11.25

03350.10 Concrete Finishes

Floor finishes				
Broom	SF		0.64	0.64
Screed	"		0.56	0.56
Darby	"		0.56	0.56
Steel float	"		0.75	0.75
Wall finishes				
Burlap rub, with cement paste	SF	0.12	0.75	0.87

03360.10 Pneumatic Concrete

Pneumatic applied concrete (gunite)				
2" thick	SF	6.33	3.33	9.66
3" thick	"	7.77	4.45	12.22
4" thick	"	9.48	5.34	14.82
Finish surface				
Minimum	SF		2.88	2.88
Maximum	"		5.76	5.76

03370.10 Curing Concrete

Sprayed membrane				
Slabs	SF	0.06	0.09	0.15
Walls	"	0.08	0.11	0.19
Curing paper				
Slabs	SF	0.08	0.11	0.19
Walls	"	0.08	0.13	0.21
Burlap				

03 CONCRETE

Cast-in-place Concrete	UNIT	MAT.	INST.	TOTAL
03370.10 **Curing Concrete** *(Cont.)*				
7.5 oz.	SF	0.07	0.15	0.22
12 oz.	"	0.10	0.16	0.26

Placing Concrete	UNIT	MAT.	INST.	TOTAL
03380.05 **Beam Concrete**				
Beams and girders				
2500# or 3000# concrete				
By crane	CY	130	96.00	226
By pump	"	130	87.00	217
By hand buggy	"	130	45.00	175
3500# or 4000# concrete				
By crane	CY	140	96.00	236
By pump	"	140	87.00	227
By hand buggy	"	140	45.00	185
03380.15 **Column Concrete**				
Columns				
2500# or 3000# concrete				
By crane	CY	140	87.00	227
By pump	"	140	80.00	220
3500# or 4000# concrete				
By crane	CY	140	87.00	227
By pump	"	140	80.00	220
03380.20 **Elevated Slab Concrete**				
Elevated slab				
2500# or 3000# concrete				
By crane	CY	140	48.00	188
By pump	"	140	37.00	177
By hand buggy	"	140	45.00	185
03380.25 **Equipment Pad Concrete**				
Equipment pad				
2500# or 3000# concrete				
By chute	CY	140	15.00	155
By pump	"	140	69.00	209
By crane	"	140	80.00	220
3500# or 4000# concrete				
By chute	CY	140	15.00	155
By pump	"	140	69.00	209

03 CONCRETE

Placing Concrete	UNIT	MAT.	INST.	TOTAL
03380.35 **Footing Concrete**				
Continuous footing				
2500# or 3000# concrete				
By chute	CY	140	15.00	155
By pump	"	140	60.00	200
By crane	"	140	69.00	209
Spread footing				
2500# or 3000# concrete				
By chute	CY	130	15.00	145
By pump	"	130	64.00	194
By crane	"	130	74.00	204
03380.50 **Grade Beam Concrete**				
Grade beam				
2500# or 3000# concrete				
By chute	CY	130	15.00	145
By crane	"	130	69.00	199
By pump	"	130	60.00	190
By hand buggy	"	130	45.00	175
3500# or 4000# concrete				
By chute	CY	140	15.00	155
By crane	"	140	69.00	209
By pump	"	140	60.00	200
By hand buggy	"	140	45.00	185
03380.53 **Pile Cap Concrete**				
Pile cap				
2500# or 3000 concrete				
By chute	CY	140	15.00	155
By crane	"	140	80.00	220
By pump	"	140	69.00	209
By hand buggy	"	140	45.00	185
03380.55 **Slab / Mat Concrete**				
Slab on grade				
2500# or 3000# concrete				
By chute	CY	140	11.25	151
By crane	"	140	40.00	180
By pump	"	140	34.25	174
By hand buggy	"	140	30.00	170
03380.58 **Sidewalks**				
Walks, cast in place with wire mesh, base not incl.				
4" thick	SF	1.93	1.50	3.43
5" thick	"	2.61	1.80	4.41
6" thick	"	3.21	2.25	5.46
03380.65 **Wall Concrete**				
Walls				
2500# or 3000# concrete				
To 4'				
By chute	CY	140	13.00	153
By crane	"	140	80.00	220

Placing Concrete	UNIT	MAT.	INST.	TOTAL
03380.65 Wall Concrete *(Cont.)*				
By pump	CY	140	74.00	214
To 8'				
By crane	CY	140	87.00	227
By pump	"	140	80.00	220
Filled block (CMU)				
3000# concrete, by pump				
4" wide	SF	0.51	3.43	3.94
6" wide	"	1.15	4.00	5.15
8" wide	"	1.80	4.80	6.60
03400.90 Precast Specialties				
Precast concrete, coping, 4' to 8' long				
12" wide	LF	10.75	6.67	17.42
10" wide	"	9.52	7.63	17.15
Splash block, 30"x12"x4"	EA	16.25	44.50	60.75
Stair unit, per riser	"	100	44.50	145
Sun screen and trellis, 8' long, 12" high				
4" thick blades	EA	110	33.50	144

Grout	UNIT	MAT.	INST.	TOTAL
03600.10 Grouting				
Grouting for bases				
Non-metallic grout				
1" deep	SF	6.46	12.50	18.96
2" deep	"	12.25	13.75	26.00
Portland cement grout (1 cement to 3 sand)				
1/2" joint thickness				
6" wide joints	LF	0.22	2.07	2.29
8" wide joints	"	0.24	2.48	2.72
1" joint thickness				
4" wide joints	LF	0.24	1.94	2.18
6" wide joints	"	0.40	2.14	2.54

Mortar, Grout And Accessories	UNIT	MAT.	INST.	TOTAL
04100.10 **Masonry Grout**				
Grout, non shrink, non-metallic, trowelable	CF	5.94	1.78	7.72
Grout door frame, hollow metal				
Single	EA	14.50	67.00	81.50
Double	"	20.50	70.00	90.50
Grout-filled concrete block (CMU)				
4" wide	SF	0.43	2.22	2.65
6" wide	"	1.12	2.42	3.54
8" wide	"	1.65	2.67	4.32
12" wide	"	2.71	2.81	5.52
Grout-filled individual CMU cells				
4" wide	LF	0.36	1.33	1.69
6" wide	"	0.48	1.33	1.81
8" wide	"	0.64	1.33	1.97
10" wide	"	0.80	1.52	2.32
12" wide	"	0.97	1.52	2.49
Bond beams or lintels, 8" deep				
6" thick	LF	0.97	2.18	3.15
8" thick	"	1.28	2.40	3.68
10" thick	"	1.61	2.67	4.28
12" thick	"	1.93	3.00	4.93
Cavity walls				
2" thick	SF	1.07	3.20	4.27
3" thick	"	1.61	3.20	4.81
4" thick	"	2.14	3.43	5.57
6" thick	"	3.21	4.00	7.21
04150.10 **Masonry Accessories**				
Foundation vents	EA	26.25	22.00	48.25
Bar reinforcing				
Horizontal				
#3 - #4	LB	0.60	2.19	2.79
#5 - #6	"	0.60	1.83	2.43
Vertical				
#3 - #4	LB	0.60	2.74	3.34
#5 - #6	"	0.60	2.19	2.79
Horizontal joint reinforcing				
Truss type				
4" wide, 6" wall	LF	0.20	0.21	0.41
6" wide, 8" wall	"	0.20	0.22	0.42
8" wide, 10" wall	"	0.25	0.23	0.48
10" wide, 12" wall	"	0.25	0.24	0.49
12" wide, 14" wall	"	0.30	0.26	0.56
Ladder type				
4" wide, 6" wall	LF	0.15	0.21	0.36
6" wide, 8" wall	"	0.17	0.22	0.39
8" wide, 10" wall	"	0.18	0.23	0.41
10" wide, 12" wall	"	0.22	0.23	0.45
Rectangular wall ties				
3/16" dia., galvanized				
2" x 6"	EA	0.38	0.91	1.29
2" x 8"	"	0.40	0.91	1.31
2" x 10"	"	0.47	0.91	1.38
2" x 12"	"	0.53	0.91	1.44

Mortar, Grout And Accessories	UNIT	MAT.	INST.	TOTAL
04150.10 **Masonry Accessories** *(Cont.)*				
4" x 6"	EA	0.44	1.09	1.53
4" x 8"	"	0.49	1.09	1.58
4" x 10"	"	0.63	1.09	1.72
4" x 12"	"	0.73	1.09	1.82
1/4" dia., galvanized				
2" x 6"	EA	0.71	0.91	1.62
2" x 8"	"	0.80	0.91	1.71
2" x 10"	"	0.91	0.91	1.82
2" x 12"	"	1.04	0.91	1.95
4" x 6"	"	0.82	1.09	1.91
4" x 8"	"	0.91	1.09	2.00
4" x 10"	"	1.04	1.09	2.13
4" x 12"	"	1.08	1.09	2.17
"Z" type wall ties, galvanized				
6" long				
1/8" dia.	EA	0.34	0.91	1.25
3/16" dia.	"	0.36	0.91	1.27
1/4" dia.	"	0.38	0.91	1.29
8" long				
1/8" dia.	EA	0.36	0.91	1.27
3/16" dia.	"	0.38	0.91	1.29
1/4" dia.	"	0.40	0.91	1.31
10" long				
1/8" dia.	EA	0.38	0.91	1.29
3/16" dia.	"	0.44	0.91	1.35
1/4" dia.	"	0.49	0.91	1.40
Dovetail anchor slots				
Galvanized steel, filled				
24 ga.	LF	1.17	1.37	2.54
20 ga.	"	2.46	1.37	3.83
16 oz. copper, foam filled	"	3.53	1.37	4.90
Dovetail anchors				
16 ga.				
3-1/2" long	EA	0.39	0.91	1.30
5-1/2" long	"	0.48	0.91	1.39
12 ga.				
3-1/2" long	EA	0.52	0.91	1.43
5-1/2" long	"	0.86	0.91	1.77
Dovetail, triangular galvanized ties, 12 ga.				
3" x 3"	EA	0.88	0.91	1.79
5" x 5"	"	0.95	0.91	1.86
7" x 7"	"	1.07	0.91	1.98
7" x 9"	"	1.14	0.91	2.05
Brick anchors				
Corrugated, 3-1/2" long				
16 ga.	EA	0.57	0.91	1.48
12 ga.	"	0.66	0.91	1.57
Non-corrugated, 3-1/2" long				
16 ga.	EA	0.47	0.91	1.38
12 ga.	"	0.85	0.91	1.76
Cavity wall anchors, corrugated, galvanized				
5" long				
16 ga.	EA	0.95	0.91	1.86

Mortar, Grout And Accessories	UNIT	MAT.	INST.	TOTAL
04150.10 **Masonry Accessories** *(Cont.)*				
12 ga.	EA	1.43	0.91	2.34
7" long				
28 ga.	EA	1.05	0.91	1.96
24 ga.	"	1.33	0.91	2.24
22 ga.	"	1.36	0.91	2.27
16 ga.	"	1.55	0.91	2.46
Mesh ties, 16 ga., 3" wide				
8" long	EA	1.28	0.91	2.19
12" long	"	1.43	0.91	2.34
20" long	"	1.96	0.91	2.87
24" long	"	2.16	0.91	3.07
04150.20 **Masonry Control Joints**				
Control joint, cross shaped PVC	LF	2.38	1.37	3.75
Closed cell joint filler				
1/2"	LF	0.41	1.37	1.78
3/4"	"	0.85	1.37	2.22
Rubber, for				
4" wall	LF	2.75	1.37	4.12
PVC, for				
4" wall	LF	1.43	1.37	2.80
04150.50 **Masonry Flashing**				
Through-wall flashing				
5 oz. coated copper	SF	4.18	4.57	8.75
0.030" elastomeric	"	1.32	3.66	4.98

Unit Masonry	UNIT	MAT.	INST.	TOTAL
04210.10 **Brick Masonry**				
Standard size brick, running bond				
Face brick, red (6.4/sf)				
Veneer	SF	5.66	9.15	14.81
Cavity wall	"	5.66	7.84	13.50
9" solid wall	"	11.25	15.75	27.00
Common brick (6.4/sf)				
Select common for veneers	SF	3.69	9.15	12.84
Back-up				
4" thick	SF	3.32	6.86	10.18
8" thick	"	6.65	11.00	17.65
Glazed brick (7.4/sf)				
Veneer	SF	15.25	9.98	25.23
Buff or gray face brick (6.4/sf)				
Veneer	SF	6.58	9.15	15.73
Cavity wall	"	6.58	7.84	14.42
Jumbo or oversize brick (3/sf)				

04 MASONRY

Unit Masonry	UNIT	MAT.	INST.	TOTAL
04210.10 **Brick Masonry** *(Cont.)*				
4" veneer	SF	4.76	5.49	10.25
4" back-up	"	4.76	4.57	9.33
8" back-up	"	5.52	7.84	13.36
Norman brick, red face, (4.5/sf)				
4" veneer	SF	7.88	6.86	14.74
Cavity wall	"	7.88	6.10	13.98
Chimney, standard brick, including flue				
16" x 16"	LF	32.00	55.00	87.00
16" x 20"	"	54.00	55.00	109
16" x 24"	"	58.00	55.00	113
20" x 20"	"	45.00	69.00	114
20" x 24"	"	61.00	69.00	130
20" x 32"	"	68.00	78.00	146
Window sill, face brick on edge	"	3.59	13.75	17.34
04210.60 **Pavers, Masonry**				
Brick walk laid on sand, sand joints				
Laid flat, (4.5 per sf)	SF	4.16	6.10	10.26
Laid on edge, (7.2 per sf)	"	6.66	9.15	15.81
Precast concrete patio blocks				
2" thick				
Natural	SF	3.58	1.83	5.41
Colors	"	4.53	1.83	6.36
Exposed aggregates, local aggregate				
Natural	SF	10.00	1.83	11.83
Colors	"	10.00	1.83	11.83
Granite or limestone aggregate	"	10.00	1.83	11.83
White tumblestone aggregate	"	10.75	1.83	12.58
Stone pavers, set in mortar				
Bluestone				
1" thick				
Irregular	SF	10.00	13.75	23.75
Snapped rectangular	"	14.75	11.00	25.75
1-1/2" thick, random rectangular	"	18.00	13.75	31.75
2" thick, random rectangular	"	20.50	15.75	36.25
Slate				
Natural cleft				
Irregular, 3/4" thick	SF	10.75	15.75	26.50
Random rectangular				
1-1/4" thick	SF	23.25	13.75	37.00
1-1/2" thick	"	26.25	15.25	41.50
Granite blocks				
3" thick, 3" to 6" wide				
4" to 12" long	SF	13.25	18.25	31.50
6" to 15" long	"	8.67	15.75	24.42
Crushed stone, white marble, 3" thick	"	1.88	0.90	2.78
04220.10 **Concrete Masonry Units**				
Hollow, load bearing				
4"	SF	1.62	4.06	5.68
6"	"	2.38	4.22	6.60
8"	"	2.73	4.57	7.30
10"	"	3.77	4.99	8.76

Unit Masonry	UNIT	MAT.	INST.	TOTAL
04220.10 **Concrete Masonry Units** *(Cont.)*				
12"	SF	4.34	5.49	9.83
Solid, load bearing				
4"	SF	2.55	4.06	6.61
6"	"	2.86	4.22	7.08
8"	"	3.91	4.57	8.48
10"	"	4.16	4.99	9.15
12"	"	6.19	5.49	11.68
Back-up block, 8" x 16"				
2"	SF	1.70	3.13	4.83
4"	"	1.78	3.23	5.01
6"	"	2.60	3.43	6.03
8"	"	2.99	3.66	6.65
10"	"	4.13	3.92	8.05
12"	"	4.75	4.22	8.97
Foundation wall, 8" x 16"				
6"	SF	2.60	3.92	6.52
8"	"	2.99	4.22	7.21
10"	"	4.13	4.57	8.70
12"	"	4.76	4.99	9.75
Solid				
6"	SF	3.15	4.22	7.37
8"	"	4.30	4.57	8.87
10"	"	4.57	4.99	9.56
12"	"	6.79	5.49	12.28
Exterior, styrofoam inserts, std weight, 8" x 16"				
6"	SF	4.57	4.22	8.79
8"	"	4.93	4.57	9.50
10"	"	6.40	4.99	11.39
12"	"	8.77	5.49	14.26
Lightweight				
6"	SF	5.09	4.22	9.31
8"	"	5.73	4.57	10.30
10"	"	6.08	4.99	11.07
12"	"	8.04	5.49	13.53
Acoustical slotted block				
4"	SF	5.31	4.99	10.30
6"	"	5.56	4.99	10.55
8"	"	6.94	5.49	12.43
Filled cavities				
4"	SF	5.69	6.10	11.79
6"	"	6.55	6.46	13.01
8"	"	8.40	6.86	15.26
Hollow, split face				
4"	SF	3.64	4.06	7.70
6"	"	4.21	4.22	8.43
8"	"	4.42	4.57	8.99
10"	"	4.95	4.99	9.94
12"	"	5.28	5.49	10.77
Split rib profile				
4"	SF	4.42	4.99	9.41
6"	"	5.13	4.99	10.12
8"	"	5.58	5.49	11.07
10"	"	6.12	5.49	11.61

Unit Masonry	UNIT	MAT.	INST.	TOTAL
04220.10 **Concrete Masonry Units** *(Cont.)*				
12"	SF	6.63	5.49	12.12
Solar screen concrete block				
4" thick				
6" x 6"	SF	4.29	12.25	16.54
8" x 8"	"	5.12	11.00	16.12
12" x 12"	"	5.24	8.44	13.68
8" thick				
8" x 16"	SF	5.24	7.84	13.08
Vertical reinforcing				
4' o.c., add 5% to labor				
2'8" o.c., add 15% to labor				
Interior partitions, add 10% to labor				
04220.90 **Bond Beams & Lintels**				
Bond beam, no grout or reinforcement				
8" x 16" x				
4" thick	LF	1.85	4.22	6.07
6" thick	"	2.83	4.39	7.22
8" thick	"	3.24	4.57	7.81
10" thick	"	4.01	4.77	8.78
12" thick	"	4.56	4.99	9.55
Beam lintel, no grout or reinforcement				
8" x 16" x				
10" thick	LF	8.85	5.49	14.34
12" thick	"	9.42	6.10	15.52
Precast masonry lintel				
6 lf, 8" high x				
4" thick	LF	7.80	9.15	16.95
6" thick	"	9.96	9.15	19.11
8" thick	"	11.25	9.98	21.23
10" thick	"	13.50	9.98	23.48
10 lf, 8" high x				
4" thick	LF	9.80	5.49	15.29
6" thick	"	12.00	5.49	17.49
8" thick	"	13.50	6.10	19.60
10" thick	"	18.25	6.10	24.35
Steel angles and plates				
Minimum	LB	1.33	0.78	2.11
Maximum	"	1.95	1.37	3.32
Various size angle lintels				
1/4" stock				
3" x 3"	LF	6.85	3.43	10.28
3" x 3-1/2"	"	7.54	3.43	10.97
3/8" stock				
3" x 4"	LF	12.00	3.43	15.43
3-1/2" x 4"	"	12.50	3.43	15.93
4" x 4"	"	13.75	3.43	17.18
5" x 3-1/2"	"	14.50	3.43	17.93
6" x 3-1/2"	"	16.25	3.43	19.68
1/2" stock				
6" x 4"	LF	18.00	3.43	21.43

Unit Masonry	UNIT	MAT.	INST.	TOTAL
04270.10 **Glass Block**				
Glass block, 4" thick				
6" x 6"	SF	38.25	18.25	56.50
8" x 8"	"	24.25	13.75	38.00
12" x 12"	"	30.75	11.00	41.75
04295.10 **Parging / Masonry Plaster**				
Parging				
1/2" thick	SF	0.44	3.66	4.10
3/4" thick	"	0.48	4.57	5.05
1" thick	"	0.63	5.49	6.12

Stone	UNIT	MAT.	INST.	TOTAL
04400.10 **Stone**				
Rubble stone				
Walls set in mortar				
8" thick	SF	17.50	13.75	31.25
12" thick	"	21.00	22.00	43.00
18" thick	"	28.00	27.50	55.50
24" thick	"	35.00	36.50	71.50
Dry set wall				
8" thick	SF	19.50	9.15	28.65
12" thick	"	22.25	13.75	36.00
18" thick	"	30.50	18.25	48.75
24" thick	"	37.25	22.00	59.25
Thresholds, 7/8" thick, 3' long, 4" to 6" wide				
Plain	EA	34.50	45.75	80.25
Beveled	"	38.25	45.75	84.00
Window sill				
6" wide, 2" thick	LF	19.25	22.00	41.25
Stools				
5" wide, 7/8" thick	LF	25.75	22.00	47.75
Granite veneer facing panels, polished				
7/8" thick				
Black	SF	48.25	22.00	70.25
Gray	"	38.00	22.00	60.00
Slate, panels				
1" thick	SF	27.50	22.00	49.50
Sills or stools				
1" thick				
6" wide	LF	12.75	22.00	34.75
10" wide	"	20.75	23.75	44.50

Stone	UNIT	MAT.	INST.	TOTAL
04520.10 **Restoration And Cleaning**				
Masonry cleaning				
Washing brick				
Smooth surface	SF	0.25	0.91	1.16
Rough surface	"	0.35	1.22	1.57
Steam clean masonry				
Smooth face				
Minimum	SF	0.53	0.77	1.30
Maximum	"	0.86	1.12	1.98
Rough face				
Minimum	SF	0.78	1.03	1.81
Maximum	"	1.15	1.55	2.70

Refractories	UNIT	MAT.	INST.	TOTAL
04550.10 **Flue Liners**				
Flue liners				
Rectangular				
8" x 12"	LF	10.50	9.15	19.65
12" x 12"	"	13.25	9.98	23.23
12" x 18"	"	23.25	11.00	34.25
16" x 16"	"	25.00	12.25	37.25
18" x 18"	"	31.00	13.00	44.00
20" x 20"	"	52.00	13.75	65.75
24" x 24"	"	62.00	15.75	77.75
Round				
18" dia.	LF	47.75	13.00	60.75
24" dia.	"	94.00	15.75	110

05 METALS

Metal Fastening	UNIT	MAT.	INST.	TOTAL
05050.10	**Structural Welding**			
Welding				
Single pass				
1/8"	LF	0.33	3.17	3.50
3/16"	"	0.55	4.23	4.78
1/4"	"	0.77	5.29	6.06
05050.90	**Metal Anchors**			
Anchor bolts, material only				
3/8" x				
8" long	EA			1.11
12" long	"			1.31
1/2" x				
8" long	EA			1.65
12" long	"			1.93
5/8" x				
8" long	EA			1.54
12" long	"			1.81
3/4" x				
8" long	EA			2.20
12" long	"			2.48
Non-drilling anchor				
1/4"	EA			0.71
3/8"	"			0.88
1/2"	"			1.35
Self-drilling anchor				
1/4"	EA			1.79
3/8"	"			2.68
1/2"	"			3.58
05050.95	**Metal Lintels**			
Lintels, steel				
Plain	LB	1.33	1.58	2.91
Galvanized	"	2.00	1.58	3.58
05120.10	**Beams, Girders, Columns, Trusses**			
Beams and girders, A-36				
Welded	TON	3,160	810	3,970
Bolted	"	3,070	730	3,800
Columns				
Pipe				
6" dia.	LB	1.61	0.80	2.41
Structural tube				
6" square				
Light sections	TON	3,770	1,610	5,380

Cold Formed Framing	UNIT	MAT.	INST.	TOTAL
05410.10 **Metal Framing**				
Furring channel, galvanized				
Beams and columns, 3/4"				
12" o.c.	SF	0.44	6.35	6.79
16" o.c.	"	0.34	5.77	6.11
Walls, 3/4"				
12" o.c.	SF	0.44	3.17	3.61
16" o.c.	"	0.34	2.64	2.98
24" o.c.	"	0.24	2.11	2.35
1-1/2"				
12" o.c.	SF	0.72	3.17	3.89
16" o.c.	"	0.55	2.64	3.19
24" o.c.	"	0.37	2.11	2.48
Stud, load bearing				
16" o.c.				
16 ga.				
2-1/2"	SF	1.33	2.82	4.15
3-5/8"	"	1.57	2.82	4.39
4"	"	1.63	2.82	4.45
6"	"	2.05	3.17	5.22
18 ga.				
2-1/2"	SF	1.08	2.82	3.90
3-5/8"	"	1.33	2.82	4.15
4"	"	1.39	2.82	4.21
6"	"	1.76	3.17	4.93
8"	"	2.12	3.17	5.29
20 ga.				
2-1/2"	SF	0.60	2.82	3.42
3-5/8"	"	0.72	2.82	3.54
4"	"	0.79	2.82	3.61
6"	"	0.96	3.17	4.13
8"	"	1.15	3.17	4.32
24" o.c.				
16 ga.				
2-1/2"	SF	0.91	2.44	3.35
3-5/8"	"	1.08	2.44	3.52
4"	"	1.15	2.44	3.59
6"	"	1.39	2.64	4.03
8"	"	1.76	2.64	4.40
18 ga.				
2-1/2"	SF	0.72	2.44	3.16
3-5/8"	"	0.84	2.44	3.28
4"	"	0.91	2.44	3.35
6"	"	1.15	2.64	3.79
8"	"	1.39	2.64	4.03
20 ga.				
2-1/2"	SF	0.44	2.44	2.88
3-5/8"	"	0.49	2.44	2.93
4"	"	0.55	2.44	2.99
6"	"	0.71	2.64	3.35
8"	"	0.88	2.64	3.52

Metal Fabrications	UNIT	MAT.	INST.	TOTAL
05520.10			**Railings**	
Railing, pipe				
1-1/4" diameter, welded steel				
2-rail				
Primed	LF	31.75	12.75	44.50
Galvanized	"	40.75	12.75	53.50
3-rail				
Primed	LF	40.75	16.00	56.75
Galvanized	"	53.00	16.00	69.00
Wall mounted, single rail, welded steel				
Primed	LF	21.25	9.77	31.02
Galvanized	"	27.50	9.77	37.27
Wall mounted, single rail, welded steel				
Primed	LF	21.75	9.77	31.52
Galvanized	"	28.50	9.77	38.27
Wall mounted, single rail, welded steel				
Primed	LF	23.75	10.50	34.25
Galvanized	"	30.75	10.50	41.25

Misc. Fabrications	UNIT	MAT.	INST.	TOTAL
05700.10			**Ornamental Metal**	
Railings, square bars, 6" o.c., shaped top rails				
Steel	LF	92.00	31.75	124
Aluminum	"	110	31.75	142
Bronze	"	230	42.25	272
Stainless steel	"	240	42.25	282
Laminated metal or wood handrails				
2-1/2" round or oval shape	LF	280	31.75	312

Fasteners And Adhesives	UNIT	MAT.	INST.	TOTAL
06050.10 **Accessories**				
Column/post base, cast aluminum				
4" x 4"	EA	19.00	14.50	33.50
6" x 6"	"	26.75	14.50	41.25
Bridging, metal, per pair				
12" o.c.	EA	2.59	5.76	8.35
16" o.c.	"	2.39	5.23	7.62
Anchors				
Bolts, threaded two ends, with nuts and washers				
1/2" dia.				
4" long	EA	3.02	3.60	6.62
7-1/2" long	"	3.52	3.60	7.12
3/4" dia.				
7-1/2" long	EA	6.68	3.60	10.28
15" long	"	10.00	3.60	13.60
Framing anchors				
10 gauge	EA	1.19	4.80	5.99
Bolts, carriage				
1/4 x 4	EA	0.79	5.76	6.55
5/16 x 6	"	1.78	6.06	7.84
3/8 x 6	"	3.60	6.06	9.66
1/2 x 6	"	5.02	6.06	11.08
Joist and beam hangers				
18 ga.				
2 x 4	EA	1.45	5.76	7.21
2 x 6	"	1.74	5.76	7.50
2 x 8	"	2.03	5.76	7.79
2 x 10	"	2.18	6.40	8.58
2 x 12	"	2.83	7.20	10.03
16 ga.				
3 x 6	EA	5.09	6.40	11.49
3 x 8	"	6.18	6.40	12.58
3 x 10	"	6.98	6.77	13.75
3 x 12	"	7.86	7.68	15.54
3 x 14	"	8.50	8.22	16.72
4 x 6	"	8.72	6.40	15.12
4 x 8	"	10.25	6.40	16.65
4 x 10	"	11.75	6.77	18.52
4 x 12	"	15.00	7.68	22.68
4 x 14	"	15.75	8.22	23.97
Rafter anchors, 18 ga., 1-1/2" wide				
5-1/4" long	EA	1.10	4.80	5.90
10-3/4" long	"	1.62	4.80	6.42
Shear plates				
2-5/8" dia.	EA	3.89	4.43	8.32
4" dia.	"	8.09	4.80	12.89
Sill anchors				
Embedded in concrete	EA	2.86	5.76	8.62
Split rings				
2-1/2" dia.	EA	2.35	6.40	8.75
4" dia.	"	4.33	7.20	11.53
Strap ties, 14 ga., 1-3/8" wide				
12" long	EA	2.94	4.80	7.74

06 WOOD AND PLASTICS

Fasteners And Adhesives	UNIT	MAT.	INST.	TOTAL
06050.10	**Accessories** *(Cont.)*			
18" long	EA	3.16	5.23	8.39
24" long	"	4.70	5.76	10.46
36" long	"	6.47	6.40	12.87
Toothed rings				
2-5/8" dia.	EA	2.72	9.60	12.32
4" dia.	"	3.16	11.50	14.66

Rough Carpentry	UNIT	MAT.	INST.	TOTAL
06110.10	**Blocking**			
Steel construction				
Walls				
2x4	LF	0.54	3.84	4.38
2x6	"	0.82	4.43	5.25
2x8	"	1.08	4.80	5.88
2x10	"	1.44	5.23	6.67
2x12	"	1.86	5.76	7.62
Ceilings				
2x4	LF	0.54	4.43	4.97
2x6	"	0.82	5.23	6.05
2x8	"	1.08	5.76	6.84
2x10	"	1.44	6.40	7.84
2x12	"	1.86	7.20	9.06
Wood construction				
Walls				
2x4	LF	0.60	3.20	3.80
2x6	"	0.92	3.60	4.52
2x8	"	1.21	3.84	5.05
2x10	"	1.62	4.11	5.73
2x12	"	2.09	4.43	6.52
Ceilings				
2x4	LF	0.60	3.60	4.20
2x6	"	0.92	4.11	5.03
2x8	"	1.21	4.43	5.64
2x10	"	1.62	4.80	6.42
2x12	"	2.09	5.23	7.32
06110.20	**Ceiling Framing**			
Ceiling joists				
12" o.c.				
2x4	SF	0.90	1.37	2.27
2x6	"	1.30	1.44	2.74
2x8	"	1.91	1.51	3.42
2x10	"	2.17	1.60	3.77
2x12	"	4.00	1.69	5.69
16" o.c.				

Rough Carpentry	UNIT	MAT.	INST.	TOTAL
06110.20 **Ceiling Framing** *(Cont.)*				
2x4	SF	0.73	1.10	1.83
2x6	"	1.09	1.15	2.24
2x8	"	1.55	1.20	2.75
2x10	"	1.74	1.25	2.99
2x12	"	3.26	1.30	4.56
24" o.c.				
2x4	SF	0.52	0.91	1.43
2x6	"	0.87	0.96	1.83
2x8	"	1.30	1.01	2.31
2x10	"	1.55	1.06	2.61
2x12	"	3.92	1.12	5.04
Headers and nailers				
2x4	LF	0.60	1.85	2.45
2x6	"	0.92	1.92	2.84
2x8	"	1.21	2.05	3.26
2x10	"	1.62	2.21	3.83
2x12	"	1.99	2.40	4.39
Sister joists for ceilings				
2x4	LF	0.60	4.11	4.71
2x6	"	0.92	4.80	5.72
2x8	"	1.21	5.76	6.97
2x10	"	1.62	7.20	8.82
2x12	"	1.99	9.60	11.59
06110.30 **Floor Framing**				
Floor joists				
12" o.c.				
2x6	SF	1.10	1.15	2.25
2x8	"	1.63	1.17	2.80
2x10	"	2.25	1.20	3.45
2x12	"	3.31	1.25	4.56
2x14	"	5.04	1.20	6.24
3x6	"	3.73	1.22	4.95
3x8	"	4.87	1.25	6.12
3x10	"	6.09	1.30	7.39
3x12	"	7.31	1.37	8.68
3x14	"	8.35	1.44	9.79
4x6	"	4.87	1.20	6.07
4x8	"	6.26	1.25	7.51
4x10	"	8.00	1.30	9.30
4x12	"	9.73	1.37	11.10
4x14	"	11.25	1.44	12.69
16" o.c.				
2x6	SF	0.95	0.96	1.91
2x8	"	1.34	0.97	2.31
2x10	"	1.63	0.99	2.62
2x12	"	2.03	1.02	3.05
2x14	"	4.52	1.06	5.58
3x6	"	3.13	0.99	4.12
3x8	"	4.00	1.02	5.02
3x10	"	5.04	1.06	6.10
3x12	"	6.09	1.10	7.19
3x14	"	7.21	1.15	8.36

06 WOOD AND PLASTICS

Rough Carpentry	UNIT	MAT.	INST.	TOTAL
06110.30 **Floor Framing** *(Cont.)*				
4x6	SF	4.00	0.99	4.99
4x8	"	5.48	1.02	6.50
4x10	"	6.78	1.06	7.84
4x12	"	8.00	1.10	9.10
4x14	"	9.57	1.15	10.72
Sister joists for floors				
2x4	LF	0.60	3.60	4.20
2x6	"	0.92	4.11	5.03
2x8	"	1.21	4.80	6.01
2x10	"	1.62	5.76	7.38
2x12	"	2.09	7.20	9.29
3x6	"	3.04	5.76	8.80
3x8	"	3.73	6.40	10.13
3x10	"	4.95	7.20	12.15
3x12	"	5.99	8.22	14.21
4x6	"	3.92	5.76	9.68
4x8	"	5.22	6.40	11.62
4x10	"	6.78	7.20	13.98
4x12	"	7.56	8.22	15.78
06110.40 **Furring**				
Furring, wood strips				
Walls				
On masonry or concrete walls				
1x2 furring				
12" o.c.	SF	0.48	1.80	2.28
16" o.c.	"	0.41	1.64	2.05
24" o.c.	"	0.40	1.51	1.91
1x3 furring				
12" o.c.	SF	0.60	1.80	2.40
16" o.c.	"	0.55	1.64	2.19
24" o.c.	"	0.42	1.51	1.93
On wood walls				
1x2 furring				
12" o.c.	SF	0.48	1.28	1.76
16" o.c.	"	0.41	1.15	1.56
24" o.c.	"	0.38	1.04	1.42
1x3 furring				
12" o.c.	SF	0.62	1.28	1.90
16" o.c.	"	0.52	1.15	1.67
24" o.c.	"	0.42	1.04	1.46
Ceilings				
On masonry or concrete ceilings				
1x2 furring				
12" o.c.	SF	0.48	3.20	3.68
16" o.c.	"	0.41	2.88	3.29
24" o.c.	"	0.38	2.61	2.99
1x3 furring				
12" o.c.	SF	0.60	3.20	3.80
16" o.c.	"	0.52	2.88	3.40
24" o.c.	"	0.42	2.61	3.03
On wood ceilings				
1x2 furring				

48

Rough Carpentry	UNIT	MAT.	INST.	TOTAL
06110.40 **Furring** *(Cont.)*				
12" o.c.	SF	0.48	2.13	2.61
16" o.c.	"	0.41	1.92	2.33
24" o.c.	"	0.38	1.74	2.12
1x3				
12" o.c.	SF	0.60	2.13	2.73
16" o.c.	"	0.52	1.92	2.44
24" o.c.	"	0.42	1.74	2.16
06110.50 **Roof Framing**				
Roof framing				
Rafters, gable end				
0-2 pitch (flat to 2-in-12)				
12" o.c.				
2x4	SF	0.87	1.20	2.07
2x6	"	1.21	1.25	2.46
2x8	"	1.74	1.30	3.04
2x10	"	2.17	1.37	3.54
2x12	"	4.00	1.44	5.44
16" o.c.				
2x6	SF	1.09	1.02	2.11
2x8	"	1.53	1.06	2.59
2x10	"	1.74	1.10	2.84
2x12	"	3.21	1.15	4.36
24" o.c.				
2x6	SF	0.60	0.87	1.47
2x8	"	1.27	0.90	2.17
2x10	"	1.48	0.92	2.40
2x12	"	2.60	0.96	3.56
4-6 pitch (4-in-12 to 6-in-12)				
12" o.c.				
2x4	SF	0.87	1.25	2.12
2x6	"	1.30	1.30	2.60
2x8	"	1.99	1.37	3.36
2x10	"	2.25	1.44	3.69
2x12	"	3.47	1.51	4.98
16" o.c.				
2x6	SF	1.09	1.06	2.15
2x8	"	1.74	1.10	2.84
2x10	"	1.99	1.15	3.14
2x12	"	2.96	1.20	4.16
24" o.c.				
2x6	SF	0.87	0.90	1.77
2x8	"	1.48	0.92	2.40
2x10	"	1.56	0.99	2.55
2x12	"	2.43	1.10	3.53
8-12 pitch (8-in-12 to 12-in-12)				
12" o.c.				
2x4	SF	0.95	1.30	2.25
2x6	"	1.48	1.37	2.85
2x8	"	2.09	1.44	3.53
2x10	"	2.43	1.51	3.94
2x12	"	3.73	1.60	5.33
16" o.c.				

Rough Carpentry	UNIT	MAT.	INST.	TOTAL
06110.50 — **Roof Framing** (Cont.)				
2x6	SF	1.21	1.10	2.31
2x8	"	1.95	1.15	3.10
2x10	"	2.17	1.20	3.37
2x12	"	3.13	1.25	4.38
24" o.c.				
2x6	SF	0.95	0.92	1.87
2x8	"	1.55	0.96	2.51
2x10	"	1.74	0.99	2.73
2x12	"	2.78	1.02	3.80
Ridge boards				
2x6	LF	0.92	2.88	3.80
2x8	"	1.21	3.20	4.41
2x10	"	1.62	3.60	5.22
2x12	"	2.09	4.11	6.20
Hip rafters				
2x6	LF	0.92	2.05	2.97
2x8	"	1.21	2.13	3.34
2x10	"	1.62	2.21	3.83
2x12	"	2.09	2.30	4.39
Jack rafters				
4-6 pitch (4-in-12 to 6-in-12)				
16" o.c.				
2x6	SF	1.13	1.69	2.82
2x8	"	1.74	1.74	3.48
2x10	"	1.99	1.85	3.84
2x12	"	2.96	1.92	4.88
24" o.c.				
2x6	SF	0.87	1.30	2.17
2x8	"	1.48	1.33	2.81
2x10	"	1.74	1.40	3.14
2x12	"	2.52	1.44	3.96
8-12 pitch (8-in-12 to 12-in-12)				
16" o.c.				
2x6	SF	1.74	1.80	3.54
2x8	"	2.17	1.85	4.02
2x10	"	3.13	1.92	5.05
2x12	"	4.34	1.98	6.32
24" o.c.				
2x6	SF	1.38	1.37	2.75
2x8	"	1.74	1.40	3.14
2x10	"	2.78	1.44	4.22
2x12	"	4.00	1.47	5.47
Sister rafters				
2x4	LF	0.60	4.11	4.71
2x6	"	0.92	4.80	5.72
2x8	"	1.21	5.76	6.97
2x10	"	1.62	7.20	8.82
2x12	"	2.09	9.60	11.69
Fascia boards				
2x4	LF	0.60	2.88	3.48
2x6	"	0.92	2.88	3.80
2x8	"	1.21	3.20	4.41
2x10	"	1.62	3.20	4.82

Rough Carpentry	UNIT	MAT.	INST.	TOTAL
06110.50 **Roof Framing** *(Cont.)*				
2x12	LF	2.09	3.60	5.69
Cant strips				
Fiber				
3x3	LF	0.48	1.64	2.12
4x4	"	0.67	1.74	2.41
Wood				
3x3	LF	2.52	1.74	4.26
06110.60 **Sleepers**				
Sleepers, over concrete				
12" o.c.				
1x2	SF	0.29	1.30	1.59
1x3	"	0.43	1.37	1.80
2x4	"	0.95	1.60	2.55
2x6	"	1.39	1.69	3.08
16" o.c.				
1x2	SF	0.26	1.15	1.41
1x3	"	0.37	1.15	1.52
2x4	"	0.79	1.37	2.16
2x6	"	1.17	1.44	2.61
06110.65 **Soffits**				
Soffit framing				
2x3	LF	0.41	4.11	4.52
2x4	"	0.51	4.43	4.94
2x6	"	0.75	4.80	5.55
2x8	"	1.06	5.23	6.29
06110.70 **Wall Framing**				
Framing wall, studs				
12" o.c.				
2x3	SF	0.53	1.06	1.59
2x4	"	0.75	1.06	1.81
2x6	"	1.09	1.15	2.24
2x8	"	1.46	1.20	2.66
16" o.c.				
2x3	SF	0.43	0.90	1.33
2x4	"	0.61	0.90	1.51
2x6	"	0.87	0.96	1.83
2x8	"	1.37	0.99	2.36
24" o.c.				
2x3	SF	0.34	0.77	1.11
2x4	"	0.46	0.77	1.23
2x6	"	0.73	0.82	1.55
2x8	"	0.95	0.84	1.79
Plates, top or bottom				
2x3	LF	0.41	1.69	2.10
2x4	"	0.51	1.80	2.31
2x6	"	0.75	1.92	2.67
2x8	"	1.06	2.05	3.11
Headers, door or window				
2x6				
Single				

Rough Carpentry	UNIT	MAT.	INST.	TOTAL
06110.70 **Wall Framing** *(Cont.)*				
3' long	EA	2.46	28.75	31.21
6' long	"	4.92	36.00	40.92
Double				
3' long	EA	4.94	32.00	36.94
6' long	"	9.89	41.25	51.14
2x8				
Single				
4' long	EA	4.51	36.00	40.51
8' long	"	9.01	44.25	53.26
Double				
4' long	EA	9.01	41.25	50.26
8' long	"	18.00	52.00	70.00
2x10				
Single				
5' long	EA	6.82	44.25	51.07
10' long	"	13.75	58.00	71.75
Double				
5' long	EA	13.75	48.00	61.75
10' long	"	27.25	58.00	85.25
2x12				
Single				
6' long	EA	9.89	44.25	54.14
12' long	"	19.50	58.00	77.50
Double				
6' long	EA	19.50	52.00	71.50
12' long	"	38.75	64.00	103
06115.10 **Floor Sheathing**				
Sub-flooring, plywood, CDX				
1/2" thick	SF	0.61	0.72	1.33
5/8" thick	"	0.88	0.82	1.70
3/4" thick	"	1.62	0.96	2.58
Structural plywood				
1/2" thick	SF	0.96	0.72	1.68
5/8" thick	"	1.54	0.82	2.36
3/4" thick	"	1.62	0.88	2.50
Board type subflooring				
1x6				
Minimum	SF	1.46	1.28	2.74
Maximum	"	1.86	1.44	3.30
1x8				
Minimum	SF	1.62	1.21	2.83
Maximum	"	1.90	1.35	3.25
1x10				
Minimum	SF	2.27	1.15	3.42
Maximum	"	2.43	1.28	3.71
Underlayment				
Hardboard, 1/4" tempered	SF	0.90	0.72	1.62
Plywood, CDX				
3/8" thick	SF	0.94	0.72	1.66
1/2" thick	"	1.12	0.76	1.88
5/8" thick	"	1.30	0.82	2.12
3/4" thick	"	1.62	0.88	2.50

Rough Carpentry	UNIT	MAT.	INST.	TOTAL
06115.20 **Roof Sheathing**				
Sheathing				
Plywood, CDX				
3/8" thick	SF	0.94	0.74	1.68
1/2" thick	"	1.12	0.76	1.88
5/8" thick	"	1.30	0.82	2.12
3/4" thick	"	1.62	0.88	2.50
Structural plywood				
3/8" thick	SF	0.59	0.74	1.33
1/2" thick	"	0.77	0.76	1.53
5/8" thick	"	0.94	0.82	1.76
3/4" thick	"	1.13	0.88	2.01
06115.30 **Wall Sheathing**				
Sheathing				
Plywood, CDX				
3/8" thick	SF	0.94	0.85	1.79
1/2" thick	"	1.12	0.88	2.00
5/8" thick	"	1.30	0.96	2.26
3/4" thick	"	1.62	1.04	2.66
Waferboard				
3/8" thick	SF	0.59	0.85	1.44
1/2" thick	"	0.77	0.88	1.65
5/8" thick	"	0.94	0.96	1.90
3/4" thick	"	1.03	1.04	2.07
Structural plywood				
3/8" thick	SF	0.94	0.85	1.79
1/2" thick	"	1.12	0.88	2.00
5/8" thick	"	1.30	0.96	2.26
3/4" thick	"	1.12	1.04	2.16
Gypsum, 1/2" thick	"	0.59	0.88	1.47
Asphalt impregnated fiberboard, 1/2" thick	"	1.03	0.88	1.91
06125.10 **Wood Decking**				
Decking, T&G solid				
Cedar				
3" thick	SF	11.75	1.44	13.19
4" thick	"	14.50	1.53	16.03
Fir				
3" thick	SF	5.10	1.44	6.54
4" thick	"	6.19	1.53	7.72
Southern yellow pine				
3" thick	SF	5.10	1.64	6.74
4" thick	"	5.39	1.77	7.16
White pine				
3" thick	SF	6.19	1.44	7.63
4" thick	"	8.38	1.53	9.91
06130.10 **Heavy Timber**				
Mill framed structures				
Beams to 20' long				
Douglas fir				
6x8	LF	8.36	8.01	16.37
6x10	"	9.87	8.28	18.15

Rough Carpentry	UNIT	MAT.	INST.	TOTAL
06130.10 **Heavy Timber** *(Cont.)*				
6x12	LF	11.75	8.90	20.65
6x14	"	14.25	9.24	23.49
6x16	"	15.50	9.61	25.11
8x10	"	13.00	8.28	21.28
8x12	"	15.50	8.90	24.40
8x14	"	17.75	9.24	26.99
8x16	"	20.25	9.61	29.86
Southern yellow pine				
6x8	LF	6.63	8.01	14.64
6x10	"	8.05	8.28	16.33
6x12	"	10.25	8.90	19.15
6x14	"	11.75	9.24	20.99
6x16	"	13.00	9.61	22.61
8x10	"	11.00	8.28	19.28
8x12	"	13.25	8.90	22.15
8x14	"	15.25	9.24	24.49
8x16	"	17.50	9.61	27.11
Columns to 12' high				
Douglas fir				
6x6	LF	6.01	12.00	18.01
8x8	"	10.25	12.00	22.25
10x10	"	18.00	13.25	31.25
12x12	"	22.25	13.25	35.50
Southern yellow pine				
6x6	LF	5.17	12.00	17.17
8x8	"	8.70	12.00	20.70
10x10	"	13.50	13.25	26.75
12x12	"	18.75	13.25	32.00
Posts, treated				
4x4	LF	2.07	2.30	4.37
6x6	"	6.01	2.88	8.89
06190.20 **Wood Trusses**				
Truss, fink, 2x4 members				
3-in-12 slope				
24' span	EA	120	69.00	189
26' span	"	130	69.00	199
28' span	"	140	73.00	213
30' span	"	150	73.00	223
34' span	"	150	78.00	228
38' span	"	150	78.00	228
5-in-12 slope				
24' span	EA	130	71.00	201
28' span	"	140	73.00	213
30' span	"	150	75.00	225
32' span	"	160	75.00	235
40' span	"	210	80.00	290
Gable, 2x4 members				
5-in-12 slope				
24' span	EA	150	71.00	221
26' span	"	160	71.00	231
28' span	"	180	73.00	253
30' span	"	190	75.00	265

Rough Carpentry	UNIT	MAT.	INST.	TOTAL
06190.20 **Wood Trusses** *(Cont.)*				
32' span	EA	200	75.00	275
36' span	"	210	78.00	288
40' span	"	230	80.00	310
King post type, 2x4 members				
4-in-12 slope				
16' span	EA	91.00	65.00	156
18' span	"	98.00	67.00	165
24' span	"	110	71.00	181
26' span	"	110	71.00	181
30' span	"	140	75.00	215
34' span	"	150	75.00	225
38' span	"	180	78.00	258
42' span	"	220	83.00	303

Finish Carpentry	UNIT	MAT.	INST.	TOTAL
06200.10 **Finish Carpentry**				
Mouldings and trim				
Apron, flat				
9/16 x 2	LF	1.99	2.88	4.87
9/16 x 3-1/2	"	4.59	3.03	7.62
Base				
Colonial				
7/16 x 2-1/4	LF	2.37	2.88	5.25
7/16 x 3	"	3.07	2.88	5.95
7/16 x 3-1/4	"	3.14	2.88	6.02
9/16 x 3	"	3.07	3.03	6.10
9/16 x 3-1/4	"	3.21	3.03	6.24
11/16 x 2-1/4	"	3.37	3.20	6.57
Ranch				
7/16 x 2-1/4	LF	2.60	2.88	5.48
7/16 x 3-1/4	"	3.07	2.88	5.95
9/16 x 2-1/4	"	2.83	3.03	5.86
9/16 x 3	"	3.07	3.03	6.10
9/16 x 3-1/4	"	3.14	3.03	6.17
Casing				
11/16 x 2-1/2	LF	2.44	2.61	5.05
11/16 x 3-1/2	"	2.76	2.74	5.50
Chair rail				
9/16 x 2-1/2	LF	2.60	2.88	5.48
9/16 x 3-1/2	"	3.60	2.88	6.48
Closet pole				
1-1/8" dia.	LF	1.76	3.84	5.60
1-5/8" dia.	"	2.60	3.84	6.44
Cove				
9/16 x 1-3/4	LF	1.99	2.88	4.87

06 WOOD AND PLASTICS

Finish Carpentry	UNIT	MAT.	INST.	TOTAL
06200.10	**Finish Carpentry** *(Cont.)*			
11/16 x 2-3/4	LF	3.07	2.88	5.95
Crown				
9/16 x 1-5/8	LF	2.60	3.84	6.44
9/16 x 2-5/8	"	2.83	4.43	7.26
11/16 x 3-5/8	"	3.07	4.80	7.87
11/16 x 4-1/4	"	4.59	5.23	9.82
11/16 x 5-1/4	"	5.14	5.76	10.90
Drip cap				
1-1/16 x 1-5/8	LF	2.76	2.88	5.64
Glass bead				
3/8 x 3/8	LF	0.99	3.60	4.59
1/2 x 9/16	"	1.22	3.60	4.82
5/8 x 5/8	"	1.30	3.60	4.90
3/4 x 3/4	"	1.53	3.60	5.13
Half round				
1/2	LF	1.15	2.30	3.45
5/8	"	1.53	2.30	3.83
3/4	"	2.07	2.30	4.37
Lattice				
1/4 x 7/8	LF	0.92	2.30	3.22
1/4 x 1-1/8	"	0.99	2.30	3.29
1/4 x 1-3/8	"	1.06	2.30	3.36
1/4 x 1-3/4	"	1.19	2.30	3.49
1/4 x 2	"	1.38	2.30	3.68
Ogee molding				
5/8 x 3/4	LF	1.83	2.88	4.71
11/16 x 1-1/8	"	4.30	2.88	7.18
11/16 x 1-3/8	"	3.37	2.88	6.25
Parting bead				
3/8 x 7/8	LF	1.53	3.60	5.13
Quarter round				
1/4 x 1/4	LF	0.54	2.30	2.84
3/8 x 3/8	"	0.76	2.30	3.06
1/2 x 1/2	"	0.99	2.30	3.29
11/16 x 11/16	"	0.99	2.50	3.49
3/4 x 3/4	"	1.83	2.50	4.33
1-1/16 x 1-1/16	"	1.45	2.61	4.06
Railings, balusters				
1-1/8 x 1-1/8	LF	4.91	5.76	10.67
1-1/2 x 1-1/2	"	5.75	5.23	10.98
Screen moldings				
1/4 x 3/4	LF	1.22	4.80	6.02
5/8 x 5/16	"	1.53	4.80	6.33
Shoe				
7/16 x 11/16	LF	1.53	2.30	3.83
Sash beads				
1/2 x 3/4	LF	1.76	4.80	6.56
1/2 x 7/8	"	1.99	4.80	6.79
1/2 x 1-1/8	"	2.15	5.23	7.38
5/8 x 7/8	"	2.15	5.23	7.38
Stop				
5/8 x 1-5/8				
Colonial	LF	1.06	3.60	4.66

Finish Carpentry	UNIT	MAT.	INST.	TOTAL
06200.10 **Finish Carpentry** *(Cont.)*				
Ranch	LF	1.06	3.60	4.66
Stools				
11/16 x 2-1/4	LF	4.68	6.40	11.08
11/16 x 2-1/2	"	4.91	6.40	11.31
11/16 x 5-1/4	"	5.06	7.20	12.26
Exterior trim, casing, select pine, 1x3	"	3.37	2.88	6.25
Douglas fir				
1x3	LF	1.60	2.88	4.48
1x4	"	1.99	2.88	4.87
1x6	"	2.60	3.20	5.80
1x8	"	3.60	3.60	7.20
Cornices, white pine, #2 or better				
1x2	LF	0.99	2.88	3.87
1x4	"	1.22	2.88	4.10
1x6	"	1.99	3.20	5.19
1x8	"	2.44	3.38	5.82
1x10	"	3.14	3.60	6.74
1x12	"	3.91	3.84	7.75
Shelving, pine				
1x8	LF	1.76	4.43	6.19
1x10	"	2.30	4.60	6.90
1x12	"	2.91	4.80	7.71
Plywood shelf, 3/4", with edge band, 12" wide	"	3.14	5.76	8.90
Adjustable shelf, and rod, 12" wide				
3' to 4' long	EA	25.00	14.50	39.50
5' to 8' long	"	47.00	19.25	66.25
Prefinished wood shelves with brackets and supports				
8" wide				
3' long	EA	74.00	14.50	88.50
4' long	"	85.00	14.50	99.50
6' long	"	120	14.50	135
10" wide				
3' long	EA	81.00	14.50	95.50
4' long	"	120	14.50	135
6' long	"	130	14.50	145
06220.10 **Millwork**				
Countertop, laminated plastic				
25" x 7/8" thick				
Minimum	LF	18.75	14.50	33.25
Average	"	35.50	19.25	54.75
Maximum	"	52.00	23.00	75.00
25" x 1-1/4" thick				
Minimum	LF	22.75	19.25	42.00
Average	"	45.50	23.00	68.50
Maximum	"	68.00	28.75	96.75
Add for cutouts	EA		36.00	36.00
Backsplash, 4" high, 7/8" thick	LF	25.00	11.50	36.50
Plywood, sanded, A-C				
1/4" thick	SF	1.61	1.92	3.53
3/8" thick	"	1.75	2.05	3.80
1/2" thick	"	1.98	2.21	4.19
A-D				

Finish Carpentry	UNIT	MAT.	INST.	TOTAL
06220.10 **Millwork** *(Cont.)*				
1/4" thick	SF	1.53	1.92	3.45
3/8" thick	"	1.75	2.05	3.80
1/2" thick	"	1.90	2.21	4.11
Base cabinet, 34-1/2" high, 24" deep, hardwood				
Minimum	LF	250	23.00	273
Average	"	280	28.75	309
Maximum	"	310	38.50	349
Wall cabinets				
Minimum	LF	75.00	19.25	94.25
Average	"	100	23.00	123
Maximum	"	130	28.75	159
06300.10 **Wood Treatment**				
Creosote preservative treatment				
8 lb/cf	BF			0.74
10 lb/cf	"			0.89
Salt preservative treatment				
Oil borne				
Minimum	BF			0.68
Maximum	"			0.96
Water borne				
Minimum	BF			0.48
Maximum	"			0.74
Fire retardant treatment				
Minimum	BF			0.96
Maximum	"			1.16
Kiln dried, softwood, add to framing costs				
1" thick	BF			0.34
2" thick	"			0.48
3" thick	"			0.61
4" thick	"			0.74

Architectural Woodwork	UNIT	MAT.	INST.	TOTAL
06420.10 **Panel Work**				
Hardboard, tempered, 1/4" thick				
Natural faced	SF	1.16	1.44	2.60
Plastic faced	"	1.75	1.64	3.39
Pegboard, natural	"	1.46	1.44	2.90
Plastic faced	"	1.75	1.64	3.39
Untempered, 1/4" thick				
Natural faced	SF	1.10	1.44	2.54
Plastic faced	"	1.90	1.64	3.54
Pegboard, natural	"	1.16	1.44	2.60
Plastic faced	"	1.68	1.64	3.32
Plywood unfinished, 1/4" thick				

Architectural Woodwork	UNIT	MAT.	INST.	TOTAL
06420.10 **Panel Work** *(Cont.)*				
Birch				
Natural	SF	1.25	1.92	3.17
Select	"	1.83	1.92	3.75
Knotty pine	"	2.41	1.92	4.33
Cedar (closet lining)				
Standard boards T&G	SF	3.00	1.92	4.92
Particle board	"	1.83	1.92	3.75
Plywood, prefinished, 1/4" thick, premium grade				
Birch veneer	SF	4.39	2.30	6.69
Cherry veneer	"	5.12	2.30	7.42
Chestnut veneer	"	9.89	2.30	12.19
Lauan veneer	"	1.90	2.30	4.20
Mahogany veneer	"	5.05	2.30	7.35
Oak veneer (red)	"	5.05	2.30	7.35
Pecan veneer	"	6.37	2.30	8.67
Rosewood veneer	"	9.89	2.30	12.19
Teak veneer	"	6.52	2.30	8.82
Walnut veneer	"	5.64	2.30	7.94
06430.10 **Stairwork**				
Risers, 1x8, 42" wide				
White oak	EA	53.00	28.75	81.75
Pine	"	47.00	28.75	75.75
Treads, 1-1/16" x 9-1/2" x 42"				
White oak	EA	63.00	36.00	99.00
06440.10 **Columns**				
Column, hollow, round wood				
12" diameter				
10' high	EA	940	89.00	1,029
12' high	"	1,150	95.00	1,245
14' high	"	1,380	110	1,490
16' high	"	1,710	130	1,840
24" diameter				
16' high	EA	3,900	130	4,030
18' high	"	4,440	140	4,580
20' high	"	5,460	140	5,600
22' high	"	5,750	150	5,900
24' high	"	6,270	150	6,420

Moisture Protection	UNIT	MAT.	INST.	TOTAL
07100.10 **Waterproofing**				
Membrane waterproofing, elastomeric				
Butyl				
1/32" thick	SF	1.47	1.80	3.27
1/16" thick	"	1.91	1.88	3.79
Neoprene				
1/32" thick	SF	2.51	1.80	4.31
1/16" thick	"	3.60	1.88	5.48
Plastic vapor barrier (polyethylene)				
4 mil	SF	0.05	0.18	0.23
6 mil	"	0.09	0.18	0.27
10 mil	"	0.12	0.22	0.34
Bituminous membrane, asphalt felt, 15 lb.				
One ply	SF	0.87	1.12	1.99
Two ply	"	1.03	1.36	2.39
Three ply	"	1.27	1.61	2.88
Bentonite waterproofing, panels				
3/16" thick	SF	2.04	1.12	3.16
1/4" thick	"	2.32	1.12	3.44
07150.10 **Dampproofing**				
Silicone dampproofing, sprayed on				
Concrete surface				
1 coat	SF	0.64	0.25	0.89
2 coats	"	1.06	0.34	1.40
Concrete block				
1 coat	SF	0.64	0.30	0.94
2 coats	"	1.06	0.41	1.47
Brick				
1 coat	SF	0.64	0.34	0.98
2 coats	"	1.06	0.45	1.51
07160.10 **Bituminous Dampproofing**				
Building paper, asphalt felt				
15 lb	SF	0.19	1.80	1.99
30 lb	"	0.37	1.88	2.25
Asphalt, troweled, cold, primer plus				
1 coat	SF	0.68	1.50	2.18
2 coats	"	1.43	2.25	3.68
3 coats	"	2.04	2.82	4.86
Fibrous asphalt, hot troweled, primer plus				
1 coat	SF	0.68	1.80	2.48
2 coats	"	1.43	2.50	3.93
3 coats	"	2.04	3.22	5.26
Asphaltic paint dampproofing, per coat				
Brush on	SF	0.35	0.64	0.99
Spray on	"	0.49	0.50	0.99
07190.10 **Vapor Barriers**				
Vapor barrier, polyethylene				
2 mil	SF	0.02	0.22	0.24
6 mil	"	0.07	0.22	0.29
8 mil	"	0.08	0.25	0.33
10 mil	"	0.09	0.25	0.34

07 THERMAL AND MOISTURE

Insulation	UNIT	MAT.	INST.	TOTAL
07210.10 **Batt Insulation**				
Ceiling, fiberglass, unfaced				
3-1/2" thick, R11	SF	0.42	0.53	0.95
6" thick, R19	"	0.56	0.60	1.16
9" thick, R30	"	1.10	0.69	1.79
Suspended ceiling, unfaced				
3-1/2" thick, R11	SF	0.42	0.50	0.92
6" thick, R19	"	0.56	0.56	1.12
9" thick, R30	"	1.10	0.64	1.74
Crawl space, unfaced				
3-1/2" thick, R11	SF	0.42	0.69	1.11
6" thick, R19	"	0.56	0.75	1.31
9" thick, R30	"	1.10	0.82	1.92
Wall, fiberglass				
Paper backed				
2" thick, R7	SF	0.33	0.47	0.80
3" thick, R8	"	0.36	0.50	0.86
4" thick, R11	"	0.59	0.53	1.12
6" thick, R19	"	0.88	0.56	1.44
Foil backed, 1 side				
2" thick, R7	SF	0.63	0.47	1.10
3" thick, R11	"	0.68	0.50	1.18
4" thick, R14	"	0.71	0.53	1.24
6" thick, R21	"	0.93	0.56	1.49
Foil backed, 2 sides				
2" thick, R7	SF	0.72	0.53	1.25
3" thick, R11	"	0.92	0.56	1.48
4" thick, R14	"	1.08	0.60	1.68
6" thick, R21	"	1.17	0.64	1.81
Unfaced				
2" thick, R7	SF	0.40	0.47	0.87
3" thick, R9	"	0.46	0.50	0.96
4" thick, R11	"	0.49	0.53	1.02
6" thick, R19	"	0.63	0.56	1.19
Mineral wool batts				
Paper backed				
2" thick, R6	SF	0.35	0.47	0.82
4" thick, R12	"	0.79	0.50	1.29
6" thick, R19	"	0.99	0.56	1.55
Fasteners, self adhering, attached to ceiling deck				
2-1/2" long	EA	0.26	0.75	1.01
4-1/2" long	"	0.29	0.82	1.11
Capped, self-locking washers	"	0.26	0.45	0.71
07210.20 **Board Insulation**				
Perlite board, roof				
1.00" thick, R2.78	SF	0.57	0.37	0.94
1.50" thick, R4.17	"	0.90	0.39	1.29
Rigid urethane				
1" thick, R6.67	SF	1.09	0.37	1.46
1.50" thick, R11.11	"	1.49	0.39	1.88
Polystyrene				
1.0" thick, R4.17	SF	0.41	0.37	0.78
1.5" thick, R6.26	"	0.63	0.39	1.02

Insulation	UNIT	MAT.	INST.	TOTAL
07210.60	**Loose Fill Insulation**			
Blown-in type				
Fiberglass				
5" thick, R11	SF	0.41	0.37	0.78
6" thick, R13	"	0.48	0.45	0.93
9" thick, R19	"	0.58	0.64	1.22
Rockwool, attic application				
6" thick, R13	SF	0.38	0.45	0.83
8" thick, R19	"	0.45	0.56	1.01
10" thick, R22	"	0.53	0.69	1.22
12" thick, R26	"	0.68	0.75	1.43
15" thick, R30	"	0.82	0.90	1.72
Poured type				
Fiberglass				
1" thick, R4	SF	0.45	0.28	0.73
2" thick, R8	"	0.84	0.32	1.16
3" thick, R12	"	1.24	0.37	1.61
4" thick, R16	"	1.63	0.45	2.08
Mineral wool				
1" thick, R3	SF	0.50	0.28	0.78
2" thick, R6	"	0.92	0.32	1.24
3" thick, R9	"	1.40	0.37	1.77
4" thick, R12	"	1.63	0.45	2.08
Vermiculite or perlite				
2" thick, R4.8	SF	0.98	0.32	1.30
3" thick, R7.2	"	1.39	0.37	1.76
4" thick, R9.6	"	1.81	0.45	2.26
Masonry, poured vermiculite or perlite				
4" block	SF	0.57	0.22	0.79
6" block	"	0.86	0.28	1.14
8" block	"	1.25	0.32	1.57
10" block	"	1.65	0.34	1.99
12" block	"	2.05	0.37	2.42
07210.70	**Sprayed Insulation**			
Foam, sprayed on				
Polystyrene				
1" thick, R4	SF	0.69	0.45	1.14
2" thick, R8	"	1.34	0.60	1.94
Urethane				
1" thick, R4	SF	0.65	0.45	1.10
2" thick, R8	"	1.24	0.60	1.84

07　THERMAL AND MOISTURE

Shingles And Tiles	UNIT	MAT.	INST.	TOTAL
07310.10 — Asphalt Shingles				
Standard asphalt shingles, strip shingles				
210 lb/square	SQ	90.00	55.00	145
235 lb/square	"	95.00	61.00	156
240 lb/square	"	99.00	69.00	168
260 lb/square	"	140	79.00	219
300 lb/square	"	150	92.00	242
385 lb/square	"	210	110	320
Roll roofing, mineral surface				
90 lb	SQ	55.00	39.25	94.25
110 lb	"	92.00	45.75	138
140 lb	"	95.00	55.00	150
07310.50 — Metal Shingles				
Aluminum, .020" thick				
Plain	SQ	280	110	390
Colors	"	310	110	420
Steel, galvanized				
26 ga.				
Plain	SQ	350	110	460
Colors	"	440	110	550
24 ga.				
Plain	SQ	400	110	510
Colors	"	510	110	620
Porcelain enamel, 22 ga.				
Minimum	SQ	870	140	1,010
Average	"	1,000	140	1,140
Maximum	"	1,120	140	1,260
07310.60 — Slate Shingles				
Slate shingles				
Pennsylvania				
Ribbon	SQ	600	280	880
Clear	"	770	280	1,050
Vermont				
Black	SQ	710	280	990
Gray	"	780	280	1,060
Green	"	800	280	1,080
Red	"	1,440	280	1,720
07310.70 — Wood Shingles				
Wood shingles, on roofs				
White cedar, #1 shingles				
4" exposure	SQ	270	180	450
5" exposure	"	240	140	380
#2 shingles				
4" exposure	SQ	190	180	370
5" exposure	"	160	140	300
Resquared and rebutted				
4" exposure	SQ	240	180	420
5" exposure	"	200	140	340
On walls				
White cedar, #1 shingles				
4" exposure	SQ	270	280	550

Shingles And Tiles	UNIT	MAT.	INST.	TOTAL
07310.70 **Wood Shingles** *(Cont.)*				
5" exposure	SQ	240	220	460
6" exposure	"	200	180	380
#2 shingles				
4" exposure	SQ	190	280	470
5" exposure	"	160	220	380
6" exposure	"	130	180	310
Add for fire retarding	"			120
07310.80 **Wood Shakes**				
Shakes, hand split, 24" red cedar, on roofs				
5" exposure	SQ	300	280	580
7" exposure	"	280	220	500
9" exposure	"	260	180	440
On walls				
6" exposure	SQ	280	280	560
8" exposure	"	270	220	490
10" exposure	"	260	180	440
Add for fire retarding	"			78.00
07460.10 **Metal Siding Panels**				
Aluminum siding panels				
Corrugated				
Plain finish				
.024"	SF	2.23	2.54	4.77
.032"	"	2.62	2.54	5.16
Painted finish				
.024"	SF	2.78	2.54	5.32
.032"	"	3.19	2.54	5.73
Steel siding panels				
Corrugated				
22 ga.	SF	2.73	4.23	6.96
24 ga.	"	2.49	4.23	6.72
Ribbed, sheets, galvanized				
22 ga.	SF	3.42	2.54	5.96
24 ga.	"	3.01	2.54	5.55
Primed				
24 ga.	SF	3.01	2.54	5.55
26 ga.	"	2.46	2.54	5.00
07460.50 **Plastic Siding**				
Horizontal vinyl siding, solid				
8" wide				
Standard	SF	1.35	2.21	3.56
Insulated	"	1.64	2.21	3.85
10" wide				
Standard	SF	1.40	2.05	3.45
Insulated	"	1.68	2.05	3.73
Vinyl moldings for doors and windows	LF	0.87	2.30	3.17

Shingles And Tiles	UNIT	MAT.	INST.	TOTAL
07460.60	**Plywood Siding**			
Rough sawn cedar, 3/8" thick	SF	2.16	1.92	4.08
Fir, 3/8" thick	"	1.19	1.92	3.11
Texture 1-11, 5/8" thick				
Cedar	SF	2.92	2.05	4.97
Fir	"	2.04	2.05	4.09
Redwood	"	3.14	1.96	5.10
Southern Yellow Pine	"	1.66	2.05	3.71
07460.80	**Wood Siding**			
Beveled siding, cedar				
A grade				
1/2 x 8	SF	5.00	2.30	7.30
3/4 x 10	"	6.43	1.92	8.35
Clear				
1/2 x 6	SF	5.44	2.88	8.32
1/2 x 8	"	5.56	2.30	7.86
3/4 x 10	"	7.45	1.92	9.37
B grade				
1/2 x 6	SF	5.26	2.88	8.14
1/2 x 8	"	5.94	2.30	8.24
3/4 x 10	"	5.60	1.92	7.52
Board and batten				
Cedar				
1x6	SF	6.84	2.88	9.72
1x8	"	6.22	2.30	8.52
1x10	"	5.62	2.05	7.67
1x12	"	5.04	1.85	6.89
Pine				
1x6	SF	1.73	2.88	4.61
1x8	"	1.69	2.30	3.99
1x10	"	1.62	2.05	3.67
1x12	"	1.49	1.85	3.34
Redwood				
1x6	SF	7.44	2.88	10.32
1x8	"	6.93	2.30	9.23
1x10	"	6.43	2.05	8.48
1x12	"	5.93	1.85	7.78
Tongue and groove				
Cedar				
1x4	SF	6.43	3.20	9.63
1x6	"	6.18	3.03	9.21
1x8	"	5.80	2.88	8.68
1x10	"	5.69	2.74	8.43
Pine				
1x4	SF	1.93	3.20	5.13
1x6	"	1.82	3.03	4.85
1x8	"	1.70	2.88	4.58
1x10	"	1.62	2.74	4.36
Redwood				
1x4	SF	6.80	3.20	10.00
1x6	"	6.55	3.03	9.58
1x8	"	6.33	2.88	9.21
1x10	"	6.04	2.74	8.78

Membrane Roofing	UNIT	MAT.	INST.	TOTAL
07510.10 **Built-up Asphalt Roofing**				
Built-up roofing, asphalt felt, including gravel				
2 ply	SQ	88.00	140	228
3 ply	"	120	180	300
4 ply	"	170	220	390
Cant strip, 4" x 4"				
Treated wood	LF	2.56	1.57	4.13
Foamglass	"	2.20	1.37	3.57
New gravel for built-up roofing, 400 lb/sq	SQ	44.50	110	155
07530.10 **Single-ply Roofing**				
Elastic sheet roofing				
Neoprene, 1/16" thick	SF	2.83	0.68	3.51
PVC				
45 mil	SF	2.34	0.68	3.02
Flashing				
Pipe flashing, 90 mil thick				
1" pipe	EA	34.00	13.75	47.75
Neoprene flashing, 60 mil thick strip				
6" wide	LF	1.72	4.58	6.30
12" wide	"	3.38	6.87	10.25

Flashing And Sheet Metal	UNIT	MAT.	INST.	TOTAL
07610.10 **Metal Roofing**				
Sheet metal roofing, copper, 16 oz, batten seam	SQ	1,800	370	2,170
Standing seam	"	1,760	340	2,100
Aluminum roofing, natural finish				
Corrugated, on steel frame				
.0175" thick	SQ	140	160	300
.0215" thick	"	180	160	340
.024" thick	"	210	160	370
.032" thick	"	260	160	420
V-beam, on steel frame				
.032" thick	SQ	270	160	430
.040" thick	"	290	160	450
.050" thick	"	370	160	530
Ridge cap				
.019" thick	LF	4.25	1.83	6.08
Corrugated galvanized steel roofing, on steel frame				
28 ga.	SQ	230	160	390
26 ga.	"	270	160	430
24 ga.	"	300	160	460
22 ga.	"	330	160	490
26 ga., factory insulated with 1" polystyrene	"	510	220	730
Ridge roll				
10" wide	LF	2.31	1.83	4.14

Flashing And Sheet Metal	UNIT	MAT.	INST.	TOTAL
07610.10 **Metal Roofing** *(Cont.)*				
20" wide	LF	4.70	2.20	6.90
07620.10 **Flashing And Trim**				
Counter flashing				
Aluminum, .032"	SF	2.09	5.50	7.59
Stainless steel, .015"	"	6.69	5.50	12.19
Copper				
16 oz.	SF	9.36	5.50	14.86
20 oz.	"	11.00	5.50	16.50
24 oz.	"	13.50	5.50	19.00
32 oz.	"	16.50	5.50	22.00
Valley flashing				
Aluminum, .032"	SF	1.74	3.43	5.17
Stainless steel, .015	"	5.56	3.43	8.99
Copper				
16 oz.	SF	9.36	3.43	12.79
20 oz.	"	11.00	4.58	15.58
24 oz.	"	13.50	3.43	16.93
32 oz.	"	16.50	3.43	19.93
Base flashing				
Aluminum, .040"	SF	2.60	4.58	7.18
Stainless steel, .018"	"	6.65	4.58	11.23
Copper				
16 oz.	SF	9.36	4.58	13.94
20 oz.	"	11.00	3.43	14.43
24 oz.	"	13.50	4.58	18.08
32 oz.	"	16.50	4.58	21.08
Flashing and trim, aluminum				
.019" thick	SF	1.28	3.92	5.20
.032" thick	"	1.57	3.92	5.49
.040" thick	"	2.69	4.23	6.92
Neoprene sheet flashing, .060" thick	"	2.14	3.43	5.57
Copper, paper backed				
2 oz.	SF	2.75	5.50	8.25
5 oz.	"	3.55	5.50	9.05
07620.20 **Gutters And Downspouts**				
Copper gutter and downspout				
Downspouts, 16 oz. copper				
Round				
3" dia.	LF	12.50	3.66	16.16
4" dia.	"	15.50	3.66	19.16
Rectangular, corrugated				
2" x 3"	LF	12.00	3.43	15.43
3" x 4"	"	14.75	3.43	18.18
Rectangular, flat surface				
2" x 3"	LF	13.75	3.66	17.41
3" x 4"	"	19.50	3.66	23.16
Lead-coated copper downspouts				
Round				
3" dia.	LF	16.25	3.43	19.68
4" dia.	"	19.75	3.92	23.67
Rectangular, corrugated				

Flashing And Sheet Metal	UNIT	MAT.	INST.	TOTAL
07620.20 — Gutters And Downspouts *(Cont.)*				
2" x 3"	LF	16.25	3.66	19.91
3" x 4"	"	19.50	3.66	23.16
Rectangular, plain				
2" x 3"	LF	11.25	3.66	14.91
3" x 4"	"	13.25	3.66	16.91
Gutters, 16 oz. copper				
Half round				
4" wide	LF	11.25	5.50	16.75
5" wide	"	13.75	6.11	19.86
Type K				
4" wide	LF	12.50	5.50	18.00
5" wide	"	13.00	6.11	19.11
Lead-coated copper gutters				
Half round				
4" wide	LF	13.50	5.50	19.00
6" wide	"	18.50	6.11	24.61
Type K				
4" wide	LF	14.75	5.50	20.25
5" wide	"	19.25	6.11	25.36
Aluminum gutter and downspout				
Downspouts				
2" x 3"	LF	1.45	3.66	5.11
3" x 4"	"	2.00	3.92	5.92
4" x 5"	"	2.31	4.23	6.54
Round				
3" dia.	LF	2.44	3.66	6.10
4" dia.	"	3.13	3.92	7.05
Gutters, stock units				
4" wide	LF	2.26	5.78	8.04
5" wide	"	2.69	6.11	8.80
Galvanized steel gutter and downspout				
Downspouts, round corrugated				
3" dia.	LF	2.09	3.66	5.75
4" dia.	"	2.80	3.66	6.46
5" dia.	"	4.18	3.92	8.10
6" dia.	"	5.54	3.92	9.46
Rectangular				
2" x 3"	LF	1.89	3.66	5.55
3" x 4"	"	2.70	3.43	6.13
4" x 4"	"	3.38	3.43	6.81
Gutters, stock units				
5" wide				
Plain	LF	1.82	6.11	7.93
Painted	"	1.98	6.11	8.09
6" wide				
Plain	LF	2.55	6.47	9.02
Painted	"	2.86	6.47	9.33

Skylights	UNIT	MAT.	INST.	TOTAL
07810.10		**Plastic Skylights**		
Single thickness, not including mounting curb				
2' x 4'	EA	410	69.00	479
4' x 4'	"	550	92.00	642
5' x 5'	"	730	140	870
6' x 8'	"	1,560	180	1,740
Double thickness, not including mounting curb				
2' x 4'	EA	540	69.00	609
4' x 4'	"	670	92.00	762
5' x 5'	"	990	140	1,130
6' x 8'	"	1,730	180	1,910

Joint Sealants	UNIT	MAT.	INST.	TOTAL
07920.10		**Caulking**		
Caulk exterior, two component				
1/4 x 1/2	LF	0.43	2.88	3.31
3/8 x 1/2	"	0.66	3.20	3.86
1/2 x 1/2	"	0.90	3.60	4.50
Caulk interior, single component				
1/4 x 1/2	LF	0.29	2.74	3.03
3/8 x 1/2	"	0.41	3.03	3.44
1/2 x 1/2	"	0.54	3.38	3.92

Metal Doors & Transoms	UNIT	MAT.	INST.	TOTAL
08110.10 **Metal Doors**				
Flush hollow metal, std. duty, 20 ga., 1-3/8" thick				
2-6 x 6-8	EA	350	64.00	414
2-8 x 6-8	"	390	64.00	454
3-0 x 6-8	"	420	64.00	484
1-3/4" thick				
2-6 x 6-8	EA	410	64.00	474
2-8 x 6-8	"	510	64.00	574
3-0 x 6-8	"	470	64.00	534
2-6 x 7-0	"	450	64.00	514
2-8 x 7-0	"	470	64.00	534
3-0 x 7-0	"	500	64.00	564
Heavy duty, 20 ga., unrated, 1-3/4"				
2-8 x 6-8	EA	450	64.00	514
3-0 x 6-8	"	490	64.00	554
2-8 x 7-0	"	520	64.00	584
3-0 x 7-0	"	500	64.00	564
3-4 x 7-0	"	520	64.00	584
18 ga., 1-3/4", unrated door				
2-0 x 7-0	EA	480	64.00	544
2-4 x 7-0	"	480	64.00	544
2-6 x 7-0	"	480	64.00	544
2-8 x 7-0	"	530	64.00	594
3-0 x 7-0	"	540	64.00	604
3-4 x 7-0	"	560	64.00	624
2", unrated door				
2-0 x 7-0	EA	530	72.00	602
2-4 x 7-0	"	530	72.00	602
2-6 x 7-0	"	530	72.00	602
2-8 x 7-0	"	580	72.00	652
3-0 x 7-0	"	600	72.00	672
3-4 x 7-0	"	610	72.00	682
08110.40 **Metal Door Frames**				
Hollow metal, stock, 18 ga., 4-3/4" x 1-3/4"				
2-0 x 7-0	EA	160	72.00	232
2-4 x 7-0	"	190	72.00	262
2-6 x 7-0	"	190	72.00	262
2-8 x 7-0	"	190	72.00	262
3-0 x 7-0	"	190	72.00	262
4-0 x 7-0	"	210	96.00	306
5-0 x 7-0	"	220	96.00	316
6-0 x 7-0	"	260	96.00	356
16 ga., 6-3/4" x 1-3/4"				
2-0 x 7-0	EA	190	79.00	269
2-4 x 7-0	"	180	79.00	259
2-6 x 7-0	"	180	79.00	259
2-8 x 7-0	"	180	79.00	259
3-0 x 7-0	"	200	79.00	279
4-0 x 7-0	"	230	110	340
6-0 x 7-0	"	260	110	370

Wood And Plastic	UNIT	MAT.	INST.	TOTAL
08210.10		**Wood Doors**		
Solid core, 1-3/8" thick				
Birch faced				
2-4 x 7-0	EA	180	72.00	252
2-8 x 7-0	"	180	72.00	252
3-0 x 7-0	"	180	72.00	252
3-4 x 7-0	"	370	72.00	442
2-4 x 6-8	"	180	72.00	252
2-6 x 6-8	"	180	72.00	252
2-8 x 6-8	"	180	72.00	252
3-0 x 6-8	"	180	72.00	252
Lauan faced				
2-4 x 6-8	EA	160	72.00	232
2-8 x 6-8	"	170	72.00	242
3-0 x 6-8	"	180	72.00	252
3-4 x 6-8	"	190	72.00	262
Tempered hardboard faced				
2-4 x 7-0	EA	200	72.00	272
2-8 x 7-0	"	210	72.00	282
3-0 x 7-0	"	240	72.00	312
3-4 x 7-0	"	250	72.00	322
Hollow core, 1-3/8" thick				
Birch faced				
2-4 x 7-0	EA	160	72.00	232
2-8 x 7-0	"	160	72.00	232
3-0 x 7-0	"	170	72.00	242
3-4 x 7-0	"	180	72.00	252
Lauan faced				
2-4 x 6-8	EA	69.00	72.00	141
2-6 x 6-8	"	74.00	72.00	146
2-8 x 6-8	"	93.00	72.00	165
3-0 x 6-8	"	97.00	72.00	169
3-4 x 6-8	"	110	72.00	182
Tempered hardboard faced				
2-4 x 7-0	EA	84.00	72.00	156
2-6 x 7-0	"	90.00	72.00	162
2-8 x 7-0	"	100	72.00	172
3-0 x 7-0	"	110	72.00	182
3-4 x 7-0	"	120	72.00	192
Solid core, 1-3/4" thick				
Birch faced				
2-4 x 7-0	EA	280	72.00	352
2-6 x 7-0	"	280	72.00	352
2-8 x 7-0	"	290	72.00	362
3-0 x 7-0	"	270	72.00	342
3-4 x 7-0	"	280	72.00	352
Lauan faced				
2-4 x 7-0	EA	190	72.00	262
2-6 x 7-0	"	220	72.00	292
2-8 x 7-0	"	230	72.00	302
3-4 x 7-0	"	240	72.00	312
3-0 x 7-0	"	260	72.00	332
Tempered hardboard faced				
2-4 x 7-0	EA	250	72.00	322

Wood And Plastic	UNIT	MAT.	INST.	TOTAL
08210.10 **Wood Doors** *(Cont.)*				
2-6 x 7-0	EA	280	72.00	352
2-8 x 7-0	"	300	72.00	372
3-0 x 7-0	"	320	72.00	392
3-4 x 7-0	"	340	72.00	412
Hollow core, 1-3/4" thick				
Birch faced				
2-4 x 7-0	EA	190	72.00	262
2-6 x 7-0	"	190	72.00	262
2-8 x 7-0	"	200	72.00	272
3-0 x 7-0	"	200	72.00	272
3-4 x 7-0	"	220	72.00	292
Lauan faced				
2-4 x 6-8	EA	110	72.00	182
2-6 x 6-8	"	130	72.00	202
2-8 x 6-8	"	110	72.00	182
3-0 x 6-8	"	120	72.00	192
3-4 x 6-8	"	120	72.00	192
Tempered hardboard				
2-4 x 7-0	EA	100	72.00	172
2-6 x 7-0	"	110	72.00	182
2-8 x 7-0	"	110	72.00	182
3-0 x 7-0	"	120	72.00	192
3-4 x 7-0	"	130	72.00	202
Add-on, louver	"	35.00	58.00	93.00
Glass	"	110	58.00	168
Exterior doors, 3-0 x 7-0 x 2-1/2", solid core				
Carved				
One face	EA	1,460	140	1,600
Two faces	"	2,020	140	2,160
Closet doors, 1-3/4" thick				
Bi-fold or bi-passing, includes frame and trim				
Paneled				
4-0 x 6-8	EA	510	96.00	606
6-0 x 6-8	"	580	96.00	676
Louvered				
4-0 x 6-8	EA	350	96.00	446
6-0 x 6-8	"	420	96.00	516
Flush				
4-0 x 6-8	EA	260	96.00	356
6-0 x 6-8	"	330	96.00	426
Primed				
4-0 x 6-8	EA	280	96.00	376
6-0 x 6-8	"	310	96.00	406
08210.90 **Wood Frames**				
Frame, interior, pine				
2-6 x 6-8	EA	100	82.00	182
2-8 x 6-8	"	110	82.00	192
3-0 x 6-8	"	120	82.00	202
5-0 x 6-8	"	120	82.00	202
6-0 x 6-8	"	130	82.00	212
2-6 x 7-0	"	120	82.00	202
2-8 x 7-0	"	140	82.00	222

Wood And Plastic	UNIT	MAT.	INST.	TOTAL
08210.90 Wood Frames *(Cont.)*				
3-0 x 7-0	EA	140	82.00	222
5-0 x 7-0	"	150	120	270
6-0 x 7-0	"	160	120	280
Exterior, custom, with threshold, including trim				
Walnut				
3-0 x 7-0	EA	420	140	560
6-0 x 7-0	"	480	140	620
Oak				
3-0 x 7-0	EA	380	140	520
6-0 x 7-0	"	430	140	570
Pine				
2-4 x 7-0	EA	160	120	280
2-6 x 7-0	"	160	120	280
2-8 x 7-0	"	170	120	290
3-0 x 7-0	"	180	120	300
3-4 x 7-0	"	200	120	320
6-0 x 7-0	"	210	190	400
08300.10 Special Doors				
Sliding glass doors				
Tempered plate glass, 1/4" thick				
6' wide				
Economy grade	EA	1,180	190	1,370
Premium grade	"	1,350	190	1,540
12' wide				
Economy grade	EA	1,650	290	1,940
Premium grade	"	2,480	290	2,770
Insulating glass, 5/8" thick				
6' wide				
Economy grade	EA	1,450	190	1,640
Premium grade	"	1,860	190	2,050
12' wide				
Economy grade	EA	1,800	290	2,090
Premium grade	"	2,890	290	3,180
1" thick				
6' wide				
Economy grade	EA	1,820	190	2,010
Premium grade	"	2,100	190	2,290
12' wide				
Economy grade	EA	2,830	290	3,120
Premium grade	"	4,140	290	4,430
Added costs				
Custom quality, add to material, 30%				
Tempered glass, 6' wide, add	SF			5.08
Residential storm door				
Minimum	EA	180	96.00	276
Average	"	240	96.00	336
Maximum	"	530	140	670

08 DOORS AND WINDOWS

Wood And Plastic	UNIT	MAT.	INST.	TOTAL
08520.10	**Aluminum Windows**			
Jalousie				
3-0 x 4-0	EA	390	79.00	469
3-0 x 5-0	"	450	79.00	529
Fixed window				
6 sf to 8 sf	SF	19.25	9.08	28.33
12 sf to 16 sf	"	17.00	7.06	24.06
Projecting window				
6 sf to 8 sf	SF	42.50	16.00	58.50
12 sf to 16 sf	"	38.25	10.50	48.75
Horizontal sliding				
6 sf to 8 sf	SF	27.75	7.94	35.69
12 sf to 16 sf	"	25.50	6.35	31.85
Double hung				
6 sf to 8 sf	SF	38.25	12.75	51.00
10 sf to 12 sf	"	34.00	10.50	44.50
Storm window, 0.5 cfm, up to				
60 u.i. (united inches)	EA	89.00	31.75	121
70 u.i.	"	92.00	31.75	124
80 u.i.	"	100	31.75	132
90 u.i.	"	100	35.25	135
100 u.i.	"	110	35.25	145
2.0 cfm, up to				
60 u.i.	EA	110	31.75	142
70 u.i.	"	120	31.75	152
80 u.i.	"	120	31.75	152
90 u.i.	"	130	35.25	165
100 u.i.	"	130	35.25	165

Wood And Plastic	UNIT	MAT.	INST.	TOTAL
08600.10	**Wood Windows**			
Double hung				
24" x 36"				
Minimum	EA	240	58.00	298
Average	"	350	72.00	422
Maximum	"	470	96.00	566
24" x 48"				
Minimum	EA	280	58.00	338
Average	"	410	72.00	482
Maximum	"	570	96.00	666
30" x 48"				
Minimum	EA	290	64.00	354
Average	"	410	82.00	492
Maximum	"	590	120	710
30" x 60"				
Minimum	EA	320	64.00	384

Wood And Plastic	UNIT	MAT.	INST.	TOTAL
08600.10 **Wood Windows** *(Cont.)*				
Average	EA	510	82.00	592
Maximum	"	630	120	750
Casement				
1 leaf, 22" x 38" high				
Minimum	EA	350	58.00	408
Average	"	430	72.00	502
Maximum	"	500	96.00	596
2 leaf, 50" x 50" high				
Minimum	EA	940	72.00	1,012
Average	"	1,230	96.00	1,326
Maximum	"	1,410	140	1,550
3 leaf, 71" x 62" high				
Minimum	EA	1,550	72.00	1,622
Average	"	1,580	96.00	1,676
Maximum	"	1,890	140	2,030
4 leaf, 95" x 75" high				
Minimum	EA	2,060	82.00	2,142
Average	"	2,350	120	2,470
Maximum	"	3,000	190	3,190
5 leaf, 119" x 75" high				
Minimum	EA	2,670	82.00	2,752
Average	"	2,880	120	3,000
Maximum	"	3,680	190	3,870
Picture window, fixed glass, 54" x 54" high				
Minimum	EA	550	72.00	622
Average	"	620	82.00	702
Maximum	"	1,100	96.00	1,196
68" x 55" high				
Minimum	EA	990	72.00	1,062
Average	"	1,140	82.00	1,222
Maximum	"	1,490	96.00	1,586
Sliding, 40" x 31" high				
Minimum	EA	330	58.00	388
Average	"	500	72.00	572
Maximum	"	600	96.00	696
52" x 39" high				
Minimum	EA	410	72.00	482
Average	"	610	82.00	692
Maximum	"	650	96.00	746
64" x 72" high				
Minimum	EA	630	72.00	702
Average	"	1,010	96.00	1,106
Maximum	"	1,110	120	1,230
Awning windows				
34" x 21" high				
Minimum	EA	330	58.00	388
Average	"	380	72.00	452
Maximum	"	440	96.00	536
40" x 21" high				
Minimum	EA	390	64.00	454
Average	"	430	82.00	512
Maximum	"	480	120	600
48" x 27" high				

Wood And Plastic	UNIT	MAT.	INST.	TOTAL
08600.10 **Wood Windows** *(Cont.)*				
Minimum	EA	410	64.00	474
Average	"	490	82.00	572
Maximum	"	570	120	690
60" x 36" high				
Minimum	EA	430	72.00	502
Average	"	760	96.00	856
Maximum	"	860	120	980
Window frame, milled				
Minimum	LF	6.09	11.50	17.59
Average	"	6.80	14.50	21.30
Maximum	"	10.25	19.25	29.50

Hardware	UNIT	MAT.	INST.	TOTAL
08710.10 **Hinges**				
Hinges, material only				
3 x 3 butts, steel, interior, plain bearing	PAIR			20.75
4 x 4 butts, steel, standard	"			30.50
5 x 4-1/2 butts, bronze/s. steel, heavy duty	"			79.00
08710.20 **Locksets**				
Latchset, heavy duty				
Cylindrical	EA	190	36.00	226
Mortise	"	200	58.00	258
Lockset, heavy duty				
Cylindrical	EA	310	36.00	346
Mortise	"	350	58.00	408
Lockset				
Privacy (bath or bedroom)	EA	220	48.00	268
Entry lock	"	240	48.00	288
08710.30 **Closers**				
Door closers				
Standard	EA	240	72.00	312
Heavy duty	"	280	72.00	352
08710.40 **Door Trim**				
Panic device				
Mortise	EA	810	140	950
Vertical rod	"	1,230	140	1,370
Labeled, rim type	"	850	140	990
Mortise	"	1,110	140	1,250
Vertical rod	"	1,180	140	1,320

08 DOORS AND WINDOWS

Hardware	UNIT	MAT.	INST.	TOTAL
08710.60 **Weatherstripping**				
Weatherstrip, head and jamb, metal strip, neoprene bulb				
Standard duty	LF	5.19	3.20	8.39
Heavy duty	"	5.77	3.60	9.37
Spring type				
Metal doors	EA	55.00	140	195
Wood doors	"	55.00	190	245
Sponge type with adhesive backing	"	51.00	58.00	109
Thresholds				
Bronze	LF	53.00	14.50	67.50
Aluminum				
Plain	LF	38.50	14.50	53.00
Vinyl insert	"	39.25	14.50	53.75
Aluminum with grit	"	37.50	14.50	52.00
Steel				
Plain	LF	29.75	14.50	44.25
Interlocking	"	39.50	48.00	87.50

Glazing	UNIT	MAT.	INST.	TOTAL
08810.10 **Glass Glazing**				
Sheet glass, 1/8" thick	SF	8.91	3.53	12.44
Plate glass, bronze or grey, 1/4" thick	"	13.00	5.77	18.77
Clear	"	10.25	5.77	16.02
Polished	"	12.00	5.77	17.77
Plexiglass				
1/8" thick	SF	5.73	5.77	11.50
1/4" thick	"	10.25	3.53	13.78
Float glass, clear				
3/16" thick	SF	6.93	5.29	12.22
1/4" thick	"	7.07	5.77	12.84
3/8" thick	"	14.25	7.94	22.19
Tinted glass, polished plate, twin ground				
3/16" thick	SF	9.55	5.29	14.84
1/4" thick	"	9.55	5.77	15.32
3/8" thick	"	15.25	7.94	23.19
Insulating glass, two lites, clear float glass				
1/2" thick	SF	13.75	10.50	24.25
5/8" thick	"	16.00	12.75	28.75
3/4" thick	"	17.50	16.00	33.50
7/8" thick	"	18.50	18.25	36.75
1" thick	"	24.75	21.25	46.00
Glass seal edge				
3/8" thick	SF	11.75	10.50	22.25
Tinted glass				
1/2" thick	SF	23.75	10.50	34.25
1" thick	"	25.50	21.25	46.75

Glazing	UNIT	MAT.	INST.	TOTAL
08810.10 — Glass Glazing *(Cont.)*				
Tempered, clear				
1" thick	SF	46.50	21.25	67.75
Plate mirror glass				
1/4" thick				
15 sf	SF	11.75	6.35	18.10
Over 15 sf	"	10.75	5.77	16.52

Glazed Curtain Walls	UNIT	MAT.	INST.	TOTAL
08910.10 — Glazed Curtain Walls				
Curtain wall, aluminum system, framing sections				
2" x 3"				
Jamb	LF	19.50	5.29	24.79
Horizontal	"	19.75	5.29	25.04
Mullion	"	26.50	5.29	31.79
2" x 4"				
Jamb	LF	26.50	7.94	34.44
Horizontal	"	27.25	7.94	35.19
Mullion	"	26.50	7.94	34.44
3" x 5-1/2"				
Jamb	LF	35.00	7.94	42.94
Horizontal	"	38.75	7.94	46.69
Mullion	"	35.25	7.94	43.19
4" corner mullion	"	46.75	10.50	57.25
Coping sections				
1/8" x 8"	LF	44.25	10.50	54.75
1/8" x 9"	"	44.50	10.50	55.00
1/8" x 12-1/2"	"	45.75	12.75	58.50
Sill section				
1/8" x 6"	LF	43.75	6.35	50.10
1/8" x 7"	"	44.25	6.35	50.60
1/8" x 8-1/2"	"	45.00	6.35	51.35
Column covers, aluminum				
1/8" x 26"	LF	43.75	16.00	59.75
1/8" x 34"	"	48.50	16.75	65.25
1/8" x 38"	"	49.00	16.75	65.75
Doors				
Aluminum framed, standard hardware				
Narrow stile				
2-6 x 7-0	EA	810	320	1,130
3-0 x 7-0	"	810	320	1,130
3-6 x 7-0	"	840	320	1,160

Support Systems	UNIT	MAT.	INST.	TOTAL
09110.10 **Metal Studs**				
Studs, non load bearing, galvanized				
2-1/2", 20 ga.				
12" o.c.	SF	0.68	1.20	1.88
16" o.c.	"	0.52	0.96	1.48
25 ga.				
12" o.c.	SF	0.46	1.20	1.66
16" o.c.	"	0.36	0.96	1.32
24" o.c.	"	0.28	0.80	1.08
3-5/8", 20 ga.				
12" o.c.	SF	0.81	1.44	2.25
16" o.c.	"	0.62	1.15	1.77
24" o.c.	"	0.47	0.96	1.43
25 ga.				
12" o.c.	SF	0.53	1.44	1.97
16" o.c.	"	0.44	1.15	1.59
24" o.c.	"	0.33	0.96	1.29
4", 20 ga.				
12" o.c.	SF	0.89	1.44	2.33
16" o.c.	"	0.68	1.15	1.83
24" o.c.	"	0.52	0.96	1.48
25 ga.				
12" o.c.	SF	0.60	1.44	2.04
16" o.c.	"	0.47	1.15	1.62
24" o.c.	"	0.35	0.96	1.31
6", 20 ga.				
12" o.c.	SF	1.14	1.80	2.94
16" o.c.	"	0.83	1.44	2.27
24" o.c.	"	0.68	1.20	1.88
25 ga.				
12" o.c.	SF	0.73	1.80	2.53
16" o.c.	"	0.58	1.44	2.02
24" o.c.	"	0.44	1.20	1.64
Load bearing studs, galvanized				
3-5/8", 16 ga.				
12" o.c.	SF	1.47	1.44	2.91
16" o.c.	"	1.36	1.15	2.51
18 ga.				
12" o.c.	SF	1.15	0.96	2.11
16" o.c.	"	1.05	1.15	2.20
4", 16 ga.				
12" o.c.	SF	1.55	1.44	2.99
16" o.c.	"	1.40	1.15	2.55
6", 16 ga.				
12" o.c.	SF	1.98	1.80	3.78
16" o.c.	"	1.78	1.44	3.22
Furring				
On beams and columns				
7/8" channel	LF	0.52	3.84	4.36
1-1/2" channel	"	0.62	4.43	5.05
On ceilings				
3/4" furring channels				
12" o.c.	SF	0.37	2.40	2.77
16" o.c.	"	0.29	2.30	2.59

Support Systems	UNIT	MAT.	INST.	TOTAL
09110.10 **Metal Studs** *(Cont.)*				
24" o.c.	SF	0.20	2.05	2.25
1-1/2" furring channels				
12" o.c.	SF	0.62	2.61	3.23
16" o.c.	"	0.47	2.40	2.87
24" o.c.	"	0.31	2.21	2.52
On walls				
3/4" furring channels				
12" o.c.	SF	0.37	1.92	2.29
16" o.c.	"	0.29	1.80	2.09
24" o.c.	"	0.20	1.69	1.89
1-1/2" furring channels				
12" o.c.	SF	0.62	2.05	2.67
16" o.c.	"	0.47	1.92	2.39
24" o.c.	"	0.31	1.80	2.11

Lath And Plaster	UNIT	MAT.	INST.	TOTAL
09205.10 **Gypsum Lath**				
Gypsum lath, 1/2" thick				
Clipped	SY	5.76	3.20	8.96
Nailed	"	5.76	3.60	9.36
09205.20 **Metal Lath**				
Diamond expanded, galvanized				
2.5 lb., on walls				
Nailed	SY	3.83	7.20	11.03
Wired	"	3.83	8.22	12.05
On ceilings				
Nailed	SY	3.83	8.22	12.05
Wired	"	3.83	9.60	13.43
3.4 lb., on walls				
Nailed	SY	5.21	7.20	12.41
Wired	"	5.21	8.22	13.43
On ceilings				
Nailed	SY	5.21	8.22	13.43
Wired	"	5.21	9.60	14.81
Flat rib				
2.75 lb., on walls				
Nailed	SY	3.63	7.20	10.83
Wired	"	3.63	8.22	11.85
On ceilings				
Nailed	SY	3.63	8.22	11.85
Wired	"	3.63	9.60	13.23
3.4 lb., on walls				
Nailed	SY	4.36	7.20	11.56
Wired	"	4.36	8.22	12.58

Lath And Plaster	UNIT	MAT.	INST.	TOTAL

09205.20 — Metal Lath *(Cont.)*

	UNIT	MAT.	INST.	TOTAL
On ceilings				
Nailed	SY	4.36	8.22	12.58
Wired	"	4.36	9.60	13.96
Stucco lath				
1.8 lb.	SY	4.51	7.20	11.71
3.6 lb.	"	5.06	7.20	12.26
Paper backed				
Minimum	SY	3.50	5.76	9.26
Maximum	"	5.65	8.22	13.87

09205.60 — Plaster Accessories

	UNIT	MAT.	INST.	TOTAL
Expansion joint, 3/4", 26 ga., galv.	LF	1.48	1.44	2.92
Plaster corner beads, 3/4", galvanized	"	0.41	1.64	2.05
Casing bead, expanded flange, galvanized	"	0.56	1.44	2.00
Expanded wing, 1-1/4" wide, galvanized	"	0.66	1.44	2.10
Joint clips for lath	EA	0.17	0.28	0.45
Metal base, galvanized, 2-1/2" high	LF	0.75	1.92	2.67
Stud clips for gypsum lath	EA	0.17	0.28	0.45
Tie wire galvanized, 18 ga., 25 lb. hank	"			47.00
Sound deadening board, 1/4"	SF	0.31	0.96	1.27

09210.10 — Plaster

	UNIT	MAT.	INST.	TOTAL
Gypsum plaster, trowel finish, 2 coats				
Ceilings	SY	4.30	16.75	21.05
Walls	"	4.30	15.75	20.05
3 coats				
Ceilings	SY	5.96	23.25	29.21
Walls	"	5.96	20.75	26.71
Vermiculite plaster				
2 coats				
Ceilings	SY	4.89	25.50	30.39
Walls	"	4.89	23.25	28.14
3 coats				
Ceilings	SY	7.68	31.50	39.18
Walls	"	7.68	28.25	35.93
Keenes cement plaster				
2 coats				
Ceilings	SY	2.09	20.75	22.84
Walls	"	2.09	17.75	19.84
3 coats				
Ceilings	SY	2.14	23.25	25.39
Walls	"	2.14	20.75	22.89
On columns, add to installation, 50%	"			
Chases, fascia, and soffits, add to installation, 50%	"			
Beams, add to installation, 50%	"			
Patch holes, average size holes				
1 sf to 5 sf				
Minimum	SF	2.47	8.94	11.41
Average	"	2.47	10.75	13.22
Maximum	"	2.47	13.50	15.97
Over 5 sf				
Minimum	SF	2.47	5.36	7.83

Lath And Plaster	UNIT	MAT.	INST.	TOTAL
09210.10 Plaster *(Cont.)*				
Average	SF	2.47	7.66	10.13
Maximum	"	2.47	8.94	11.41
Patch cracks				
Minimum	SF	2.47	1.78	4.25
Average	"	2.47	2.68	5.15
Maximum	"	2.47	5.36	7.83
09220.10 Portland Cement Plaster				
Stucco, portland, gray, 3 coat, 1" thick				
Sand finish	SY	8.03	23.25	31.28
Trowel finish	"	8.03	24.50	32.53
White cement				
Sand finish	SY	9.17	24.50	33.67
Trowel finish	"	9.17	26.75	35.92
Scratch coat				
For ceramic tile	SY	2.91	5.36	8.27
For quarry tile	"	2.91	5.36	8.27
Portland cement plaster				
2 coats, 1/2"	SY	5.78	10.75	16.53
3 coats, 7/8"	"	6.90	13.50	20.40
09250.10 Gypsum Board				
Drywall, plasterboard, 3/8" clipped to				
Metal furred ceiling	SF	0.41	0.64	1.05
Columns and beams	"	0.41	1.44	1.85
Walls	"	0.41	0.57	0.98
Nailed or screwed to				
Wood or metal framed ceiling	SF	0.41	0.57	0.98
Columns and beams	"	0.41	1.28	1.69
Walls	"	0.41	0.52	0.93
1/2", clipped to				
Metal furred ceiling	SF	0.42	0.64	1.06
Columns and beams	"	0.38	1.44	1.82
Walls	"	0.38	0.57	0.95
Nailed or screwed to				
Wood or metal framed ceiling	SF	0.38	0.57	0.95
Columns and beams	"	0.38	1.28	1.66
Walls	"	0.38	0.52	0.90
5/8", clipped to				
Metal furred ceiling	SF	0.42	0.72	1.14
Columns and beams	"	0.42	1.60	2.02
Walls	"	0.42	0.64	1.06
Nailed or screwed to				
Wood or metal framed ceiling	SF	0.42	0.72	1.14
Columns and beams	"	0.42	1.60	2.02
Walls	"	0.42	0.64	1.06
Vinyl faced, clipped to metal studs				
1/2"	SF	1.19	0.72	1.91
5/8"	"	1.13	0.72	1.85
Add for				
Fire resistant	SF			0.12
Water resistant	"			0.19
Water and fire resistant	"			0.24

Lath And Plaster	UNIT	MAT.	INST.	TOTAL
09250.10 **Gypsum Board** *(Cont.)*				
Taping and finishing joints				
Minimum	SF	0.04	0.38	0.42
Average	"	0.07	0.48	0.55
Maximum	"	0.10	0.57	0.67
Casing bead				
Minimum	LF	0.16	1.64	1.80
Average	"	0.18	1.92	2.10
Maximum	"	0.22	2.88	3.10
Corner bead				
Minimum	LF	0.18	1.64	1.82
Average	"	0.22	1.92	2.14
Maximum	"	0.27	2.88	3.15

Tile	UNIT	MAT.	INST.	TOTAL
09310.10 **Ceramic Tile**				
Glazed wall tile, 4-1/4" x 4-1/4"				
Minimum	SF	2.32	3.92	6.24
Average	"	3.68	4.57	8.25
Maximum	"	13.25	5.49	18.74
Base, 4-1/4" high				
Minimum	LF	4.47	6.86	11.33
Average	"	5.20	6.86	12.06
Maximum	"	6.87	6.86	13.73
Unglazed floor tile				
Portland cem., cushion edge, face mtd				
1" x 1"	SF	8.87	4.99	13.86
2" x 2"	"	9.38	4.57	13.95
4" x 4"	"	8.73	4.57	13.30
6" x 6"	"	3.12	3.92	7.04
12" x 12"	"	2.75	3.43	6.18
16" x 16"	"	2.38	3.05	5.43
18" x 18"	"	2.31	2.74	5.05
Adhesive bed, with white grout				
1" x 1"	SF	7.38	4.99	12.37
2" x 2"	"	7.81	4.57	12.38
4" x 4"	"	7.81	4.57	12.38
6" x 6"	"	2.60	3.92	6.52
12" x 12"	"	2.28	3.43	5.71
16" x 16"	"	1.98	3.05	5.03
18" x 18"	"	1.92	2.74	4.66
Organic adhesive bed, thin set, back mounted				
1" x 1"	SF	7.38	4.99	12.37
2" x 2"	"	8.59	4.57	13.16
For group 2 colors, add to material, 10%				
For group 3 colors, add to material, 20%				

Tile	UNIT	MAT.	INST.	TOTAL
09310.10 Ceramic Tile *(Cont.)*				
For abrasive surface, add to material, 25%				
Porcelain floor tile				
1" x 1"	SF	9.90	4.99	14.89
2" x 2"	"	9.05	4.77	13.82
4" x 4"	"	8.41	4.57	12.98
6" x 6"	"	3.02	3.92	6.94
12" x 12"	"	2.72	3.43	6.15
16" x 16"	"	2.16	3.05	5.21
18" x 18"	"	2.04	2.74	4.78
Unglazed wall tile				
Organic adhesive, face mounted cushion edge				
1" x 1"				
Minimum	SF	4.65	4.57	9.22
Average	"	6.09	4.99	11.08
Maximum	"	9.09	5.49	14.58
2" x 2"				
Minimum	SF	5.37	4.22	9.59
Average	"	6.09	4.57	10.66
Maximum	"	9.96	4.99	14.95
Back mounted				
1" x 1"				
Minimum	SF	4.65	4.57	9.22
Average	"	6.09	4.99	11.08
Maximum	"	9.09	5.49	14.58
2" x 2"				
Minimum	SF	5.37	4.22	9.59
Average	"	6.09	4.57	10.66
Maximum	"	9.96	4.99	14.95
For glazed finish, add to material, 25%				
For glazed mosaic, add to material, 100%				
For metallic colors, add to material, 125%				
For exterior wall use, add to total, 25%				
For exterior soffit, add to total, 25%				
For portland cement bed, add to total, 25%				
For dry set portland cement bed, add to total, 10%				
Ceramic accessories				
Towel bar, 24" long				
Minimum	EA	18.00	22.00	40.00
Average	"	22.25	27.50	49.75
Maximum	"	59.00	36.50	95.50
Soap dish				
Minimum	EA	8.47	36.50	44.97
Average	"	11.50	45.75	57.25
Maximum	"	30.25	55.00	85.25
09330.10 Quarry Tile				
Floor				
4 x 4 x 1/2"	SF	6.57	7.32	13.89
6 x 6 x 1/2"	"	6.44	6.86	13.30
6 x 6 x 3/4"	"	7.99	6.86	14.85
12 x 12x 3/4"	"	11.25	6.10	17.35
16x1 6 x 3/4"	"	7.83	5.49	13.32
18 x 18 x 3/4"	"	5.52	4.57	10.09

Tile	UNIT	MAT.	INST.	TOTAL
09330.10 **Quarry Tile** *(Cont.)*				
Medallion				
36" dia.	EA	350	140	490
48" dia.	"	410	140	550
Wall, applied to 3/4" portland cement bed				
4 x 4 x 1/2"	SF	5.84	11.00	16.84
6 x 6 x 3/4"	"	6.53	9.15	15.68
Cove base				
5 x 6 x 1/2" straight top	LF	6.66	9.15	15.81
6 x 6 x 3/4" round top	"	6.18	9.15	15.33
Moldings				
2 x 12	LF	10.50	5.49	15.99
4 x 12	"	16.50	5.49	21.99
Stair treads 6 x 6 x 3/4"	"	9.13	13.75	22.88
Window sill 6 x 8 x 3/4"	"	8.33	11.00	19.33
For abrasive surface, add to material, 25%				
09410.10 **Terrazzo**				
Floors on concrete, 1-3/4" thick, 5/8" topping				
Gray cement	SF	5.67	7.66	13.33
White cement	"	5.92	7.66	13.58
Sand cushion, 3" thick, 5/8" top, 1/4"				
Gray cement	SF	6.69	8.94	15.63
White cement	"	6.96	8.94	15.90
Monolithic terrazzo, 3-1/2" base slab, 5/8" topping	"	4.75	6.70	11.45
Terrazzo wainscot, cast-in-place, 1/2" thick	"	5.77	13.50	19.27
Base, cast in place, terrazzo cove type, 6" high	LF	7.26	7.66	14.92
Curb, cast in place, 6" wide x 6" high, polished top	"	6.60	26.75	33.35
For venetian type terrazzo, add to material, 10%				
For abrasive heavy duty terrazzo, add to material, 15%				
Divider strips				
Zinc	LF			1.43
Brass	"			2.66
Stairs, cast-in-place, topping on concrete or metal				
1-1/2" thick treads, 12" wide	LF	5.61	26.75	32.36
Combined tread and riser	"	8.42	67.00	75.42
Precast terrazzo, thin set				
Terrazzo tiles, non-slip surface				
9" x 9" x 1" thick	SF	16.00	7.66	23.66
12" x 12"				
1" thick	SF	17.25	7.15	24.40
1-1/2" thick	"	18.00	7.66	25.66
18" x 18" x 1-1/2" thick	"	23.50	7.66	31.16
24" x 24" x 1-1/2" thick	"	30.25	6.31	36.56
For white cement, add to material, 10%				
For venetian type terrazzo, add to material, 25%				
Terrazzo wainscot				
12" x 12" x 1" thick	SF	8.83	13.50	22.33
18" x 18" x 1-1/2" thick	"	14.50	15.25	29.75
Base				
6" high				
Straight	LF	12.75	4.12	16.87
Coved	"	15.00	4.12	19.12
8" high				

Tile	UNIT	MAT.	INST.	TOTAL
09410.10 Terrazzo *(Cont.)*				
Straight	LF	14.25	4.47	18.72
Coved	"	16.75	4.47	21.22
Terrazzo curbs				
8" wide x 8" high	LF	33.00	21.50	54.50
6" wide x 6" high	"	29.75	17.75	47.50
Precast terrazzo stair treads, 12" wide				
1-1/2" thick				
Diamond pattern	LF	39.75	9.75	49.50
Non-slip surface	"	41.75	9.75	51.50
2" thick				
Diamond pattern	LF	41.75	9.75	51.50
Non-slip surface	"	44.00	10.75	54.75
Stair risers, 1" thick to 6" high				
Straight sections	LF	13.25	5.36	18.61
Cove sections	"	15.75	5.36	21.11
Combined tread and riser				
Straight sections				
1-1/2" tread, 3/4" riser	LF	57.00	15.25	72.25
3" tread, 1" riser	"	69.00	15.25	84.25
Curved sections				
2" tread, 1" riser	LF	73.00	17.75	90.75
3" tread, 1" riser	"	76.00	17.75	93.75
Stair stringers, notched for treads and risers				
1" thick	LF	34.50	13.50	48.00
2" thick	"	36.00	17.75	53.75
Landings, structural, nonslip				
1-1/2" thick	SF	32.75	8.94	41.69
3" thick	"	45.75	10.75	56.50

Acoustical Treatment	UNIT	MAT.	INST.	TOTAL
09510.10 Ceilings And Walls				
Acoustical panels, suspension system not included				
Fiberglass panels				
5/8" thick				
2' x 2'	SF	1.61	0.82	2.43
2' x 4'	"	1.34	0.64	1.98
3/4" thick				
2' x 2'	SF	2.14	0.82	2.96
2' x 4'	"	2.07	0.64	2.71
Glass cloth faced fiberglass panels				
3/4" thick	SF	3.05	0.96	4.01
1" thick	"	3.41	0.96	4.37
Mineral fiber panels				
5/8" thick				
2' x 2'	SF	1.37	0.82	2.19

Acoustical Treatment	UNIT	MAT.	INST.	TOTAL
09510.10 **Ceilings And Walls** *(Cont.)*				
2' x 4'	SF	1.37	0.64	2.01
3/4" thick				
2' x 2'	SF	2.14	0.82	2.96
2' x 4'	"	2.07	0.64	2.71
Wood fiber panels				
1/2" thick				
2' x 2'	SF	1.77	0.82	2.59
2' x 4'	"	1.77	0.64	2.41
5/8" thick				
2' x 2'	SF	2.03	0.82	2.85
2' x 4'	"	2.03	0.64	2.67
Acoustical tiles, suspension system not included				
Fiberglass tile, 12" x 12"				
5/8" thick	SF	2.00	1.04	3.04
3/4" thick	"	2.32	1.28	3.60
Glass cloth faced fiberglass tile				
3/4" thick	SF	3.73	1.28	5.01
3" thick	"	4.17	1.44	5.61
Mineral fiber tile, 12" x 12"				
5/8" thick				
Standard	SF	1.05	1.15	2.20
Vinyl faced	"	2.08	1.15	3.23
3/4" thick				
Standard	SF	1.53	1.15	2.68
Vinyl faced	"	2.66	1.15	3.81
Ceiling suspension systems				
T bar system				
2' x 4'	SF	1.15	0.57	1.72
2' x 2'	"	1.25	0.64	1.89
Concealed Z bar suspension system, 12" module	"	1.18	0.96	2.14
For 1-1/2" carrier channels, 4' o.c., add	"			0.38
Carrier channel for recessed light fixtures	"			0.69

Flooring	UNIT	MAT.	INST.	TOTAL
09550.10 **Wood Flooring**				
Wood strip flooring, unfinished				
Fir floor				
C and better				
Vertical grain	SF	3.52	1.92	5.44
Flat grain	"	4.40	1.92	6.32
Oak floor				
Minimum	SF	3.72	2.74	6.46
Average	"	5.13	2.74	7.87
Maximum	"	7.42	2.74	10.16
Maple floor				

Flooring	UNIT	MAT.	INST.	TOTAL
09550.10 **Wood Flooring** *(Cont.)*				
25/32" x 2-1/4"				
Minimum	SF	5.32	2.74	8.06
Maximum	"	7.54	2.74	10.28
33/32" x 3-1/4"				
Minimum	SF	7.42	2.74	10.16
Maximum	"	8.38	2.74	11.12
Added costs				
For factory finish, add to material, 10%				
For random width floor, add to total, 20%				
For simulated pegs, add to total, 10%				
Wood block industrial flooring				
Creosoted				
2" thick	SF	4.18	1.51	5.69
2-1/2" thick	"	4.34	1.80	6.14
3" thick	"	4.51	1.92	6.43
Parquet, 5/16", white oak				
Finished	SF	10.00	2.88	12.88
Unfinished	"	4.84	2.88	7.72
Gym floor, 2 ply felt, 25/32" maple, finished, in mastic	"	8.54	3.20	11.74
Over wood sleepers	"	8.69	3.60	12.29
Finishing, sand, fill, finish, and wax	"	0.66	1.44	2.10
Refinish sand, seal, and 2 coats of polyurethane	"	1.16	1.92	3.08
Clean and wax floors	"	0.24	0.28	0.52
09630.10 **Unit Masonry Flooring**				
Clay brick				
9 x 4-1/2 x 3" thick				
Glazed	SF	8.03	4.80	12.83
Unglazed	"	7.70	4.80	12.50
8 x 4 x 3/4" thick				
Glazed	SF	7.26	5.00	12.26
Unglazed	"	6.93	5.00	11.93
For herringbone pattern, add to labor, 15%				
09660.10 **Resilient Tile Flooring**				
Solid vinyl tile, 1/8" thick, 12" x 12"				
Marble patterns	SF	4.66	1.44	6.10
Solid colors	"	6.05	1.44	7.49
Travertine patterns	"	6.79	1.44	8.23
Conductive resilient flooring, vinyl tile				
1/8" thick, 12" x 12"	SF	7.38	1.64	9.02
09665.10 **Resilient Sheet Flooring**				
Vinyl sheet flooring				
Minimum	SF	3.83	0.57	4.40
Average	"	6.19	0.68	6.87
Maximum	"	10.50	0.96	11.46
Cove, to 6"	LF	2.28	1.15	3.43
Fluid applied resilient flooring				
Polyurethane, poured in place, 3/8" thick	SF	10.50	4.80	15.30
Vinyl sheet goods, backed				
0.070" thick	SF	3.90	0.72	4.62
0.093" thick	"	6.05	0.72	6.77

Flooring	UNIT	MAT.	INST.	TOTAL
09665.10 **Resilient Sheet Flooring** *(Cont.)*				
0.125" thick	SF	6.98	0.72	7.70
0.250" thick	"	8.03	0.72	8.75
09678.10 **Resilient Base And Accessories**				
Wall base, vinyl				
4" high	LF	1.28	1.92	3.20
6" high	"	1.74	1.92	3.66

Carpet	UNIT	MAT.	INST.	TOTAL
09682.10 **Carpet Padding**				
Carpet padding				
Foam rubber, waffle type, 0.3" thick	SY	6.74	2.88	9.62
Jute padding				
Minimum	SY	4.57	2.61	7.18
Average	"	5.95	2.88	8.83
Maximum	"	8.98	3.20	12.18
Sponge rubber cushion				
Minimum	SY	5.42	2.61	8.03
Average	"	7.22	2.88	10.10
Maximum	"	10.25	3.20	13.45
Urethane cushion, 3/8" thick				
Minimum	SY	5.42	2.61	8.03
Average	"	6.32	2.88	9.20
Maximum	"	8.24	3.20	11.44
09685.10 **Carpet**				
Carpet, acrylic				
24 oz., light traffic	SY	17.75	6.40	24.15
28 oz., medium traffic	"	21.25	6.40	27.65
Nylon				
15 oz., light traffic	SY	24.50	6.40	30.90
28 oz., medium traffic	"	32.00	6.40	38.40
Nylon				
28 oz., medium traffic	SY	30.50	6.40	36.90
35 oz., heavy traffic	"	37.25	6.40	43.65
Wool				
30 oz., medium traffic	SY	51.00	6.40	57.40
36 oz., medium traffic	"	53.00	6.40	59.40
42 oz., heavy traffic	"	71.00	6.40	77.40
Carpet tile				
Foam backed				
Minimum	SF	4.07	1.15	5.22
Average	"	4.71	1.28	5.99
Maximum	"	7.47	1.44	8.91
Tufted loop or shag				

Carpet	UNIT	MAT.	INST.	TOTAL
09685.10	**Carpet** *(Cont.)*			
Minimum	SF	4.41	1.15	5.56
Average	"	5.32	1.28	6.60
Maximum	"	8.56	1.44	10.00
Clean and vacuum carpet				
Minimum	SY	0.36	0.22	0.58
Average	"	0.56	0.38	0.94
Maximum	"	0.77	0.57	1.34

Paint	UNIT	MAT.	INST.	TOTAL
09905.10	**Painting Preparation**			
Dropcloths				
Minimum	SF	0.16	0.03	0.19
Average	"	0.18	0.04	0.22
Maximum	"	0.37	0.05	0.42
Masking				
Paper and tape				
Minimum	LF	0.04	0.48	0.52
Average	"	0.06	0.60	0.66
Maximum	"	0.07	0.80	0.87
Doors				
Minimum	EA	0.05	6.01	6.06
Average	"	0.06	8.01	8.07
Maximum	"	0.07	10.75	10.82
Windows				
Minimum	EA	0.05	6.01	6.06
Average	"	0.06	8.01	8.07
Maximum	"	0.07	10.75	10.82
Sanding				
Walls and flat surfaces				
Minimum	SF		0.32	0.32
Average	"		0.40	0.40
Maximum	"		0.48	0.48
Doors and windows				
Minimum	EA		8.01	8.01
Average	"		12.00	12.00
Maximum	"		16.00	16.00
Trim				
Minimum	LF		0.60	0.60
Average	"		0.80	0.80
Maximum	"		1.06	1.06
Puttying				
Minimum	SF	0.01	0.73	0.74
Average	"	0.02	0.96	0.98
Maximum	"	0.03	1.20	1.23

Paint	UNIT	MAT.	INST.	TOTAL

09910.05 — Ext. Painting, Sitework

	UNIT	MAT.	INST.	TOTAL
Concrete Block				
Roller				
First Coat				
Minimum	SF	0.19	0.24	0.43
Average	"	0.21	0.32	0.53
Maximum	"	0.22	0.48	0.70
Second Coat				
Minimum	SF	0.19	0.20	0.39
Average	"	0.21	0.26	0.47
Maximum	"	0.22	0.40	0.62
Spray				
First Coat				
Minimum	SF	0.15	0.13	0.28
Average	"	0.17	0.16	0.33
Maximum	"	0.18	0.18	0.36
Second Coat				
Minimum	SF	0.15	0.08	0.23
Average	"	0.17	0.10	0.27
Maximum	"	0.18	0.15	0.33
Fences, Chain Link				
Roller				
First Coat				
Minimum	SF	0.13	0.34	0.47
Average	"	0.14	0.40	0.54
Maximum	"	0.15	0.45	0.60
Second Coat				
Minimum	SF	0.13	0.20	0.33
Average	"	0.14	0.24	0.38
Maximum	"	0.15	0.30	0.45
Spray				
First Coat				
Minimum	SF	0.10	0.15	0.25
Average	"	0.11	0.17	0.28
Maximum	"	0.13	0.20	0.33
Second Coat				
Minimum	SF	0.10	0.11	0.21
Average	"	0.11	0.13	0.24
Maximum	"	0.13	0.15	0.28
Fences, Wood or Masonry				
Brush				
First Coat				
Minimum	SF	0.19	0.50	0.69
Average	"	0.21	0.60	0.81
Maximum	"	0.22	0.80	1.02
Second Coat				
Minimum	SF	0.19	0.30	0.49
Average	"	0.21	0.36	0.57
Maximum	"	0.22	0.48	0.70
Roller				
First Coat				
Minimum	SF	0.19	0.26	0.45
Average	"	0.21	0.32	0.53
Maximum	"	0.22	0.36	0.58

Paint	UNIT	MAT.	INST.	TOTAL
09910.05 **Ext. Painting, Sitework** *(Cont.)*				
Second Coat				
Minimum	SF	0.19	0.18	0.37
Average	"	0.21	0.22	0.43
Maximum	"	0.22	0.30	0.52
Spray				
First Coat				
Minimum	SF	0.15	0.17	0.32
Average	"	0.17	0.21	0.38
Maximum	"	0.18	0.30	0.48
Second Coat				
Minimum	SF	0.15	0.12	0.27
Average	"	0.17	0.15	0.32
Maximum	"	0.18	0.20	0.38
09910.15 **Ext. Painting, Buildings**				
Decks, Wood, Stained				
Brush				
First Coat				
Minimum	SF	0.15	0.24	0.39
Average	"	0.17	0.26	0.43
Maximum	"	0.18	0.30	0.48
Second Coat				
Minimum	SF	0.15	0.17	0.32
Average	"	0.17	0.18	0.35
Maximum	"	0.18	0.20	0.38
Roller				
First Coat				
Minimum	SF	0.15	0.17	0.32
Average	"	0.17	0.18	0.35
Maximum	"	0.18	0.20	0.38
Second Coat				
Minimum	SF	0.15	0.15	0.30
Average	"	0.17	0.16	0.33
Maximum	"	0.18	0.18	0.36
Spray				
First Coat				
Minimum	SF	0.13	0.15	0.28
Average	"	0.14	0.16	0.30
Maximum	"	0.15	0.18	0.33
Second Coat				
Minimum	SF	0.13	0.13	0.26
Average	"	0.14	0.14	0.28
Maximum	"	0.15	0.16	0.31
Doors, Wood				
Brush				
First Coat				
Minimum	SF	0.15	0.73	0.88
Average	"	0.17	0.96	1.13
Maximum	"	0.18	1.20	1.38
Second Coat				
Minimum	SF	0.15	0.60	0.75
Average	"	0.17	0.68	0.85
Maximum	"	0.18	0.80	0.98

Paint	UNIT	MAT.	INST.	TOTAL
09910.15 — **Ext. Painting, Buildings** *(Cont.)*				
Roller				
First Coat				
Minimum	SF	0.15	0.32	0.47
Average	"	0.17	0.40	0.57
Maximum	"	0.18	0.60	0.78
Second Coat				
Minimum	SF	0.15	0.24	0.39
Average	"	0.17	0.26	0.43
Maximum	"	0.18	0.40	0.58
Spray				
First Coat				
Minimum	SF	0.13	0.15	0.28
Average	"	0.14	0.18	0.32
Maximum	"	0.15	0.24	0.39
Second Coat				
Minimum	SF	0.13	0.12	0.25
Average	"	0.14	0.13	0.27
Maximum	"	0.15	0.16	0.31
Gutters and Downspouts				
Brush				
First Coat				
Minimum	LF	0.19	0.60	0.79
Average	"	0.21	0.68	0.89
Maximum	"	0.22	0.80	1.02
Second Coat				
Minimum	LF	0.19	0.40	0.59
Average	"	0.21	0.48	0.69
Maximum	"	0.22	0.60	0.82
Siding, Wood				
Roller				
First Coat				
Minimum	SF	0.13	0.17	0.30
Average	"	0.14	0.20	0.34
Maximum	"	0.15	0.21	0.36
Second Coat				
Minimum	SF	0.13	0.20	0.33
Average	"	0.14	0.21	0.35
Maximum	"	0.15	0.24	0.39
Spray				
First Coat				
Minimum	SF	0.13	0.16	0.29
Average	"	0.14	0.17	0.31
Maximum	"	0.15	0.18	0.33
Second Coat				
Minimum	SF	0.13	0.12	0.25
Average	"	0.14	0.16	0.30
Maximum	"	0.15	0.24	0.39
Stucco				
Roller				
First Coat				
Minimum	SF	0.19	0.21	0.40
Average	"	0.21	0.25	0.46
Maximum	"	0.22	0.30	0.52

Paint	UNIT	MAT.	INST.	TOTAL
09910.15 **Ext. Painting, Buildings** *(Cont.)*				
Second Coat				
Minimum	SF	0.19	0.17	0.36
Average	"	0.21	0.20	0.41
Maximum	"	0.22	0.24	0.46
Spray				
First Coat				
Minimum	SF	0.15	0.15	0.30
Average	"	0.17	0.17	0.34
Maximum	"	0.18	0.20	0.38
Second Coat				
Minimum	SF	0.15	0.12	0.27
Average	"	0.17	0.13	0.30
Maximum	"	0.18	0.16	0.34
Trim				
Brush				
First Coat				
Minimum	LF	0.19	0.20	0.39
Average	"	0.21	0.24	0.45
Maximum	"	0.22	0.30	0.52
Second Coat				
Minimum	LF	0.19	0.15	0.34
Average	"	0.21	0.20	0.41
Maximum	"	0.22	0.30	0.52
Walls				
Roller				
First Coat				
Minimum	SF	0.15	0.17	0.32
Average	"	0.17	0.17	0.34
Maximum	"	0.18	0.19	0.37
Second Coat				
Minimum	SF	0.15	0.15	0.30
Average	"	0.17	0.16	0.33
Maximum	"	0.18	0.18	0.36
Spray				
First Coat				
Minimum	SF	0.11	0.07	0.18
Average	"	0.13	0.09	0.22
Maximum	"	0.14	0.12	0.26
Second Coat				
Minimum	SF	0.11	0.06	0.17
Average	"	0.13	0.08	0.21
Maximum	"	0.14	0.10	0.24
Windows				
Brush				
First Coat				
Minimum	SF	0.13	0.80	0.93
Average	"	0.14	0.96	1.10
Maximum	"	0.15	1.20	1.35
Second Coat				
Minimum	SF	0.13	0.68	0.81
Average	"	0.14	0.80	0.94
Maximum	"	0.15	0.96	1.11

Paint	UNIT	MAT.	INST.	TOTAL
09910.25 **Ext. Painting, Misc.**				
Shakes				
Spray				
First Coat				
Minimum	SF	0.14	0.20	0.34
Average	"	0.15	0.21	0.36
Maximum	"	0.17	0.24	0.41
Second Coat				
Minimum	SF	0.14	0.18	0.32
Average	"	0.15	0.20	0.35
Maximum	"	0.17	0.21	0.38
Shingles, Wood				
Roller				
First Coat				
Minimum	SF	0.15	0.26	0.41
Average	"	0.17	0.30	0.47
Maximum	"	0.18	0.34	0.52
Second Coat				
Minimum	SF	0.15	0.18	0.33
Average	"	0.17	0.20	0.37
Maximum	"	0.18	0.21	0.39
Spray				
First Coat				
Minimum	LF	0.13	0.18	0.31
Average	"	0.14	0.20	0.34
Maximum	"	0.15	0.21	0.36
Second Coat				
Minimum	LF	0.13	0.14	0.27
Average	"	0.14	0.15	0.29
Maximum	"	0.15	0.16	0.31
Shutters and Louvres				
Brush				
First Coat				
Minimum	EA	0.19	9.61	9.80
Average	"	0.21	12.00	12.21
Maximum	"	0.22	16.00	16.22
Second Coat				
Minimum	EA	0.19	6.01	6.20
Average	"	0.21	7.39	7.60
Maximum	"	0.22	9.61	9.83
Spray				
First Coat				
Minimum	EA	0.14	3.20	3.34
Average	"	0.15	3.84	3.99
Maximum	"	0.17	4.80	4.97
Second Coat				
Minimum	EA	0.14	2.40	2.54
Average	"	0.15	3.20	3.35
Maximum	"	0.17	3.84	4.01
Stairs, metal				
Brush				
First Coat				
Minimum	SF	0.19	0.53	0.72
Average	"	0.21	0.60	0.81

Paint	UNIT	MAT.	INST.	TOTAL
09910.25 **Ext. Painting, Misc.** *(Cont.)*				
Maximum	SF	0.22	0.68	0.90
Second Coat				
Minimum	SF	0.19	0.30	0.49
Average	"	0.21	0.34	0.55
Maximum	"	0.22	0.40	0.62
Spray				
First Coat				
Minimum	SF	0.14	0.26	0.40
Average	"	0.15	0.34	0.49
Maximum	"	0.17	0.36	0.53
Second Coat				
Minimum	SF	0.14	0.20	0.34
Average	"	0.15	0.24	0.39
Maximum	"	0.17	0.30	0.47
09910.35 **Int. Painting, Buildings**				
Acoustical Ceiling				
Roller				
First Coat				
Minimum	SF	0.19	0.30	0.49
Average	"	0.21	0.40	0.61
Maximum	"	0.22	0.60	0.82
Second Coat				
Minimum	SF	0.19	0.24	0.43
Average	"	0.21	0.30	0.51
Maximum	"	0.22	0.40	0.62
Spray				
First Coat				
Minimum	SF	0.15	0.13	0.28
Average	"	0.17	0.16	0.33
Maximum	"	0.18	0.20	0.38
Second Coat				
Minimum	SF	0.15	0.10	0.25
Average	"	0.17	0.12	0.29
Maximum	"	0.18	0.13	0.31
Cabinets and Casework				
Brush				
First Coat				
Minimum	SF	0.19	0.48	0.67
Average	"	0.21	0.53	0.74
Maximum	"	0.22	0.60	0.82
Second Coat				
Minimum	SF	0.19	0.40	0.59
Average	"	0.21	0.43	0.64
Maximum	"	0.22	0.48	0.70
Spray				
First Coat				
Minimum	SF	0.15	0.24	0.39
Average	"	0.17	0.28	0.45
Maximum	"	0.18	0.34	0.52
Second Coat				
Minimum	SF	0.15	0.19	0.34
Average	"	0.17	0.20	0.37

Paint	UNIT	MAT.	INST.	TOTAL
09910.35 **Int. Painting, Buildings** *(Cont.)*				
Maximum	SF	0.18	0.26	0.44
Ceilings				
Roller				
First Coat				
Minimum	SF	0.15	0.20	0.35
Average	"	0.17	0.21	0.38
Maximum	"	0.18	0.24	0.42
Second Coat				
Minimum	SF	0.15	0.16	0.31
Average	"	0.17	0.18	0.35
Maximum	"	0.18	0.20	0.38
Spray				
First Coat				
Minimum	SF	0.13	0.12	0.25
Average	"	0.14	0.13	0.27
Maximum	"	0.15	0.15	0.30
Second Coat				
Minimum	SF	0.13	0.09	0.22
Average	"	0.14	0.10	0.24
Maximum	"	0.15	0.12	0.27
Doors, Wood				
Brush				
First Coat				
Minimum	SF	0.19	0.68	0.87
Average	"	0.21	0.87	1.08
Maximum	"	0.22	1.06	1.28
Second Coat				
Minimum	SF	0.14	0.53	0.67
Average	"	0.15	0.60	0.75
Maximum	"	0.17	0.68	0.85
Spray				
First Coat				
Minimum	SF	0.14	0.14	0.28
Average	"	0.15	0.17	0.32
Maximum	"	0.17	0.21	0.38
Second Coat				
Minimum	SF	0.14	0.11	0.25
Average	"	0.15	0.12	0.27
Maximum	"	0.17	0.15	0.32
Trim				
Brush				
First Coat				
Minimum	LF	0.19	0.19	0.38
Average	"	0.21	0.21	0.42
Maximum	"	0.22	0.26	0.48
Second Coat				
Minimum	LF	0.19	0.14	0.33
Average	"	0.21	0.18	0.39
Maximum	"	0.22	0.26	0.48
Walls				
Roller				
First Coat				
Minimum	SF	0.15	0.17	0.32

Paint	UNIT	MAT.	INST.	TOTAL
09910.35 **Int. Painting, Buildings** *(Cont.)*				
Average	SF	0.17	0.17	0.34
Maximum	"	0.18	0.20	0.38
Second Coat				
Minimum	SF	0.15	0.15	0.30
Average	"	0.17	0.16	0.33
Maximum	"	0.18	0.18	0.36
Spray				
First Coat				
Minimum	SF	0.13	0.07	0.20
Average	"	0.14	0.09	0.23
Maximum	"	0.15	0.12	0.27
Second Coat				
Minimum	SF	0.13	0.06	0.19
Average	"	0.14	0.08	0.22
Maximum	"	0.15	0.10	0.25
09955.10 **Wall Covering**				
Vinyl wall covering				
Medium duty	SF	1.10	0.68	1.78
Heavy duty	"	2.26	0.80	3.06
Over pipes and irregular shapes				
Lightweight, 13 oz.	SF	1.89	0.96	2.85
Medium weight, 25 oz.	"	2.26	1.06	3.32
Heavy weight, 34 oz.	"	2.77	1.20	3.97
Cork wall covering				
1' x 1' squares				
1/4" thick	SF	5.66	1.20	6.86
1/2" thick	"	7.19	1.20	8.39
3/4" thick	"	8.10	1.20	9.30
Wall fabrics				
Natural fabrics, grass cloths				
Minimum	SF	1.65	0.73	2.38
Average	"	1.83	0.80	2.63
Maximum	"	6.16	0.96	7.12
Flexible gypsum coated wall fabric, fire resistant	"	1.85	0.48	2.33
Vinyl corner guards				
3/4" x 3/4" x 8'	EA	8.70	6.01	14.71
2-3/4" x 2-3/4" x 4'	"	5.14	6.01	11.15
09980.15 **Paint**				
Paint, enamel				
600 sf per gal.	GAL			54.00
550 sf per gal.	"			50.00
500 sf per gal.	"			36.00
450 sf per gal.	"			33.75
350 sf per gal.	"			32.50
Filler, 60 sf per gal.	"			38.50
Latex, 400 sf per gal.	"			36.00
Aluminum				
400 sf per gal.	GAL			48.00
500 sf per gal.	"			77.00
Red lead, 350 sf per gal.	"			67.00
Primer				

Paint	UNIT	MAT.	INST.	TOTAL
09980.15 **Paint** *(Cont.)*				
400 sf per gal.	GAL			32.50
300 sf per gal.	"			32.50
Latex base, interior, white	"			36.00
Sealer and varnish				
400 sf per gal.	GAL			33.75
425 sf per gal.	"			48.00
600 sf per gal.	"			62.00

Specialties	UNIT	MAT.	INST.	TOTAL
10185.10 **Shower Stalls**				
Shower receptors				
Precast, terrazzo				
32" x 32"	EA	690	53.00	743
32" x 48"	"	720	63.00	783
Concrete				
32" x 32"	EA	280	53.00	333
48" x 48"	"	310	70.00	380
Shower door, trim and hardware				
Economy, 24" wide, chrome, tempered glass	EA	310	63.00	373
Porcelain enameled steel, flush	"	560	63.00	623
Baked enameled steel, flush	"	330	63.00	393
Aluminum, tempered glass, 48" wide, sliding	"	690	79.00	769
Folding	"	670	79.00	749
Aluminum and tempered glass, molded plastic				
Complete with receptor and door				
32" x 32"	EA	840	160	1,000
36" x 36"	"	950	160	1,110
40" x 40"	"	990	180	1,170
10210.10 **Vents And Wall Louvers**				
Block vent, 8"x16"x4" alum., w/screen, mill finish	EA	170	21.25	191
Standard	"	96.00	19.75	116
Vents w/screen, 4" deep, 8" wide, 5" high				
Modular	EA	110	19.75	130
Aluminum gable louvers	SF	20.50	10.50	31.00
Vent screen aluminum, 4" wide, continuous	LF	5.92	2.11	8.03
Aluminum louvers				
Residential use, fixed type, with screen				
8" x 8"	EA	20.75	31.75	52.50
12" x 12"	"	23.00	31.75	54.75
12" x 18"	"	27.50	31.75	59.25
14" x 24"	"	39.50	31.75	71.25
18" x 24"	"	44.50	31.75	76.25
30" x 24"	"	61.00	35.25	96.25
10290.10 **Pest Control**				
Termite control				
Under slab spraying				
Minimum	SF	1.21	0.11	1.32
Average	"	1.21	0.22	1.43
Maximum	"	1.73	0.45	2.18
10350.10 **Flagpoles**				
Installed in concrete base				
Fiberglass				
25' high	EA	1,640	380	2,020
50' high	"	4,330	960	5,290
Aluminum				
25' high	EA	1,590	380	1,970
50' high	"	3,150	960	4,110
Bonderized steel				
25' high	EA	1,780	440	2,220
50' high	"	3,560	1,150	4,710

10   SPECIALTIES

Specialties	UNIT	MAT.	INST.	TOTAL
10350.10 **Flagpoles** *(Cont.)*				
Freestanding tapered, fiberglass				
30' high	EA	1,950	410	2,360
40' high	"	2,540	520	3,060
10800.10 **Bath Accessories**				
Grab bar, 1-1/2" dia., stainless steel, wall mounted				
24" long	EA	57.00	28.75	85.75
36" long	"	64.00	30.25	94.25
1" dia., stainless steel				
12" long	EA	35.50	25.00	60.50
24" long	"	48.25	28.75	77.00
36" long	"	64.00	32.00	96.00
Medicine cabinet, 16 x 22, baked enamel, lighted	"	160	23.00	183
With mirror, lighted	"	230	38.50	269
Mirror, 1/4" plate glass, up to 10 sf	SF	12.50	5.76	18.26
Mirror, stainless steel frame				
18"x24"	EA	95.00	19.25	114
18"x32"	"	110	23.00	133
24"x30"	"	120	28.75	149
24"x60"	"	430	58.00	488
Soap dish, stainless steel, wall mounted	"	160	38.50	199
Toilet tissue dispenser, stainless, wall mounted				
Single roll	EA	79.00	14.50	93.50
Towel bar, stainless steel				
18" long	EA	98.00	23.00	121
24" long	"	130	26.25	156
30" long	"	140	28.75	169
36" long	"	150	32.00	182
Toothbrush and tumbler holder	"	62.00	19.25	81.25

Architectural Equipment	UNIT	MAT.	INST.	TOTAL
11010.10 **Maintenance Equipment**				
Vacuum cleaning system				
3 valves				
1.5 hp	EA	1,000	640	1,640
2.5 hp	"	1,200	820	2,020
5 valves	"	1,870	1,150	3,020
7 valves	"	2,490	1,440	3,930
11450.10 **Residential Equipment**				
Compactor, 4 to 1 compaction	EA	1,680	150	1,830
Dishwasher, built-in				
2 cycles	EA	830	290	1,120
4 or more cycles	"	2,230	290	2,520
Disposal				
Garbage disposer	EA	230	200	430
Heaters, electric, built-in				
Ceiling type	EA	470	200	670
Wall type				
Minimum	EA	240	150	390
Maximum	"	820	200	1,020
Hood for range, 2-speed, vented				
30" wide	EA	650	200	850
42" wide	"	1,200	200	1,400
Ice maker, automatic				
30 lb per day	EA	2,200	84.00	2,284
50 lb per day	"	2,790	290	3,080
Folding access stairs, disappearing metal stair				
8' long	EA	1,150	84.00	1,234
11' long	"	1,200	84.00	1,284
12' long	"	1,280	84.00	1,364
Wood frame, wood stair				
22" x 54" x 8'9" long	EA	220	59.00	279
25" x 54" x 10' long	"	270	59.00	329
Ranges electric				
Built-in, 30", 1 oven	EA	2,410	200	2,610
2 oven	"	2,790	200	2,990
Counter top, 4 burner, standard	"	1,390	150	1,540
With grill	"	3,480	150	3,630
Free standing, 21", 1 oven	"	1,250	200	1,450
30", 1 oven	"	2,440	120	2,560
2 oven	"	3,970	120	4,090
Water softener				
30 grains per gallon	EA	1,360	200	1,560
70 grains per gallon	"	1,720	290	2,010

Casework	UNIT	MAT.	INST.	TOTAL
12302.10 — **Wood Casework**				
Kitchen base cabinet, standard, 24" deep, 35" high				
12"wide	EA	220	58.00	278
18" wide	"	260	58.00	318
24" wide	"	330	64.00	394
27" wide	"	370	64.00	434
36" wide	"	450	72.00	522
48" wide	"	540	72.00	612
Drawer base, 24" deep, 35" high				
15"wide	EA	280	58.00	338
18" wide	"	300	58.00	358
24" wide	"	480	64.00	544
27" wide	"	550	64.00	614
30" wide	"	640	64.00	704
Sink-ready, base cabinet				
30" wide	EA	300	64.00	364
36" wide	"	310	64.00	374
42" wide	"	340	64.00	404
60" wide	"	410	72.00	482
Corner cabinet, 36" wide	"	560	72.00	632
Wall cabinet, 12" deep, 12" high				
30" wide	EA	280	58.00	338
36" wide	"	300	58.00	358
15" high				
30" wide	EA	330	64.00	394
36" wide	"	500	64.00	564
24" high				
30" wide	EA	370	64.00	434
36" wide	"	380	64.00	444
30" high				
12" wide	EA	210	72.00	282
18" wide	"	250	72.00	322
24" wide	"	260	72.00	332
27" wide	"	310	72.00	382
30" wide	"	350	82.00	432
36" wide	"	360	82.00	442
Corner cabinet, 30" high				
24" wide	EA	390	96.00	486
30" wide	"	470	96.00	566
36" wide	"	510	96.00	606
Wardrobe	"	1,040	140	1,180
Vanity with top, laminated plastic				
24" wide	EA	860	140	1,000
30" wide	"	960	140	1,100
36" wide	"	1,110	190	1,300
48" wide	"	1,240	230	1,470
12390.10 — **Countertops**				
Stainless steel, counter top, with backsplash	SF	260	14.50	275
Acid-proof, kemrock surface	"	100	9.60	110

Casework		UNIT	MAT.	INST.	TOTAL
12500.10	**Window Treatment**				
Drapery tracks, wall or ceiling mounted					
Basic traverse rod					
50 to 90"		EA	57.00	28.75	85.75
84 to 156"		"	76.00	32.00	108
136 to 250"		"	110	32.00	142
165 to 312"		"	170	36.00	206
Traverse rod with stationary curtain rod					
30 to 50"		EA	86.00	28.75	115
50 to 90"		"	98.00	28.75	127
84 to 156"		"	140	32.00	172
136 to 250"		"	170	36.00	206
Double traverse rod					
30 to 50"		EA	100	28.75	129
50 to 84"		"	130	28.75	159
84 to 156"		"	140	32.00	172
136 to 250"		"	170	36.00	206
12510.10	**Blinds**				
Venetian blinds					
2" slats		SF	39.50	1.44	40.94
1" slats		"	42.25	1.44	43.69

13 SPECIAL CONSTRUCTION

Construction	UNIT	MAT.	INST.	TOTAL
13056.10	**Vaults**			
Floor safes				
1.0 cf	EA	930	48.00	978
1.3 cf	"	1,030	72.00	1,102
13121.10	**Pre-engineered Buildings**			
Pre-engineered metal building, 40'x100'				
14' eave height	SF	9.33	5.37	14.70
16' eave height	"	10.50	6.19	16.69
13200.10	**Storage Tanks**			
Oil storage tank, underground, single wall, no excv.				
Steel				
500 gals	EA	4,040	330	4,370
1,000 gals	"	5,470	450	5,920
Fiberglass, double wall				
550 gals	EA	11,370	450	11,820
1,000 gals	"	14,620	450	15,070
Above ground				
Steel, single wall				
275 gals	EA	2,290	270	2,560
500 gals	"	5,720	450	6,170
1,000 gals	"	7,810	530	8,340
Fill cap	"	140	63.00	203
Vent cap	"	140	63.00	203
Level indicator	"	220	63.00	283

Recycling Systems	UNIT	MAT.	INST.	TOTAL
13250.10	**Gray Water Recycling System**			
Residential, small commercial, 150 Gallons				
Min.	EA			4,040
Ave.	"			4,580
Max.	"			5,110
250 Gallons				
Min.	EA			4,710
Ave.	"			5,250
Max.	"			5,790
350 Gallons				
Min.	EA			5,110
Ave.	"			5,520
Max.	"			5,920
450 Gallons				
Min.	EA			5,380
Ave.	"			5,790
Max.	"			6,190
550 Gallons				

Recycling Systems	UNIT	MAT.	INST.	TOTAL
13250.10 **Gray Water Recycling System** *(Cont.)*				
Min.	EA			6,600
Ave.	"			5,450
Max.	"			7,000

Lifts	UNIT	MAT.	INST.	TOTAL
14410.10 **Personnel Lifts**				
Electrically operated, 1 or 2 person lift				
With attached foot platforms				
3 stops	EA			12,390
Residential stair climber, per story	"	5,750	490	6,240
14410.20 **Wheelchair Lifts**				
600 lb, Residential	EA	6,830	590	7,420

Basic Materials	UNIT	MAT.	INST.	TOTAL
15100.10 Specialties				
Wall penetration				
Concrete wall, 6" thick				
2" dia.	EA		15.00	15.00
4" dia.	"		22.50	22.50
12" thick				
2" dia.	EA		20.50	20.50
4" dia.	"		32.25	32.25
15120.10 Backflow Preventers				
Backflow preventer, flanged, cast iron, with valves				
3" pipe	EA	4,030	320	4,350
4" pipe	"	4,700	350	5,050
Threaded				
3/4" pipe	EA	750	39.50	790
2" pipe	"	1,320	63.00	1,383
15140.11 Pipe Hangers, Light				
A band, black iron				
1/2"	EA	1.03	4.51	5.54
1"	"	1.11	4.67	5.78
1-1/4"	"	1.23	4.85	6.08
1-1/2"	"	1.28	5.26	6.54
2"	"	1.36	5.74	7.10
2-1/2"	"	2.03	6.31	8.34
3"	"	2.48	7.01	9.49
4"	"	3.26	7.89	11.15
Copper				
1/2"	EA	1.67	4.51	6.18
3/4"	"	1.94	4.67	6.61
1"	"	1.94	4.67	6.61
1-1/4"	"	2.09	4.85	6.94
1-1/2"	"	2.24	5.26	7.50
2"	"	2.37	5.74	8.11
2-1/2"	"	4.79	6.31	11.10
3"	"	4.99	7.01	12.00
4"	"	5.51	7.89	13.40
2 hole clips, galvanized				
3/4"	EA	0.27	4.21	4.48
1"	"	0.30	4.35	4.65
1-1/4"	"	0.39	4.51	4.90
1-1/2"	"	0.48	4.67	5.15
2"	"	0.63	4.85	5.48
2-1/2"	"	1.14	5.05	6.19
3"	"	1.66	5.26	6.92
4"	"	3.56	5.74	9.30
Perforated strap				
3/4"				
Galvanized, 20 ga.	LF	0.44	3.15	3.59
Copper, 22 ga.	"	0.69	3.15	3.84
J-Hooks				
1/2"	EA	0.79	2.87	3.66
3/4"	"	0.84	2.87	3.71
1"	"	0.86	3.00	3.86

Basic Materials	UNIT	MAT.	INST.	TOTAL

15140.11 — Pipe Hangers, Light (Cont.)

	UNIT	MAT.	INST.	TOTAL
1-1/4"	EA	0.91	3.08	3.99
1-1/2"	"	0.93	3.15	4.08
2"	"	0.97	3.15	4.12
3"	"	1.12	3.32	4.44
4"	"	1.21	3.32	4.53
PVC coated hangers, galvanized, 28 ga.				
1-1/2" x 12"	EA	1.29	4.21	5.50
2" x 12"	"	1.41	4.51	5.92
3" x 12"	"	1.58	4.85	6.43
4" x 12"	"	1.76	5.26	7.02
Copper, 30 ga.				
1-1/2" x 12"	EA	1.99	4.21	6.20
2" x 12"	"	2.36	4.51	6.87
3" x 12"	"	2.61	4.85	7.46
4" x 12"	"	2.87	5.26	8.13
Wire hook hangers				
Black wire, 1/2" x				
4"	EA	0.44	3.15	3.59
6"	"	0.50	3.32	3.82
Copper wire hooks				
1/2" x				
4"	EA	0.58	3.15	3.73
6"	"	0.66	3.32	3.98

15240.10 — Vibration Control

	UNIT	MAT.	INST.	TOTAL
Vibration isolator, in-line, stainless connector				
1/2"	EA	100	35.00	135
3/4"	"	120	37.25	157
1"	"	120	39.50	160
1-1/4"	"	170	42.00	212
1-1/2"	"	190	45.00	235
2"	"	230	48.50	279
2-1/2"	"	340	53.00	393
3"	"	400	57.00	457
4"	"	510	63.00	573

Insulation	UNIT	MAT.	INST.	TOTAL

15290.10 — Ductwork Insulation

	UNIT	MAT.	INST.	TOTAL
Fiberglass duct insulation, plain blanket				
1-1/2" thick	SF	0.22	0.78	1.00
2" thick	"	0.30	1.05	1.35
With vapor barrier				
1-1/2" thick	SF	0.26	0.78	1.04
2" thick	"	0.33	1.05	1.38
Rigid with vapor barrier				

15 MECHANICAL

Insulation	UNIT	MAT.	INST.	TOTAL
15290.10	**Ductwork Insulation** *(Cont.)*			
2" thick	SF	1.45	2.10	3.55

Facility Water Distribution	UNIT	MAT.	INST.	TOTAL
15410.05	**C.I. Pipe, Above Ground**			
No hub pipe				
1-1/2" pipe	LF	10.75	4.51	15.26
2" pipe	"	9.45	5.26	14.71
3" pipe	"	13.00	6.31	19.31
4" pipe	"	17.00	10.50	27.50
No hub fittings, 1-1/2" pipe				
1/4 bend	EA	9.85	21.00	30.85
1/8 bend	"	8.25	21.00	29.25
Sanitary tee	"	13.75	31.50	45.25
Sanitary cross	"	18.50	31.50	50.00
Plug	"			5.44
Coupling	"			19.50
Wye	"	17.00	31.50	48.50
Tapped tee	"	18.00	21.00	39.00
P-trap	"	15.50	21.00	36.50
Tapped cross	"	20.50	21.00	41.50
2" pipe				
1/4 bend	EA	11.50	25.25	36.75
1/8 bend	"	9.18	25.25	34.43
Sanitary tee	"	15.50	42.00	57.50
Sanitary cross	"	26.25	42.00	68.25
Plug	"			5.44
Coupling	"			17.25
Wye	"	14.50	53.00	67.50
Double wye	"	22.50	53.00	75.50
2x1-1/2" wye & 1/8 bend	"	27.25	39.50	66.75
Double wye & 1/8 bend	"	22.50	53.00	75.50
Test tee less 2" plug	"	14.00	25.25	39.25
Tapped tee				
2"x2"	EA	18.25	25.25	43.50
2"x1-1/2"	"	17.25	25.25	42.50
P-trap				
2"x2"	EA	16.50	25.25	41.75
Tapped cross				
2"x1-1/2"	EA	23.50	25.25	48.75
3" pipe				
1/4 bend	EA	15.50	31.50	47.00
1/8 bend	"	13.00	31.50	44.50
Sanitary tee	"	19.00	39.50	58.50
3"x2" sanitary tee	"	17.25	39.50	56.75
3"x1-1/2" sanitary tee	"	18.00	39.50	57.50

110

Facility Water Distribution	UNIT	MAT.	INST.	TOTAL
15410.05	**C.I. Pipe, Above Ground** *(Cont.)*			
Sanitary cross	EA	40.50	53.00	93.50
3x2" sanitary cross	"	36.00	53.00	89.00
Plug	"			8.07
Coupling	"			19.75
Wye	"	20.75	53.00	73.75
3x2" wye	"	15.50	53.00	68.50
Double wye	"	41.50	53.00	94.50
3x2" double wye	"	35.25	53.00	88.25
3x2" wye & 1/8 bend	"	19.50	45.00	64.50
3x1-1/2" wye & 1/8 bend	"	19.50	45.00	64.50
Double wye & 1/8 bend	"	41.50	53.00	94.50
3x2" double wye & 1/8 bend	"	35.25	53.00	88.25
3x2" reducer	"	7.87	28.75	36.62
Test tee, less 3" plug	"	21.75	31.50	53.25
Plug	"			8.07
3x3" tapped tee	"	49.50	31.50	81.00
3x2" tapped tee	"	26.75	31.50	58.25
3x1-1/2" tapped tee	"	23.00	31.50	54.50
P-trap	"	36.50	31.50	68.00
3x2" tapped cross	"	33.75	31.50	65.25
3x1-1/2" tapped cross	"	31.75	31.50	63.25
Closet flange, 3-1/2" deep	"	23.00	15.75	38.75
4" pipe				
1/4 bend	EA	22.50	31.50	54.00
1/8 bend	"	16.50	31.50	48.00
Sanitary tee	"	29.50	53.00	82.50
4x3" sanitary tee	"	27.25	53.00	80.25
4x2" sanitary tee	"	22.50	53.00	75.50
Sanitary cross	"	77.00	63.00	140
4x3" sanitary cross	"	62.00	63.00	125
4x2" sanitary cross	"	52.00	63.00	115
Plug	"			12.50
Coupling	"			19.25
Wye	"	33.75	53.00	86.75
4x3" wye	"	29.50	53.00	82.50
4x2" wye	"	21.75	53.00	74.75
Double wye	"	85.00	63.00	148
4x3" double wye	"	53.00	63.00	116
4x2" double wye	"	46.75	63.00	110
Wye & 1/8 bend	"	46.00	53.00	99.00
4x3" wye & 1/8 bend	"	33.50	53.00	86.50
4x2" wye & 1/8 bend	"	26.00	53.00	79.00
Double wye & 1/8 bend	"	120	63.00	183
4x3" double wye & 1/8 bend	"	78.00	63.00	141
4x2" double wye & 1/8 bend	"	74.00	63.00	137
4x3" reducer	"	12.25	31.50	43.75
4x2" reducer	"	12.25	31.50	43.75
Test tee, less 4" plug	"	36.75	31.50	68.25
Plug	"			12.50
4x2" tapped tee	"	27.00	31.50	58.50
4x1-1/2" tapped tee	"	23.75	31.50	55.25

Facility Water Distribution	UNIT	MAT.	INST.	TOTAL
15410.05	**C.I. Pipe, Above Ground** (Cont.)			
P-trap	EA	63.00	31.50	94.50
4x2" tapped cross	"	48.25	31.50	79.75
4x1-1/2" tapped cross	"	37.75	31.50	69.25
Closet flange				
3" deep	EA	25.25	31.50	56.75
8" deep	"	65.00	31.50	96.50
15410.06	**C.I. Pipe, Below Ground**			
No hub pipe				
1-1/2" pipe	LF	8.66	3.15	11.81
2" pipe	"	8.89	3.50	12.39
3" pipe	"	12.25	3.94	16.19
4" pipe	"	16.00	5.26	21.26
Fittings, 1-1/2"				
1/4 bend	EA	10.25	18.00	28.25
1/8 bend	"	8.63	18.00	26.63
Plug	"			5.44
Wye	"	14.50	25.25	39.75
Wye & 1/8 bend	"	15.50	18.00	33.50
P-trap	"	17.00	18.00	35.00
2"				
1/4 bend	EA	11.25	21.00	32.25
1/8 bend	"	9.67	21.00	30.67
Plug	"			5.44
Double wye	"	22.50	39.50	62.00
Wye & 1/8 bend	"	15.75	31.50	47.25
Double wye & 1/8 bend	"	38.75	39.50	78.25
P-trap	"	16.50	21.00	37.50
3"				
1/4 bend	EA	15.50	25.25	40.75
1/8 bend	"	13.00	25.25	38.25
Plug	"			8.07
Wye	"	20.75	39.50	60.25
3x2" wye	"	15.50	39.50	55.00
Wye & 1/8 bend	"	25.00	39.50	64.50
Double wye & 1/8 bend	"	60.00	39.50	99.50
3x2" double wye & 1/8 bend	"	45.25	39.50	84.75
3x2" reducer	"	7.87	25.25	33.12
P-trap	"	36.50	25.25	61.75
4"				
1/4 bend	EA	22.50	25.25	47.75
1/8 bend	"	16.50	25.25	41.75
Plug	"			12.50
Wye	"	33.75	39.50	73.25
4x3" wye	"	29.50	39.50	69.00
4x2" wye	"	21.75	39.50	61.25
Double wye	"	85.00	53.00	138
4x3" double wye	"	53.00	53.00	106
4x2" double wye	"	46.75	53.00	99.75
Wye & 1/8 bend	"	46.00	39.50	85.50
4x3" wye & 1/8 bend	"	33.50	39.50	73.00

Facility Water Distribution	UNIT	MAT.	INST.	TOTAL
15410.06 **C.I. Pipe, Below Ground** *(Cont.)*				
4x2" wye & 1/8 bend	EA	26.00	39.50	65.50
Double wye & 1/8 bend	"	120	53.00	173
4x3" double wye & 1/8 bend	"	78.00	53.00	131
4x2" double wye & 1/8 bend	"	74.00	53.00	127
4x3" reducer	"	12.25	25.25	37.50
4x2" reducer	"	12.25	25.25	37.50
15410.10 **Copper Pipe**				
Type "K" copper				
1/2"	LF	3.77	1.97	5.74
3/4"	"	7.03	2.10	9.13
1"	"	9.20	2.25	11.45
DWV, copper				
1-1/4"	LF	10.25	2.63	12.88
1-1/2"	"	13.00	2.87	15.87
2"	"	17.00	3.15	20.15
3"	"	29.00	3.50	32.50
4"	"	50.00	3.94	53.94
6"	"	200	4.51	205
Refrigeration tubing, copper, sealed				
1/8"	LF	0.73	2.52	3.25
3/16"	"	0.85	2.63	3.48
1/4"	"	1.02	2.74	3.76
Type "L" copper				
1/4"	LF	1.52	1.85	3.37
3/8"	"	2.33	1.85	4.18
1/2"	"	2.71	1.97	4.68
3/4"	"	4.33	2.10	6.43
1"	"	6.50	2.25	8.75
Type "M" copper				
1/2"	LF	1.91	1.97	3.88
3/4"	"	3.12	2.10	5.22
1"	"	5.06	2.25	7.31
Type "K" tube, coil, material only				
1/4" x 60'	EA			110
1/2" x 60'	"			240
1/2" x 100'	"			390
3/4" x 60'	"			440
3/4" x 100'	"			730
1" x 60'	"			570
1" x 100'	"			950
Type "L" tube, coil				
1/4" x 60'	EA			120
3/8" x 60'	"			190
1/2" x 60'	"			250
1/2" x 100'	"			420
3/4" x 60'	"			400
3/4" x 100'	"			670
1" x 60'	"			580
1" x 100'	"			970

Facility Water Distribution	UNIT	MAT.	INST.	TOTAL
15410.11 — **Copper Fittings**				
Coupling, with stop				
1/4"	EA	0.95	21.00	21.95
3/8"	"	1.24	25.25	26.49
1/2"	"	0.99	27.50	28.49
5/8"	"	2.87	31.50	34.37
3/4"	"	1.97	35.00	36.97
1"	"	4.06	37.25	41.31
Reducing coupling				
1/4" x 1/8"	EA	2.54	25.25	27.79
3/8" x 1/4"	"	2.79	27.50	30.29
1/2" x				
3/8"	EA	2.10	31.50	33.60
1/4"	"	2.54	31.50	34.04
1/8"	"	2.80	31.50	34.30
3/4" x				
3/8"	EA	4.50	35.00	39.50
1/2"	"	3.56	35.00	38.56
1" x				
3/8"	EA	8.08	39.50	47.58
1" x 1/2"	"	7.82	39.50	47.32
1" x 3/4"	"	6.59	39.50	46.09
Slip coupling				
1/4"	EA	0.78	21.00	21.78
1/2"	"	1.31	25.25	26.56
3/4"	"	2.74	31.50	34.24
1"	"	5.82	35.00	40.82
Coupling with drain				
1/2"	EA	10.00	31.50	41.50
3/4"	"	14.75	35.00	49.75
1"	"	18.25	39.50	57.75
Reducer				
3/8" x 1/4"	EA	2.85	25.25	28.10
1/2" x 3/8"	"	2.29	25.25	27.54
3/4" x				
1/4"	EA	4.65	28.75	33.40
3/8"	"	4.86	28.75	33.61
1/2"	"	5.06	28.75	33.81
1" x				
1/2"	EA	6.99	31.50	38.49
3/4"	"	5.36	31.50	36.86
Female adapters				
1/4"	EA	7.40	25.25	32.65
3/8"	"	7.58	28.75	36.33
1/2"	"	3.60	31.50	35.10
3/4"	"	4.94	35.00	39.94
1"	"	11.50	35.00	46.50
Increasing female adapters				
1/8" x				
3/8"	EA	7.27	25.25	32.52
1/2"	"	6.78	25.25	32.03
1/4" x 1/2"	"	7.10	27.50	34.60
3/8" x 1/2"	"	7.62	28.75	36.37
1/2" X				

Facility Water Distribution	UNIT	MAT.	INST.	TOTAL
15410.11 **Copper Fittings** *(Cont.)*				
3/4"	EA	8.08	31.50	39.58
1"	"	16.25	31.50	47.75
3/4" X				
1"	EA	17.25	35.00	52.25
1-1/4"	"	29.25	35.00	64.25
1" x				
1-1/4"	EA	31.00	35.00	66.00
1-1/2"	"	34.00	35.00	69.00
Reducing female adapters				
3/8" x 1/4"	EA	6.55	28.75	35.30
1/2" x				
1/4"	EA	5.63	31.50	37.13
3/8"	"	5.63	31.50	37.13
3/4" x 1/2"	"	7.86	35.00	42.86
1" x				
1/2"	EA	21.00	35.00	56.00
3/4"	"	16.75	35.00	51.75
Female fitting adapters				
1/2"	EA	10.00	31.50	41.50
3/4"	"	13.00	31.50	44.50
3/4" x 1/2"	"	15.50	33.25	48.75
1"	"	17.25	35.00	52.25
Male adapters				
1/4"	EA	11.25	28.75	40.00
3/8"	"	5.63	28.75	34.38
Increasing male adapters				
3/8" x 1/2"	EA	7.68	28.75	36.43
1/2" x				
3/4"	EA	6.66	31.50	38.16
1"	"	15.00	31.50	46.50
3/4" x				
1"	EA	14.75	33.25	48.00
1-1/4"	"	18.75	33.25	52.00
1" x 1-1/4"	"	18.75	35.00	53.75
Reducing male adapters				
1/2" x				
1/4"	EA	9.76	31.50	41.26
3/8"	"	8.10	31.50	39.60
3/4" x 1/2"	"	9.25	33.25	42.50
1" x				
1/2"	EA	25.50	35.00	60.50
3/4"	"	20.25	35.00	55.25
Fitting x male adapters				
1/2"	EA	14.00	31.50	45.50
3/4"	"	18.00	33.25	51.25
1"	"	18.25	35.00	53.25
90 ells				
1/8"	EA	2.12	25.25	27.37
1/4"	"	3.37	25.25	28.62
3/8"	"	3.20	28.75	31.95
1/2"	"	1.06	31.50	32.56
3/4"	"	2.40	33.25	35.65
1"	"	5.90	35.00	40.90

Facility Water Distribution	UNIT	MAT.	INST.	TOTAL
15410.11 — **Copper Fittings** *(Cont.)*				
Reducing 90 ell				
3/8" x 1/4"	EA	5.49	28.75	34.24
1/2" x				
1/4"	EA	7.86	31.50	39.36
3/8"	"	7.86	31.50	39.36
3/4" x 1/2"	"	6.90	33.25	40.15
1" x				
1/2"	EA	12.00	35.00	47.00
3/4"	"	11.25	35.00	46.25
Street ells, copper				
1/4"	EA	5.69	25.25	30.94
3/8"	"	3.92	28.75	32.67
1/2"	"	1.58	31.50	33.08
3/4"	"	3.33	33.25	36.58
1"	"	8.62	35.00	43.62
Female, 90 ell				
1/2"	EA	1.07	31.50	32.57
3/4"	"	2.42	33.25	35.67
1"	"	5.96	35.00	40.96
Female increasing, 90 ell				
3/8" x 1/2"	EA	12.00	28.75	40.75
1/2" x				
3/4"	EA	8.33	31.50	39.83
1"	"	17.00	31.50	48.50
3/4" x 1"	"	15.25	33.25	48.50
1" x 1-1/4"	"	39.00	35.00	74.00
Female reducing, 90 ell				
1/2" x 3/8"	EA	13.25	31.50	44.75
3/4" x 1/2"	"	14.75	33.25	48.00
1" x				
1/2"	EA	20.75	35.00	55.75
3/4"	"	22.25	35.00	57.25
Male, 90 ell				
1/4"	EA	10.25	25.25	35.50
3/8"	"	11.00	28.75	39.75
1/2"	"	5.66	31.50	37.16
3/4"	"	12.75	33.25	46.00
1"	"	14.75	35.00	49.75
Male, increasing 90 ell				
1/2" x				
3/4"	EA	20.25	31.50	51.75
1"	"	39.75	31.50	71.25
3/4" x 1"	"	38.00	33.25	71.25
1" x 1-1/4"	"	35.00	35.00	70.00
Male, reducing 90 ell				
1/2" x 3/8"	EA	11.50	31.50	43.00
3/4" x 1/2"	"	20.25	33.25	53.50
1" x				
1/2"	EA	38.00	35.00	73.00
3/4"	"	36.25	35.00	71.25
Drop ear ells				
1/2"	EA	7.30	31.50	38.80
Female drop ear ells				

Facility Water Distribution	UNIT	MAT.	INST.	TOTAL
15410.11 **Copper Fittings** *(Cont.)*				
1/2"	EA	7.30	31.50	38.80
1/2" x 3/8"	"	12.75	31.50	44.25
3/4"	"	21.25	33.25	54.50
Female flanged sink ell				
1/2"	EA	13.25	31.50	44.75
45 ells				
1/4"	EA	6.09	25.25	31.34
3/8"	"	4.94	28.75	33.69
45 street ell				
1/4"	EA	6.89	25.25	32.14
3/8"	"	7.45	28.75	36.20
1/2"	"	2.21	31.50	33.71
3/4"	"	3.33	33.25	36.58
1"	"	8.79	35.00	43.79
Tee				
1/8"	EA	5.17	25.25	30.42
1/4"	"	5.45	25.25	30.70
3/8"	"	4.17	28.75	32.92
Caps				
1/4"	EA	0.94	25.25	26.19
3/8"	"	1.50	28.75	30.25
Test caps				
1/2"	EA	0.88	31.50	32.38
3/4"	"	0.99	33.25	34.24
1"	"	1.77	35.00	36.77
Flush bushing				
1/4" x 1/8"	EA	1.79	25.25	27.04
1/2" x				
1/4"	EA	2.43	31.50	33.93
3/8"	"	2.15	31.50	33.65
3/4" x				
3/8"	EA	4.51	33.25	37.76
1/2"	"	3.99	33.25	37.24
1" x				
1/2"	EA	6.88	35.00	41.88
3/4"	"	6.11	35.00	41.11
Female flush bushing				
1/2" x				
1/2" x 1/8"	EA	5.43	31.50	36.93
1/4"	"	5.71	31.50	37.21
Union				
1/4"	EA	33.75	25.25	59.00
3/8"	"	46.50	28.75	75.25
Female				
1/2"	EA	16.25	31.50	47.75
3/4"	"	16.25	33.25	49.50
Male				
1/2"	EA	17.50	31.50	49.00
3/4"	"	23.25	33.25	56.50
1"	"	51.00	35.00	86.00
45 degree wye				
1/2"	EA	21.75	31.50	53.25
3/4"	"	31.25	33.25	64.50

Facility Water Distribution	UNIT	MAT.	INST.	TOTAL
15410.11 **Copper Fittings** *(Cont.)*				
1"	EA	42.00	35.00	77.00
1" x 3/4" x 3/4"	"	58.00	35.00	93.00
Twin ells				
1" x 3/4" x 3/4"	EA	15.25	35.00	50.25
1" x 1" x 1"	"	15.25	35.00	50.25
90 union ells, male				
1/2"	EA	23.50	31.50	55.00
3/4"	"	39.00	33.25	72.25
1"	"	58.00	35.00	93.00
DWV fittings, coupling with stop				
1-1/4"	EA	4.82	37.25	42.07
1-1/2"	"	6.01	39.50	45.51
1-1/2" x 1-1/4"	"	9.72	39.50	49.22
2"	"	8.32	42.00	50.32
2" x 1-1/4"	"	11.25	42.00	53.25
2" x 1-1/2"	"	11.25	42.00	53.25
3"	"	16.00	53.00	69.00
3" x 1-1/2"	"	38.50	53.00	91.50
3" x 2"	"	36.75	53.00	89.75
4"	"	51.00	63.00	114
Slip coupling				
1-1/2"	EA	9.34	39.50	48.84
2"	"	11.00	42.00	53.00
3"	"	20.25	53.00	73.25
90 ells				
1-1/2"	EA	11.50	39.50	51.00
1-1/2" x 1-1/4"	"	31.25	39.50	70.75
2"	"	20.75	42.00	62.75
2" x 1-1/2"	"	42.00	42.00	84.00
3"	"	55.00	53.00	108
4"	"	180	63.00	243
Street, 90 elbows				
1-1/2"	EA	14.50	39.50	54.00
2"	"	31.75	42.00	73.75
3"	"	81.00	53.00	134
4"	"	200	63.00	263
Female, 90 elbows				
1-1/2"	EA	14.25	39.50	53.75
2"	"	27.75	42.00	69.75
Male, 90 elbows				
1-1/2"	EA	25.25	39.50	64.75
2"	"	52.00	42.00	94.00
90 with side inlet				
3" x 3" x 1"	EA	76.00	53.00	129
3" x 3" x 1-1/2"	"	79.00	53.00	132
3" x 3" x 2"	"	79.00	53.00	132
45 ells				
1-1/4"	EA	9.48	37.25	46.73
1-1/2"	"	7.82	39.50	47.32
2"	"	18.00	42.00	60.00
3"	"	38.25	53.00	91.25
4"	"	170	63.00	233
Street, 45 ell				

Facility Water Distribution	UNIT	MAT.	INST.	TOTAL
15410.11 **Copper Fittings** *(Cont.)*				
1-1/2"	EA	12.50	39.50	52.00
2"	"	22.50	42.00	64.50
3"	"	65.00	53.00	118
60 ell				
1-1/2"	EA	19.75	39.50	59.25
2"	"	36.25	42.00	78.25
3"	"	81.00	53.00	134
22-1/2 ell				
1-1/2"	EA	24.00	39.50	63.50
2"	"	30.75	42.00	72.75
3"	"	54.00	53.00	107
11-1/4 ell				
1-1/2"	EA	26.50	39.50	66.00
2"	"	37.25	42.00	79.25
3"	"	75.00	53.00	128
Wye				
1-1/4"	EA	40.25	37.25	77.50
1-1/2"	"	43.75	39.50	83.25
2"	"	57.00	42.00	99.00
2" x 1-1/2" x 1-1/2"	"	63.00	42.00	105
2" x 1-1/2" x 2"	"	70.00	42.00	112
2" x 1-1/2" x 2-1/2"	"	70.00	42.00	112
3"	"	140	53.00	193
3" x 3" x 1-1/2"	"	130	53.00	183
3" x 3" x 2"	"	130	53.00	183
4"	"	280	63.00	343
4" x 4" x 2"	"	200	63.00	263
4" x 4" x 3"	"	200	63.00	263
Sanitary tee				
1-1/4"	EA	20.25	37.25	57.50
1-1/2"	"	25.25	39.50	64.75
2"	"	29.50	42.00	71.50
2" x 1-1/2" x 1-1/2"	"	47.25	42.00	89.25
2" x 1-1/2" x 2"	"	48.25	42.00	90.25
2" x 2" x 1-1/2"	"	28.00	42.00	70.00
3"	"	110	53.00	163
3" x 3" x 1-1/2"	"	84.00	53.00	137
3" x 3" x 2"	"	84.00	53.00	137
4"	"	280	63.00	343
4" x 4" x 3"	"	230	63.00	293
Female sanitary tee				
1-1/2"	EA	49.25	39.50	88.75
Long turn tee				
1-1/2"	EA	48.75	39.50	88.25
2"	"	110	42.00	152
3" x 1-1/2"	"	140	53.00	193
Double wye				
1-1/2"	EA	72.00	39.50	112
2"	"	130	42.00	172
2" x 2" x 1-1/2" x 1-1/2"	"	100	42.00	142
3"	"	200	53.00	253
3" x 3" x 1-1/2" x 1-1/2"	"	200	53.00	253
3" x 3" x 2" x 2"	"	200	53.00	253

Facility Water Distribution	UNIT	MAT.	INST.	TOTAL
15410.11 **Copper Fittings** *(Cont.)*				
4" x 4" x 1-1/2" x 1-1/2"	EA	220	63.00	283
Double sanitary tee				
1-1/2"	EA	49.75	39.50	89.25
2"	"	110	42.00	152
2" x 2" x 1-1/2"	"	100	42.00	142
3"	"	130	53.00	183
3" x 3" x 1-1/2" x 1-1/2"	"	170	53.00	223
3" x 3" x 2" x 2"	"	140	53.00	193
4" x 4" x 1-1/2" x 1-1/2"	"	310	63.00	373
Long				
2" x 1-1/2"	EA	130	42.00	172
Twin elbow				
1-1/2"	EA	65.00	39.50	105
2"	"	100	42.00	142
2" x 1-1/2" x 1-1/2"	"	90.00	42.00	132
Spigot adapter, manoff				
1-1/2" x 2"	EA	42.75	39.50	82.25
1-1/2" x 3"	"	52.00	39.50	91.50
2"	"	21.25	42.00	63.25
2" x 3"	"	49.75	42.00	91.75
2" x 4"	"	71.00	42.00	113
3"	"	73.00	53.00	126
3" x 4"	"	130	53.00	183
4"	"	110	63.00	173
No-hub adapters				
1-1/2" x 2"	EA	26.00	39.50	65.50
2"	"	24.50	42.00	66.50
2" x 3"	"	56.00	42.00	98.00
3"	"	49.25	53.00	102
3" x 4"	"	100	53.00	153
4"	"	110	63.00	173
Fitting reducers				
1-1/2" x 1-1/4"	EA	9.09	39.50	48.59
2" x 1-1/2"	"	14.50	42.00	56.50
3" x 1-1/2"	"	40.25	53.00	93.25
3" x 2"	"	36.00	53.00	89.00
Slip joint (Desanco)				
1-1/4"	EA	15.50	37.25	52.75
1-1/2"	"	16.00	39.50	55.50
1-1/2" x 1-1/4"	"	16.50	39.50	56.00
Street x slip joint (Desanco)				
1-1/2"	EA	19.75	39.50	59.25
1-1/2" x 1-1/4"	"	21.00	39.50	60.50
Flush bushing				
1-1/2" x 1-1/4"	EA	11.50	39.50	51.00
2" x 1-1/2"	"	19.50	42.00	61.50
3" x 1-1/2"	"	35.00	53.00	88.00
3" x 2"	"	35.00	53.00	88.00
Male hex trap bushing				
1-1/4" x 1-1/2"	EA	16.50	37.25	53.75
1-1/2"	"	12.25	39.50	51.75
1-1/2" x 2"	"	18.50	39.50	58.00
2"	"	14.25	42.00	56.25

Facility Water Distribution	UNIT	MAT.	INST.	TOTAL
15410.11 **Copper Fittings** *(Cont.)*				
Round trap bushing				
1-1/2"	EA	14.00	39.50	53.50
2"	"	15.00	42.00	57.00
Female adapter				
1-1/4"	EA	16.50	37.25	53.75
1-1/2"	"	25.75	39.50	65.25
1-1/2" x 2"	"	65.00	39.50	105
2"	"	35.00	42.00	77.00
2" x 1-1/2"	"	57.00	42.00	99.00
3"	"	140	53.00	193
Fitting x female adapter				
1-1/2"	EA	34.50	39.50	74.00
2"	"	46.00	42.00	88.00
Male adapters				
1-1/4"	EA	14.25	37.25	51.50
1-1/4" x 1-1/2"	"	33.50	37.25	70.75
1-1/2"	"	16.50	39.50	56.00
1-1/2" x 2"	"	62.00	39.50	102
2"	"	27.75	42.00	69.75
2" x 1-1/2"	"	64.00	42.00	106
3"	"	140	53.00	193
Male x slip joint adapters				
1-1/2" x 1-1/4"	EA	26.25	39.50	65.75
Dandy cleanout				
1-1/2"	EA	46.25	39.50	85.75
2"	"	55.00	42.00	97.00
3"	"	190	53.00	243
End cleanout, flush pattern				
1-1/2" x 1"	EA	28.25	39.50	67.75
2" x 1-1/2"	"	33.75	42.00	75.75
3" x 2-1/2"	"	71.00	53.00	124
Copper caps				
1-1/2"	EA	9.57	39.50	49.07
2"	"	17.75	42.00	59.75
Closet flanges				
3"	EA	41.75	53.00	94.75
4"	"	74.00	63.00	137
Drum traps, with cleanout				
1-1/2" x 3" x 6"	EA	160	39.50	200
P-trap, swivel, with cleanout				
1-1/2"	EA	99.00	39.50	139
P-trap, solder union				
1-1/2"	EA	41.50	39.50	81.00
2"	"	73.00	42.00	115
With cleanout				
1-1/2"	EA	45.50	39.50	85.00
2"	"	81.00	42.00	123
2" x 1-1/2"	"	81.00	42.00	123
Swivel joint, with cleanout				
1-1/2" x 1-1/4"	EA	58.00	39.50	97.50
1-1/2"	"	74.00	39.50	114
2" x 1-1/2"	"	91.00	42.00	133
Estabrook TY, with inlets				

Facility Water Distribution	UNIT	MAT.	INST.	TOTAL
15410.11 **Copper Fittings** *(Cont.)*				
3", with 1-1/2" inlet	EA	130	53.00	183
Fine thread adapters				
1/2"	EA	3.33	31.50	34.83
1/2" x 1/2" IPS	"	3.76	31.50	35.26
1/2" x 3/4" IPS	"	6.14	31.50	37.64
1/2" x male	"	2.32	31.50	33.82
1/2" x female	"	4.74	31.50	36.24
Copper pipe fittings				
1/2"				
90 deg ell	EA	1.47	14.00	15.47
45 deg ell	"	1.86	14.00	15.86
Tee	"	2.46	18.00	20.46
Cap	"	1.00	7.01	8.01
Coupling	"	1.07	14.00	15.07
Union	"	7.46	15.75	23.21
3/4"				
90 deg ell	EA	3.21	15.75	18.96
45 deg ell	"	3.76	15.75	19.51
Tee	"	5.38	21.00	26.38
Cap	"	1.96	7.43	9.39
Coupling	"	2.19	15.75	17.94
Union	"	11.00	18.00	29.00
1"				
90 deg ell	EA	7.46	21.00	28.46
45 deg ell	"	9.72	21.00	30.72
Tee	"	12.25	25.25	37.50
Cap	"	3.64	10.50	14.14
Coupling	"	5.38	21.00	26.38
Union	"	14.50	21.00	35.50
15410.14 **Brass I.P.S. Fittings**				
Fittings, iron pipe size, 45 deg ell				
1/8"	EA	3.93	25.25	29.18
1/4"	"	3.56	25.25	28.81
3/8"	"	3.94	28.75	32.69
1/2"	"	4.54	31.50	36.04
3/4"	"	7.15	33.25	40.40
1"	"	12.00	35.00	47.00
90 deg ell				
1/8"	EA	4.04	25.25	29.29
1/4"	"	4.11	25.25	29.36
3/8"	"	4.11	28.75	32.86
1/2"	"	4.11	31.50	35.61
3/4"	"	8.59	33.25	41.84
1"	"	13.25	35.00	48.25
90 deg ell, reducing				
1/4" x 1/8"	EA	15.00	25.25	40.25
3/8" x 1/8"	"	15.00	28.75	43.75
3/8" x 1/4"	"	16.50	28.75	45.25
1/2" x 1/4"	"	13.50	31.50	45.00
1/2" x 3/8"	"	12.25	31.50	43.75
3/4" x 1/2"	"	16.75	33.25	50.00
1" x 3/8"	"	82.00	35.00	117

Facility Water Distribution	UNIT	MAT.	INST.	TOTAL
15410.14 **Brass I.P.S. Fittings** *(Cont.)*				
1" x 1/2"	EA	29.50	35.00	64.50
1" x 3/4"	"	29.25	35.00	64.25
Street ell, 45 deg				
1/2"	EA	7.62	31.50	39.12
3/4"	"	10.75	33.25	44.00
90 deg				
1/8"	EA	6.13	25.25	31.38
1/4"	"	6.13	25.25	31.38
3/8"	"	6.13	28.75	34.88
1/2"	"	6.13	31.50	37.63
3/4"	"	10.75	33.25	44.00
1"	"	17.75	35.00	52.75
Tee, 1/8"	"	5.60	25.25	30.85
1/4"	"	5.60	25.25	30.85
3/8"	"	5.60	28.75	34.35
1/2"	"	5.60	31.50	37.10
3/4"	"	10.25	33.25	43.50
1"	"	18.00	35.00	53.00
Tee, reducing, 3/8" x				
1/4"	EA	17.25	28.75	46.00
1/2"	"	8.45	28.75	37.20
1/2" x				
1/4"	EA	8.45	31.50	39.95
3/8"	"	8.45	31.50	39.95
3/4"	"	11.50	31.50	43.00
3/4" x				
1/4"	EA	13.25	33.25	46.50
1/2"	"	13.25	33.25	46.50
1"	"	24.00	33.25	57.25
1" x				
1/2"	EA	44.50	35.00	79.50
3/4"	"	44.50	35.00	79.50
Tee, reducing				
1/2" x 3/8" x 1/2"	EA	8.78	31.50	40.28
3/4" x 1/2" x 1/2"	"	8.78	33.25	42.03
3/4" x 1/2" x 3/4"	"	10.00	33.25	43.25
1" x 1/2" x 1/2"	"	43.00	35.00	78.00
1" x 1/2" x 3/4"	"	43.00	35.00	78.00
1" x 3/4" x 1/2"	"	41.25	35.00	76.25
1" x 3/4" x 3/4"	"	16.50	35.00	51.50
Union				
1/8"	EA	10.75	25.25	36.00
1/4"	"	10.75	25.25	36.00
3/8"	"	10.75	28.75	39.50
1/2"	"	10.75	31.50	42.25
3/4"	"	15.50	33.25	48.75
1"	"	20.50	35.00	55.50
Brass face bushing				
3/8" x 1/4"	EA	9.70	28.75	38.45
1/2" x 3/8"	"	9.70	31.50	41.20
3/4" x 1/2"	"	12.00	33.25	45.25
1" x 3/4"	"	26.25	35.00	61.25
Hex bushing, 1/4" x 1/8"	"	3.38	25.25	28.63

Facility Water Distribution	UNIT	MAT.	INST.	TOTAL
15410.14 **Brass I.P.S. Fittings** *(Cont.)*				
1/2" x				
1/4"	EA	3.38	31.50	34.88
3/8"	"	3.38	31.50	34.88
5/8" x				
1/8"	EA	6.26	31.50	37.76
1/4"	"	6.26	31.50	37.76
3/4" x				
1/8"	EA	5.55	33.25	38.80
1/4"	"	5.55	33.25	38.80
3/8"	"	4.54	33.25	37.79
1/2"	"	4.54	33.25	37.79
1" x				
1/4"	EA	8.57	35.00	43.57
3/8"	"	8.57	35.00	43.57
1/2"	"	7.09	35.00	42.09
3/4"	"	7.09	35.00	42.09
Caps				
1/8"	EA	18.25	25.25	43.50
1/4"	"	13.50	25.25	38.75
3/8"	"	13.50	28.75	42.25
1/2"	"	5.02	31.50	36.52
3/4"	"	7.90	33.25	41.15
1"	"	14.25	35.00	49.25
Couplings				
1/8"	EA	7.17	25.25	32.42
1/4"	"	7.17	25.25	32.42
3/8"	"	7.17	28.75	35.92
1/2"	"	6.29	31.50	37.79
3/4"	"	9.01	33.25	42.26
1"	"	12.75	35.00	47.75
Couplings, reducing, 1/4" x 1/8"	"	18.00	25.25	43.25
3/8" x				
1/8"	EA	23.50	28.75	52.25
1/4"	"	7.69	28.75	36.44
1/2" x				
1/8"	EA	25.00	31.50	56.50
1/4"	"	6.72	31.50	38.22
3/8"	"	6.72	31.50	38.22
3/4" x				
1/4"	EA	11.25	33.25	44.50
3/8"	"	10.00	33.25	43.25
1/2"	"	9.98	33.25	43.23
1" x				
1/2"	EA	16.50	33.25	49.75
3/4"	"	16.50	33.25	49.75
Square head plug, solid				
1/8"	EA	5.06	25.25	30.31
1/4"	"	5.06	25.25	30.31
3/8"	"	5.06	28.75	33.81
1/2"	"	5.25	31.50	36.75
3/4"	"	6.02	33.25	39.27
Cored				
1/2"	EA	3.14	31.50	34.64

Facility Water Distribution	UNIT	MAT.	INST.	TOTAL
15410.14 **Brass I.P.S. Fittings** *(Cont.)*				
3/4"	EA	3.60	33.25	36.85
1"	"	13.25	35.00	48.25
Countersunk				
1/2"	EA	19.50	31.50	51.00
3/4"	"	19.50	33.25	52.75
Locknut				
3/4"	EA	7.11	33.25	40.36
1"	"	8.87	35.00	43.87
Close standard red nipple, 1/8"	"	2.31	25.25	27.56
1/8" x				
1-1/2"	EA	1.43	25.25	26.68
2"	"	1.56	25.25	26.81
2-1/2"	"	1.81	25.25	27.06
3"	"	1.91	25.25	27.16
3-1/2"	"	2.11	25.25	27.36
4"	"	2.38	25.25	27.63
4-1/2"	"	2.42	25.25	27.67
5"	"	2.81	25.25	28.06
5-1/2"	"	3.02	25.25	28.27
6"	"	3.31	25.25	28.56
1/4" x close	"	3.63	25.25	28.88
1/4" x				
1-1/2"	EA	6.30	25.25	31.55
2"	"	6.68	25.25	31.93
2-1/2"	"	6.98	25.25	32.23
3"	"	7.32	25.25	32.57
3-1/2"	"	8.17	25.25	33.42
4"	"	8.50	25.25	33.75
4-1/2"	"	9.07	25.25	34.32
5"	"	9.35	25.25	34.60
5-1/2"	"	10.25	25.25	35.50
6"	"	10.50	25.25	35.75
3/8" x close	"	4.26	28.75	33.01
3/8" x				
1-1/2"	EA	4.98	28.75	33.73
2"	"	5.48	28.75	34.23
2-1/2"	"	6.60	28.75	35.35
3"	"	8.04	28.75	36.79
3-1/2"	"	8.79	28.75	37.54
4"	"	11.25	28.75	40.00
4-1/2"	"	11.50	28.75	40.25
5"	"	12.25	28.75	41.00
5-1/2"	"	13.25	28.75	42.00
6"	"	14.75	28.75	43.50
1/2" x close	"	5.61	31.50	37.11
1/2" x				
1-1/2"	EA	6.99	31.50	38.49
2"	"	8.51	31.50	40.01
2-1/2"	"	12.50	31.50	44.00
3"	"	11.00	31.50	42.50
3-1/2"	"	12.25	31.50	43.75
4"	"	12.75	31.50	44.25
4-1/2"	"	13.75	31.50	45.25

Facility Water Distribution	UNIT	MAT.	INST.	TOTAL
15410.14 **Brass I.P.S. Fittings** *(Cont.)*				
5"	EA	14.25	31.50	45.75
5-1/2"	"	17.25	31.50	48.75
6"	"	16.00	31.50	47.50
7-1/2"	"	48.50	31.50	80.00
8"	"	48.50	31.50	80.00
3/4" x close	"	17.25	33.25	50.50
3/4" x				
1-1/2"	EA	10.50	33.25	43.75
2"	"	11.00	33.25	44.25
2-1/2"	"	12.25	33.25	45.50
3"	"	13.25	33.25	46.50
3-1/2"	"	14.50	33.25	47.75
4"	"	15.50	33.25	48.75
4-1/2"	"	16.50	33.25	49.75
5"	"	17.25	33.25	50.50
5-1/2"	"	19.50	33.25	52.75
6"	"	20.25	33.25	53.50
1" x close	"	17.50	35.00	52.50
1" x				
2"	EA	20.00	35.00	55.00
2-1/2"	"	20.25	35.00	55.25
3"	"	21.00	35.00	56.00
3-1/2"	"	20.00	35.00	55.00
4"	"	21.50	35.00	56.50
4-1/2"	"	30.00	35.00	65.00
5"	"	30.25	35.00	65.25
5-1/2"	"	30.25	35.00	65.25
6"	"	32.00	35.00	67.00
15410.15 **Brass Fittings**				
Compression fittings, union				
3/8"	EA	1.81	10.50	12.31
1/2"	"	3.91	10.50	14.41
5/8"	"	4.98	10.50	15.48
Union elbow				
3/8"	EA	5.80	10.50	16.30
1/2"	"	9.66	10.50	20.16
5/8"	"	13.00	10.50	23.50
Union tee				
3/8"	EA	5.44	10.50	15.94
1/2"	"	7.87	10.50	18.37
5/8"	"	11.00	10.50	21.50
Male connector				
3/8"	EA	3.66	10.50	14.16
1/2"	"	4.33	10.50	14.83
5/8"	"	2.26	10.50	12.76
Female connector				
3/8"	EA	3.09	10.50	13.59
1/2"	"	4.18	10.50	14.68
5/8"	"	4.56	10.50	15.06
Brass flare fittings, union				
3/8"	EA	2.35	10.25	12.60
1/2"	"	3.24	10.25	13.49

Facility Water Distribution	UNIT	MAT.	INST.	TOTAL
15410.15 **Brass Fittings** *(Cont.)*				
5/8"	EA	4.18	10.25	14.43
90 deg elbow union				
3/8"	EA	5.04	10.25	15.29
1/2"	"	7.39	10.25	17.64
5/8"	"	13.00	10.25	23.25
Three way tee				
3/8"	EA	7.20	17.00	24.20
1/2"	"	7.44	17.00	24.44
5/8"	"	9.95	17.00	26.95
Cross				
3/8"	EA	11.75	22.50	34.25
1/2"	"	23.50	22.50	46.00
5/8"	"	49.50	22.50	72.00
Male connector, half union				
3/8"	EA	1.63	10.25	11.88
1/2"	"	2.86	10.25	13.11
5/8"	"	4.42	10.25	14.67
Female connector, half union				
3/8"	EA	2.24	10.25	12.49
1/2"	"	2.07	10.25	12.32
5/8"	"	3.99	10.25	14.24
Long forged nut				
3/8"	EA	1.77	10.25	12.02
1/2"	"	2.56	10.25	12.81
5/8"	"	8.86	10.25	19.11
Short forged nut				
3/8"	EA	1.43	10.25	11.68
1/2"	"	1.94	10.25	12.19
5/8"	"	2.37	10.25	12.62
Nut, material only				
1/8"	EA			0.33
1/4"	"			0.33
5/16"	"			0.40
3/8"	"			0.48
1/2"	"			0.70
5/8"	"			1.49
Sleeve				
1/8"	EA	0.25	12.75	13.00
1/4"	"	0.08	12.75	12.83
5/16"	"	0.23	12.75	12.98
3/8"	"	0.34	12.75	13.09
1/2"	"	0.40	12.75	13.15
5/8"	"	0.58	12.75	13.33
Tee				
1/4"	EA	3.67	18.00	21.67
5/16"	"	5.67	18.00	23.67
Male tee				
5/16" x 1/8"	EA	7.57	18.00	25.57
Female union				
1/8" x 1/8"	EA	1.92	15.75	17.67
1/4" x 3/8"	"	3.52	15.75	19.27
3/8" x 1/4"	"	3.24	15.75	18.99
3/8" x 1/2"	"	3.67	15.75	19.42

15 MECHANICAL

Facility Water Distribution	UNIT	MAT.	INST.	TOTAL
15410.15	**Brass Fittings** *(Cont.)*			
5/8" x 1/2"	EA	5.73	18.00	23.73
Male union, 1/4"				
1/4" x 1/4"	EA	1.73	15.75	17.48
3/8"	"	2.29	15.75	18.04
1/2"	"	3.45	15.75	19.20
5/16" x				
1/8"	EA	1.88	15.75	17.63
1/4"	"	2.14	15.75	17.89
3/8"	"	3.11	15.75	18.86
3/8" x				
1/8"	EA	1.98	15.75	17.73
1/4"	"	2.29	15.75	18.04
1/2"	"	3.20	15.75	18.95
5/8" x				
3/8"	EA	4.57	18.00	22.57
1/2"	"	3.94	18.00	21.94
Female elbow, 1/4" x 1/4"	"	4.40	18.00	22.40
5/16" x				
1/8"	EA	4.74	18.00	22.74
1/4"	"	6.93	18.00	24.93
3/8" x				
3/8"	EA	3.82	18.00	21.82
1/2"	"	3.20	18.00	21.20
Male elbow, 1/8" x 1/8"	"	4.29	18.00	22.29
3/16" x 1/4"	"	3.91	18.00	21.91
1/4" x				
1/8"	EA	2.42	18.00	20.42
1/4"	"	2.89	18.00	20.89
3/8"	"	2.46	18.00	20.46
5/16" x				
1/8"	EA	2.54	18.00	20.54
1/4"	"	2.82	18.00	20.82
3/8"	"	4.86	18.00	22.86
3/8" x				
1/8"	EA	2.43	18.00	20.43
1/4"	"	3.35	18.00	21.35
3/8"	"	2.54	18.00	20.54
1/2"	"	3.38	18.00	21.38
1/2" x				
1/4"	EA	5.32	21.00	26.32
3/8"	"	4.95	21.00	25.95
1/2"	"	4.11	21.00	25.11
5/8" x				
3/8"	EA	5.35	21.00	26.35
1/2"	"	5.66	21.00	26.66
3/4"	"	11.75	21.00	32.75
Union				
1/8"	EA	2.44	18.00	20.44
3/16"	"	2.72	18.00	20.72
1/4"	"	2.07	18.00	20.07
5/16"	"	2.25	18.00	20.25

15 MECHANICAL

Facility Water Distribution	UNIT	MAT.	INST.	TOTAL
15410.15	**Brass Fittings** *(Cont.)*			
3/8"	EA	2.67	18.00	20.67
Reducing union				
3/8" x 1/4"	EA	3.05	21.00	24.05
5/8" x				
3/8"	EA	5.18	21.00	26.18
1/2"	"	5.31	21.00	26.31
15410.17	**Chrome Plated Fittings**			
Fittings				
90 ell				
3/8"	EA	27.50	15.75	43.25
1/2"	"	35.50	15.75	51.25
45 ell				
3/8"	EA	35.50	15.75	51.25
1/2"	"	46.75	15.75	62.50
Tee				
3/8"	EA	34.00	21.00	55.00
1/2"	"	40.50	21.00	61.50
Coupling				
3/8"	EA	21.50	15.75	37.25
1/2"	"	21.50	15.75	37.25
Union				
3/8"	EA	35.50	15.75	51.25
1/2"	"	36.75	15.75	52.50
Tee				
1/2" x 3/8" x 3/8"	EA	40.50	21.00	61.50
1/2" x 3/8" x 1/2"	"	41.25	21.00	62.25
15410.30	**PVC/CPVC Pipe**			
PVC schedule 40				
1/2" pipe	LF	0.50	2.63	3.13
3/4" pipe	"	0.69	2.87	3.56
1" pipe	"	0.88	3.15	4.03
1-1/4" pipe	"	1.13	3.50	4.63
1-1/2" pipe	"	1.69	3.94	5.63
2" pipe	"	2.14	4.51	6.65
2-1/2" pipe	"	3.46	5.26	8.72
3" pipe	"	4.41	6.31	10.72
4" pipe	"	6.30	7.89	14.19
Fittings, 1/2"				
90 deg ell	EA	0.48	7.89	8.37
45 deg ell	"	0.66	7.89	8.55
Tee	"	0.49	9.02	9.51
Reducing insert	"	0.50	10.50	11.00
Threaded	"	1.18	7.89	9.07
Male adapter	"	0.47	10.50	10.97
Female adapter	"	0.50	7.89	8.39
Coupling	"	0.37	7.89	8.26
Union	"	3.94	12.75	16.69
Cap	"	0.46	10.50	10.96
Flange	"	8.33	12.75	21.08
3/4"				
90 deg elbow	EA	0.47	10.50	10.97

129

Facility Water Distribution	UNIT	MAT.	INST.	TOTAL
15410.30 **PVC/CPVC Pipe** *(Cont.)*				
45 deg elbow	EA	1.12	10.50	11.62
Tee	"	0.64	12.75	13.39
Reducing insert	"	0.47	9.02	9.49
Threaded	"	0.71	10.50	11.21
1"				
90 deg elbow	EA	0.82	12.75	13.57
45 deg elbow	"	1.21	12.75	13.96
Tee	"	1.10	14.00	15.10
Reducing insert	"	0.82	12.75	13.57
Threaded	"	1.10	14.00	15.10
Male adapter	"	0.77	15.75	16.52
Female adapter	"	0.66	15.75	16.41
Coupling	"	0.60	15.75	16.35
Union	"	5.92	21.00	26.92
Cap	"	0.66	12.75	13.41
Flange	"	8.39	21.00	29.39
1-1/4"				
90 deg elbow	EA	1.43	18.00	19.43
45 deg elbow	"	1.69	18.00	19.69
Tee	"	1.65	21.00	22.65
Reducing insert	"	0.99	21.00	21.99
Threaded	"	1.65	21.00	22.65
Female adapter	"	1.04	21.00	22.04
Coupling	"	0.88	21.00	21.88
Union	"	13.50	25.25	38.75
Cap	"	0.93	21.00	21.93
Flange	"	8.50	25.25	33.75
1-1/2"				
90 deg elbow	EA	1.59	18.00	19.59
45 deg elbow	"	2.35	18.00	20.35
Tee	"	2.20	21.00	23.20
Reducing insert	"	1.10	21.00	22.10
Threaded	"	1.98	21.00	22.98
Male adapter	"	1.32	21.00	22.32
Female adapter	"	1.32	21.00	22.32
Coupling	"	0.99	21.00	21.99
Union	"	18.50	31.50	50.00
Cap	"	1.04	21.00	22.04
Flange	"	14.25	31.50	45.75
2"				
90 deg elbow	EA	2.53	21.00	23.53
45 deg elbow	"	3.19	21.00	24.19
Tee	"	3.35	25.25	28.60
Reducing insert	"	2.13	25.25	27.38
Threaded	"	2.84	25.25	28.09
Male adapter	"	1.76	25.25	27.01
Female adapter	"	1.81	25.25	27.06
Coupling	"	1.47	25.25	26.72
Union	"	25.50	39.50	65.00
Cap	"	1.37	25.25	26.62
Flange	"	15.25	39.50	54.75
2-1/2"				
90 deg elbow	EA	7.62	39.50	47.12

Facility Water Distribution	UNIT	MAT.	INST.	TOTAL
15410.30 — PVC/CPVC Pipe *(Cont.)*				
45 deg elbow	EA	11.00	39.50	50.50
Tee	"	9.82	42.00	51.82
Reducing insert	"	3.06	42.00	45.06
Threaded	"	4.38	42.00	46.38
Male adapter	"	5.10	42.00	47.10
Female adapter	"	4.22	42.00	46.22
Coupling	"	3.04	42.00	45.04
Union	"	34.00	53.00	87.00
Cap	"	4.07	39.50	43.57
Flange	"	20.25	53.00	73.25
3"				
90 deg elbow	EA	8.23	53.00	61.23
45 deg elbow	"	10.75	53.00	63.75
Tee	"	13.00	57.00	70.00
Reducing insert	"	3.94	53.00	56.94
Threaded	"	5.10	53.00	58.10
Male adapter	"	6.25	53.00	59.25
Female adapter	"	5.04	53.00	58.04
Coupling	"	4.66	53.00	57.66
Union	"	35.50	63.00	98.50
Cap	"	4.07	53.00	57.07
Flange	"	18.75	63.00	81.75
4"				
90 deg elbow	EA	14.75	63.00	77.75
45 deg elbow	"	19.25	63.00	82.25
Tee	"	21.75	70.00	91.75
Reducing insert	"	8.94	63.00	71.94
Threaded	"	11.50	63.00	74.50
Male adapter	"	7.95	63.00	70.95
Female adapter	"	8.55	63.00	71.55
Coupling	"	6.79	63.00	69.79
Union	"	43.50	79.00	123
Cap	"	9.21	63.00	72.21
Flange	"	25.25	79.00	104
PVC schedule 80 pipe				
1-1/2" pipe	LF	2.15	3.94	6.09
2" pipe	"	2.91	4.51	7.42
3" pipe	"	6.00	6.31	12.31
4" pipe	"	7.84	7.89	15.73
Fittings, 1-1/2"				
90 deg elbow	EA	6.69	21.00	27.69
45 deg elbow	"	14.75	21.00	35.75
Tee	"	23.00	31.50	54.50
Reducing insert	"	4.24	21.00	25.24
Threaded	"	5.06	21.00	26.06
Male adapter	"	8.06	21.00	29.06
Female adapter	"	8.71	21.00	29.71
Coupling	"	9.30	21.00	30.30
Union	"	16.75	31.50	48.25
Cap	"	4.74	21.00	25.74
Flange	"	11.00	31.50	42.50
2"				
90 deg elbow	EA	8.10	25.25	33.35

Facility Water Distribution	UNIT	MAT.	INST.	TOTAL
15410.30 **PVC/CPVC Pipe** *(Cont.)*				
45 deg elbow	EA	19.00	25.25	44.25
Tee	"	28.75	39.50	68.25
Reducing insert	"	6.03	25.25	31.28
Threaded	"	6.09	25.25	31.34
Male adapter	"	11.00	25.25	36.25
Female adapter	"	15.25	25.25	40.50
2-1/2"				
90 deg elbow	EA	19.00	39.50	58.50
45 deg elbow	"	40.00	39.50	79.50
Tee	"	31.25	53.00	84.25
Reducing insert	"	10.50	39.50	50.00
Threaded	"	13.00	39.50	52.50
Male adapter	"	13.25	39.50	52.75
Female adapter	"	24.00	39.50	63.50
Coupling	"	13.00	39.50	52.50
Union	"	36.00	53.00	89.00
Cap	"	15.25	39.50	54.75
Flange	"	19.25	53.00	72.25
3"				
90 deg elbow	EA	17.00	53.00	70.00
45 deg elbow	"	48.75	53.00	102
Tee	"	39.25	63.00	102
Reducing insert	"	16.75	53.00	69.75
Threaded	"	24.25	53.00	77.25
Male adapter	"	14.75	53.00	67.75
Female adapter	"	27.00	53.00	80.00
Coupling	"	14.75	53.00	67.75
Union	"	46.00	63.00	109
Cap	"	19.25	53.00	72.25
Flange	"	22.00	63.00	85.00
4"				
90 deg elbow	EA	43.50	63.00	107
45 deg elbow	"	88.00	63.00	151
Tee	"	45.25	79.00	124
Reducing insert	"	23.00	63.00	86.00
Threaded	"	37.25	63.00	100
Male adapter	"	26.00	63.00	89.00
Coupling	"	18.50	63.00	81.50
Union	"	43.50	79.00	123
Cap	"	23.50	63.00	86.50
Flange	"	30.00	79.00	109
CPVC schedule 40				
1/2" pipe	LF	0.61	2.63	3.24
3/4" pipe	"	0.82	2.87	3.69
1" pipe	"	1.19	3.15	4.34
1-1/4" pipe	"	1.57	3.50	5.07
1-1/2" pipe	"	1.90	3.94	5.84
2" pipe	"	2.53	4.51	7.04
Fittings, CPVC, schedule 80				
1/2", 90 deg ell	EA	3.44	6.31	9.75
Tee	"	10.50	10.50	21.00
3/4", 90 deg ell	"	4.48	6.31	10.79
Tee	"	15.50	10.50	26.00

Facility Water Distribution	UNIT	MAT.	INST.	TOTAL
15410.30	**PVC/CPVC Pipe** *(Cont.)*			
1", 90 deg ell	EA	7.09	7.01	14.10
Tee	"	16.50	11.50	28.00
1-1/4", 90 deg ell	"	13.00	7.01	20.01
Tee	"	15.50	11.50	27.00
1-1/2", 90 deg ell	"	14.25	12.75	27.00
Tee	"	17.50	15.75	33.25
2", 90 deg ell	"	15.50	12.75	28.25
Tee	"	19.75	15.75	35.50
15410.33	**ABS DWV Pipe**			
Schedule 40 ABS				
1-1/2" pipe	LF	1.58	3.15	4.73
2" pipe	"	2.11	3.50	5.61
3" pipe	"	4.32	4.51	8.83
4" pipe	"	6.12	6.31	12.43
Fittings				
1/8 bend				
1-1/2"	EA	2.30	12.75	15.05
2"	"	3.39	15.75	19.14
3"	"	8.14	21.00	29.14
4"	"	14.50	25.25	39.75
Tee, sanitary				
1-1/2"	EA	3.33	21.00	24.33
2"	"	5.14	25.25	30.39
3"	"	14.00	31.50	45.50
4"	"	25.75	39.50	65.25
Tee, sanitary reducing				
2 x 1-1/2 x 1-1/2	EA	4.73	25.25	29.98
2 x 1-1/2 x 2	"	4.87	26.25	31.12
2 x 2 x 1-1/2	"	4.52	28.75	33.27
3 x 3 x 1-1/2	"	8.20	31.50	39.70
3 x 3 x 2	"	10.25	35.00	45.25
4 x 4 x 1-1/2	"	25.50	39.50	65.00
4 x 4 x 2	"	23.75	45.00	68.75
4 x 4 x 3	"	20.75	48.50	69.25
Wye				
1-1/2"	EA	4.87	18.00	22.87
2"	"	6.81	25.25	32.06
3"	"	15.50	31.50	47.00
4"	"	33.50	39.50	73.00
Reducer				
2 x 1-1/2	EA	3.27	15.75	19.02
3 x 1-1/2	"	8.42	21.00	29.42
3 x 2	"	7.17	21.00	28.17
4 x 2	"	14.50	25.25	39.75
4 x 3	"	14.75	25.25	40.00
P-trap				
1-1/2"	EA	7.59	21.00	28.59
2"	"	10.25	23.50	33.75
3"	"	39.25	27.50	66.75
4"	"	80.00	31.50	112
Double sanitary, tee				
1-1/2"	EA	7.37	25.25	32.62

Facility Water Distribution	UNIT	MAT.	INST.	TOTAL
15410.33 **ABS DWV Pipe** *(Cont.)*				
2"	EA	10.75	31.50	42.25
3"	"	29.25	39.50	68.75
4"	"	47.00	53.00	100
Long sweep, 1/4 bend				
1-1/2"	EA	3.83	12.75	16.58
2"	"	4.87	15.75	20.62
3"	"	11.75	21.00	32.75
4"	"	21.75	31.50	53.25
Wye, standard				
1-1/2"	EA	4.92	21.00	25.92
2"	"	6.81	25.25	32.06
3"	"	15.75	31.50	47.25
4"	"	33.50	39.50	73.00
Wye, reducing				
2 x 1-1/2 x 1-1/2	EA	9.12	21.00	30.12
2 x 2 x 1-1/2	"	8.70	25.25	33.95
4 x 4 x 2	"	18.50	39.50	58.00
4 x 4 x 3	"	25.50	42.00	67.50
Double wye				
1-1/2"	EA	11.25	25.25	36.50
2"	"	13.25	31.50	44.75
3"	"	34.00	39.50	73.50
4"	"	69.00	53.00	122
2 x 2 x 1-1/2 x 1-1/2	"	13.25	31.50	44.75
3 x 3 x 2 x 2	"	28.00	39.50	67.50
4 x 4 x 3 x 3	"	65.00	53.00	118
Combination wye and 1/8 bend				
1-1/2"	EA	7.79	21.00	28.79
2"	"	9.39	25.25	34.64
3"	"	20.25	31.50	51.75
4"	"	41.25	39.50	80.75
2 x 2 x 1-1/2	"	8.07	25.25	33.32
3 x 3 x 1-1/2	"	19.25	31.50	50.75
3 x 3 x 2	"	13.50	31.50	45.00
4 x 4 x 2	"	26.75	39.50	66.25
4 x 4 x 3	"	32.75	39.50	72.25
15410.80 **Steel Pipe**				
Black steel, extra heavy pipe, threaded				
1/2" pipe	LF	2.81	2.52	5.33
3/4" pipe	"	3.64	2.52	6.16
Fittings, malleable iron, threaded, 1/2" pipe				
90 deg ell	EA	3.37	21.00	24.37
45 deg ell	"	4.56	21.00	25.56
Tee	"	3.66	31.50	35.16
Reducing tee	"	8.22	31.50	39.72
Cap	"	2.85	12.75	15.60
Coupling	"	3.81	25.25	29.06
Union	"	16.00	21.00	37.00
Nipple, 4" long	"	3.00	21.00	24.00
3/4" pipe				
90 deg ell	EA	3.95	21.00	24.95
45 deg ell	"	6.26	31.50	37.76

Facility Water Distribution	UNIT	MAT.	INST.	TOTAL
15410.80 **Steel Pipe** *(Cont.)*				
Tee	EA	5.31	31.50	36.81
Reducing tee	"	9.18	21.00	30.18
Cap	"	3.81	12.75	16.56
Coupling	"	4.50	21.00	25.50
Union	"	18.00	21.00	39.00
Nipple, 4" long	"	3.47	21.00	24.47
Cast iron fittings				
1/2" pipe				
90 deg. ell	EA	4.23	21.00	25.23
45 deg. ell	"	8.59	21.00	29.59
Tee	"	5.59	31.50	37.09
Reducing tee	"	10.50	31.50	42.00
3/4" pipe				
90 deg. ell	EA	4.52	21.00	25.52
45 deg. ell	"	5.59	21.00	26.59
Tee	"	6.99	31.50	38.49
Reducing tee	"	9.05	31.50	40.55
15410.82 **Galvanized Steel Pipe**				
Galvanized pipe				
1/2" pipe	LF	3.01	6.31	9.32
3/4" pipe	"	3.92	7.89	11.81
90 degree ell, 150 lb malleable iron, galvanized				
1/2"	EA	2.02	12.75	14.77
3/4"	"	2.68	15.75	18.43
45 degree ell, 150 lb m.i., galv.				
1/2"	EA	3.23	12.75	15.98
3/4"	"	4.38	15.75	20.13
Tees, straight, 150 lb m.i., galv.				
1/2"	EA	2.68	15.75	18.43
3/4"	"	4.47	18.00	22.47
Tees, reducing, out, 150 lb m.i., galv.				
1/2"	EA	4.64	15.75	20.39
3/4"	"	5.37	18.00	23.37
Couplings, straight, 150 lb m.i., galv.				
1/2"	EA	2.48	12.75	15.23
3/4"	"	2.98	14.00	16.98
Couplings, reducing, 150 lb m.i., galv				
1/2"	EA	2.89	12.75	15.64
3/4"	"	3.23	14.00	17.23
Caps, 150 lb m.i., galv.				
1/2"	EA	2.06	6.31	8.37
3/4"	"	2.72	6.64	9.36
Unions, 150 lb m.i., galv.				
1/2"	EA	11.50	15.75	27.25
3/4"	"	13.00	18.00	31.00
Nipples, galvanized steel, 4" long				
1/2"	EA	2.98	7.89	10.87
3/4"	"	3.97	8.42	12.39
90 degree reducing ell, 150 lb m.i., galv.				
3/4" x 1/2"	EA	3.23	12.75	15.98
1" x 3/4"	"	4.38	14.00	18.38
Square head plug (C.I.)				

Facility Water Distribution	UNIT	MAT.	INST.	TOTAL
15410.82 **Galvanized Steel Pipe** *(Cont.)*				
1/2"	EA	2.04	7.01	9.05
3/4"	"	4.55	7.89	12.44
15430.23 **Cleanouts**				
Cleanout, wall				
2"	EA	240	42.00	282
3"	"	340	42.00	382
4"	"	340	53.00	393
Floor				
2"	EA	220	53.00	273
3"	"	290	53.00	343
4"	"	300	63.00	363
15430.25 **Hose Bibbs**				
Hose bibb				
1/2"	EA	10.50	21.00	31.50
3/4"	"	11.00	21.00	32.00
15430.60 **Valves**				
Gate valve, 125 lb, bronze, soldered				
1/2"	EA	34.50	15.75	50.25
3/4"	"	41.25	15.75	57.00
Threaded				
1/4", 125 lb	EA	32.25	25.25	57.50
1/2"				
125 lb	EA	31.00	25.25	56.25
150 lb	"	41.50	25.25	66.75
300 lb	"	78.00	25.25	103
3/4"				
125 lb	EA	36.25	25.25	61.50
150 lb	"	49.25	25.25	74.50
300 lb	"	94.00	25.25	119
Check valve, bronze, soldered, 125 lb				
1/2"	EA	65.00	15.75	80.75
3/4"	"	81.00	15.75	96.75
Threaded				
1/2"				
125 lb	EA	76.00	21.00	97.00
150 lb	"	70.00	21.00	91.00
200 lb	"	73.00	21.00	94.00
3/4"				
125 lb	EA	56.00	25.25	81.25
150 lb	"	88.00	25.25	113
200 lb	"	97.00	25.25	122
Vertical check valve, bronze, 125 lb, threaded				
1/2"	EA	86.00	25.25	111
3/4"	"	130	28.75	159
Globe valve, bronze, soldered, 125 lb				
1/2"	EA	80.00	18.00	98.00
3/4"	"	99.00	19.75	119
Threaded				
1/2"				
125 lb	EA	77.00	21.00	98.00

Facility Water Distribution	UNIT	MAT.	INST.	TOTAL
15430.60 **Valves** *(Cont.)*				
150 lb	EA	100	21.00	121
300 lb	"	190	21.00	211
3/4"				
125 lb	EA	110	25.25	135
150 lb	"	120	25.25	145
300 lb	"	230	25.25	255
Ball valve, bronze, 250 lb, threaded				
1/2"	EA	20.50	25.25	45.75
3/4"	"	30.50	25.25	55.75
Angle valve, bronze, 150 lb, threaded				
1/2"	EA	100	22.50	123
3/4"	"	140	25.25	165
Balancing valve, meter connections, circuit setter				
1/2"	EA	90.00	25.25	115
3/4"	"	95.00	28.75	124
Balancing valve, straight type				
1/2"	EA	24.50	25.25	49.75
3/4"	"	29.75	25.25	55.00
Angle type				
1/2"	EA	33.00	25.25	58.25
3/4"	"	45.75	25.25	71.00
Square head cock, 125 lb, bronze body				
1/2"	EA	19.25	21.00	40.25
3/4"	"	23.00	25.25	48.25
Radiator temp control valve, with control and sensor				
1/2" valve	EA	130	39.50	170
Pressure relief valve, 1/2", bronze				
Low pressure	EA	30.75	25.25	56.00
High pressure	"	36.00	25.25	61.25
Pressure and temperature relief valve				
Bronze, 3/4"	EA	110	25.25	135
Cast iron, 3/4"				
High pressure	EA	53.00	25.25	78.25
Temperature relief	"	72.00	25.25	97.25
Pressure & temp relief valve	"	86.00	25.25	111
Pressure reducing valve, bronze, threaded, 250 lb				
1/2"	EA	180	39.50	220
3/4"	"	180	39.50	220
Solar water temperature regulating valve				
3/4"	EA	700	53.00	753
Tempering valve, threaded				
3/4"	EA	380	21.00	401
Thermostatic mixing valve, threaded				
1/2"	EA	130	22.50	153
3/4"	"	130	25.25	155
Sweat connection				
1/2"	EA	150	22.50	173
3/4"	"	180	25.25	205
Mixing valve, sweat connection				
1/2"	EA	79.00	22.50	102
3/4"	"	79.00	25.25	104
Liquid level gauge, aluminum body				

Facility Water Distribution	UNIT	MAT.	INST.	TOTAL
15430.60 — Valves *(Cont.)*				
3/4"	EA	400	25.25	425
125 psi, pvc body				
3/4"	EA	470	25.25	495
150 psi, crs body				
3/4"	EA	380	25.25	405
175 psi, bronze body, 1/2"	"	760	22.50	783
15430.65 — Vacuum Breakers				
Vacuum breaker, atmospheric, threaded connection				
3/4"	EA	55.00	25.25	80.25
Anti-siphon, brass				
3/4"	EA	60.00	25.25	85.25
15430.68 — Strainers				
Strainer, Y pattern, 125 psi, cast iron body, threaded				
3/4"	EA	13.75	22.50	36.25
250 psi, brass body, threaded				
3/4"	EA	36.00	25.25	61.25
Cast iron body, threaded				
3/4"	EA	21.00	25.25	46.25
15430.70 — Drains, Roof & Floor				
Floor drain, cast iron, with cast iron top				
2"	EA	180	53.00	233
3"	"	180	53.00	233
4"	"	390	53.00	443
Roof drain, cast iron				
2"	EA	280	53.00	333
3"	"	290	53.00	343
4"	"	370	53.00	423
15430.80 — Traps				
Bucket trap, threaded				
3/4"	EA	230	39.50	270
Inverted bucket steam trap, threaded				
3/4"	EA	280	39.50	320
With stainless interior				
1/2"	EA	170	39.50	210
3/4"	"	200	39.50	240
Brass interior				
3/4"	EA	310	39.50	350
Cast steel body, threaded, high temperature				
3/4"	EA	800	39.50	840
Float trap, 15 psi				
3/4"	EA	200	39.50	240
Float and thermostatic trap, 15 psi				
3/4"	EA	210	39.50	250
Steam trap, cast iron body, threaded, 125 psi				
3/4"	EA	250	39.50	290
Thermostatic trap, low pressure, angle type, 25 psi				
1/2"	EA	77.00	39.50	117
3/4"	"	130	39.50	170
Cast iron body, threaded, 125 psi				

Facility Water Distribution	UNIT	MAT.	INST.	TOTAL
15430.80 **Traps** *(Cont.)*				
3/4"	EA	170	39.50	210

Plumbing Fixtures	UNIT	MAT.	INST.	TOTAL
15440.10 **Baths**				
Bath tub, 5' long				
Minimum	EA	580	210	790
Average	"	1,270	320	1,590
Maximum	"	2,900	630	3,530
6' long				
Minimum	EA	650	210	860
Average	"	1,330	320	1,650
Maximum	"	3,760	630	4,390
Square tub, whirlpool, 4'x4'				
Minimum	EA	2,000	320	2,320
Average	"	2,830	630	3,460
Maximum	"	8,640	790	9,430
5'x5'				
Minimum	EA	2,000	320	2,320
Average	"	2,830	630	3,460
Maximum	"	8,800	790	9,590
6'x6'				
Minimum	EA	2,430	320	2,750
Average	"	3,560	630	4,190
Maximum	"	10,200	790	10,990
For trim and rough-in				
Minimum	EA	210	210	420
Average	"	300	320	620
Maximum	"	860	630	1,490
15440.12 **Disposals & Accessories**				
Disposal, continuous feed				
Minimum	EA	79.00	130	209
Average	"	220	160	380
Maximum	"	420	210	630
Batch feed, 1/2 hp				
Minimum	EA	300	130	430
Average	"	600	160	760
Maximum	"	1,040	210	1,250
Hot water dispenser				
Minimum	EA	220	130	350
Average	"	350	160	510
Maximum	"	560	210	770
Epoxy finish faucet	"	310	130	440
Lock stop assembly	"	67.00	79.00	146
Mounting gasket	"	7.74	53.00	60.74

Plumbing Fixtures	UNIT	MAT.	INST.	TOTAL
15440.12 **Disposals & Accessories** *(Cont.)*				
Tailpipe gasket	EA	1.13	53.00	54.13
Stopper assembly	"	26.50	63.00	89.50
Switch assembly, on/off	"	30.25	110	140
Tailpipe gasket washer	"	1.21	31.50	32.71
Stop gasket	"	2.66	35.00	37.66
Tailpipe flange	"	0.30	31.50	31.80
Tailpipe	"	3.44	39.50	42.94
15440.15 **Faucets**				
Kitchen				
Minimum	EA	91.00	110	201
Average	"	250	130	380
Maximum	"	310	160	470
Bath				
Minimum	EA	91.00	110	201
Average	"	270	130	400
Maximum	"	410	160	570
Lavatory, domestic				
Minimum	EA	97.00	110	207
Average	"	310	130	440
Maximum	"	510	160	670
Washroom				
Minimum	EA	120	110	230
Average	"	300	130	430
Maximum	"	560	160	720
Handicapped				
Minimum	EA	130	130	260
Average	"	400	160	560
Maximum	"	620	210	830
Shower				
Minimum	EA	120	110	230
Average	"	350	130	480
Maximum	"	560	160	720
For trim and rough-in				
Minimum	EA	85.00	130	215
Average	"	130	160	290
Maximum	"	220	320	540
15440.18 **Hydrants**				
Wall hydrant				
8" thick	EA	400	110	510
12" thick	"	470	130	600
15440.20 **Lavatories**				
Lavatory, counter top, porcelain enamel on cast iron				
Minimum	EA	210	130	340
Average	"	320	160	480
Maximum	"	570	210	780
Wall hung, china				
Minimum	EA	290	130	420
Average	"	340	160	500
Maximum	"	850	210	1,060
Handicapped				

Plumbing Fixtures	UNIT	MAT.	INST.	TOTAL
15440.20 Lavatories *(Cont.)*				
Minimum	EA	470	160	630
Average	"	540	210	750
Maximum	"	910	320	1,230
For trim and rough-in				
Minimum	EA	240	160	400
Average	"	410	210	620
Maximum	"	510	320	830
15440.30 Showers				
Shower, fiberglass, 36"x34"x84"				
Minimum	EA	630	450	1,080
Average	"	880	630	1,510
Maximum	"	1,270	630	1,900
Steel, 1 piece, 36"x36"				
Minimum	EA	580	450	1,030
Average	"	880	630	1,510
Maximum	"	1,040	630	1,670
Receptor, molded stone, 36"x36"				
Minimum	EA	240	210	450
Average	"	410	320	730
Maximum	"	630	530	1,160
For trim and rough-in				
Minimum	EA	240	290	530
Average	"	410	350	760
Maximum	"	510	630	1,140
15440.40 Sinks				
Service sink, 24"x29"				
Minimum	EA	700	160	860
Average	"	870	210	1,080
Maximum	"	1,280	320	1,600
Kitchen sink, single, stainless steel, single bowl				
Minimum	EA	310	130	440
Average	"	350	160	510
Maximum	"	640	210	850
Double bowl				
Minimum	EA	350	160	510
Average	"	390	210	600
Maximum	"	680	320	1,000
Porcelain enamel, cast iron, single bowl				
Minimum	EA	220	130	350
Average	"	290	160	450
Maximum	"	450	210	660
Double bowl				
Minimum	EA	300	160	460
Average	"	420	210	630
Maximum	"	600	320	920
Mop sink, 24"x36"x10"				
Minimum	EA	530	130	660
Average	"	640	160	800
Maximum	"	860	210	1,070
Washing machine box				
Minimum	EA	190	160	350

Plumbing Fixtures	UNIT	MAT.	INST.	TOTAL
15440.40 **Sinks** (Cont.)				
Average	EA	280	210	490
Maximum	"	340	320	660
For trim and rough-in				
Minimum	EA	310	210	520
Average	"	480	320	800
Maximum	"	620	420	1,040
15440.60 **Water Closets**				
Water closet flush tank, floor mounted				
Minimum	EA	360	160	520
Average	"	710	210	920
Maximum	"	1,130	320	1,450
Handicapped				
Minimum	EA	490	210	700
Average	"	890	320	1,210
Maximum	"	1,680	630	2,310
For trim and rough-in				
Minimum	EA	230	160	390
Average	"	270	210	480
Maximum	"	360	320	680
15440.70 **Domestic Water Heaters**				
Water heater, electric				
6 gal	EA	450	110	560
10 gal	"	460	110	570
15 gal	"	450	110	560
20 gal	"	630	130	760
30 gal	"	650	130	780
40 gal	"	710	130	840
52 gal	"	800	160	960
Oil fired				
20 gal	EA	1,430	320	1,750
50 gal	"	2,230	450	2,680
Tankless water heater, natural gas				
Min.	EA	770	420	1,190
Ave.	"	880	630	1,510
Max.	"	990	1,260	2,250
Propane				
Min.	EA	660	420	1,080
Ave.	"	770	630	1,400
Max.	"	880	1,260	2,140
15440.75 **Solar Water Heaters**				
Hydronic system, 100-120 Gallons including material				
Min.	EA			13,670
Ave.	"			14,400
Max.	"			15,130
Direct-Solar, 100-120 Gallons				
Min.	EA			9,080
Ave.	"			9,500
Max.	"			9,920
Indirect-Solar tank, 50-80 Gallons				
Min.	EA			1,810

Plumbing Fixtures	UNIT	MAT.	INST.	TOTAL
15440.75 **Solar Water Heaters** *(Cont.)*				
Ave.	EA			2,240
Max.	"			2,660
100-120 Gallons				
Min.	EA			3,030
Ave.	"			3,450
Max.	"			3,870
Solar water collector panel, 3 x 8				
Min.	EA	970	79.00	1,049
Ave.	"	1,000	90.00	1,090
Max.	"	1,030	110	1,140
4 x 7				
Min.	EA	1,090	79.00	1,169
Ave.	"	1,120	90.00	1,210
Max.	"	1,150	110	1,260
4 x 8				
Min.	EA	1,150	79.00	1,229
Ave.	"	1,180	90.00	1,270
Max.	"	1,210	110	1,320
4 x 10				
Min.	EA	1,330	90.00	1,420
Ave.	"	1,450	110	1,560
Max.	"	1,570	130	1,700
Passive tube tank system, 12 Tube				
Min.	EA	730	79.00	809
Ave.	"	910	90.00	1,000
Max.	"	1,090	110	1,200
24 Tube				
Min.	EA	1,090	79.00	1,169
Ave.	"	1,270	90.00	1,360
Max.	"	1,450	110	1,560
27 Tube				
Min.	EA	1,210	79.00	1,289
Ave.	"	1,510	90.00	1,600
Max.	"	1,810	110	1,920
15440.76 **Solar Water Heaters, Pools**				
Solar Water Heater, 1000 BTU/SF panel, 4x8	EA	130	79.00	209
4X10	"	150	90.00	240
4X12	"	170	110	280
Panel Mounting Kit, 4x8	"	36.25	31.50	67.75
4X10	"	54.00	35.00	89.00
4X12	"	67.00	39.50	107
15450.40 **Storage Tanks**				
Hot water storage tank, cement lined				
10 gallon	EA	510	210	720
70 gallon	"	1,620	320	1,940

HVAC	UNIT	MAT.	INST.	TOTAL
15555.10 **Boilers**				
Cast iron, gas fired, hot water				
115 mbh	EA	3,150	2,230	5,380
175 mbh	"	3,750	2,430	6,180
235 mbh	"	4,800	2,670	7,470
Steam				
115 mbh	EA	3,470	2,230	5,700
175 mbh	"	4,180	2,430	6,610
235 mbh	"	4,950	2,670	7,620
Electric, hot water				
115 mbh	EA	5,630	1,340	6,970
175 mbh	"	6,230	1,340	7,570
235 mbh	"	7,110	1,340	8,450
Steam				
115 mbh	EA	7,110	1,340	8,450
175 mbh	"	8,700	1,340	10,040
235 mbh	"	9,500	1,340	10,840
Oil fired, hot water				
115 mbh	EA	4,150	1,780	5,930
175 mbh	"	5,270	2,060	7,330
235 mbh	"	7,280	2,430	9,710
Steam				
115 mbh	EA	4,150	1,780	5,930
175 mbh	"	5,270	2,060	7,330
235 mbh	"	6,720	2,430	9,150
15610.10 **Furnaces**				
Electric, hot air				
40 mbh	EA	850	320	1,170
60 mbh	"	920	330	1,250
80 mbh	"	1,000	350	1,350
100 mbh	"	1,130	370	1,500
125 mbh	"	1,380	380	1,760
Gas fired hot air				
40 mbh	EA	850	320	1,170
60 mbh	"	910	330	1,240
80 mbh	"	1,050	350	1,400
100 mbh	"	1,090	370	1,460
125 mbh	"	1,200	380	1,580
Oil fired hot air				
40 mbh	EA	1,140	320	1,460
60 mbh	"	1,890	330	2,220
80 mbh	"	1,900	350	2,250
100 mbh	"	1,930	370	2,300
125 mbh	"	2,000	380	2,380

Central Cooling Equipment	UNIT	MAT.	INST.	TOTAL
15670.10 **Condensing Units**				
Air cooled condenser, single circuit				
3 ton	EA	1,730	110	1,840
5 ton	"	2,590	110	2,700
With low ambient dampers				
3 ton	EA	1,880	160	2,040
5 ton	"	2,970	160	3,130
15780.20 **Rooftop Units**				
Packaged, single zone rooftop unit, with roof curb				
2 ton	EA	4,130	630	4,760
3 ton	"	4,340	630	4,970
4 ton	"	4,740	790	5,530
15830.10 **Radiation Units**				
Baseboard radiation unit				
1.7 mbh/lf	LF	90.00	25.25	115
2.1 mbh/lf	"	120	31.50	152
15830.70 **Unit Heaters**				
Steam unit heater, horizontal				
12,500 btuh, 200 cfm	EA	560	110	670
17,000 btuh, 300 cfm	"	740	110	850

Air Handling	UNIT	MAT.	INST.	TOTAL
15855.10 **Air Handling Units**				
Air handling unit, medium pressure, single zone				
1500 cfm	EA	4,840	390	5,230
3000 cfm	"	6,360	700	7,060
Rooftop air handling units				
4950 cfm	EA	13,910	700	14,610
7370 cfm	"	17,640	900	18,540
15870.20 **Exhaust Fans**				
Belt drive roof exhaust fans				
640 cfm, 2618 fpm	EA	1,140	79.00	1,219
940 cfm, 2604 fpm	"	1,480	79.00	1,559

Air Distribution	UNIT	MAT.	INST.	TOTAL
15890.10 **Metal Ductwork**				
Rectangular duct				
Galvanized steel				
Minimum	LB	0.88	5.74	6.62
Average	"	1.10	7.01	8.11
Maximum	"	1.68	10.50	12.18
Aluminum				
Minimum	LB	2.29	12.75	15.04
Average	"	3.05	15.75	18.80
Maximum	"	3.79	21.00	24.79
Fittings				
Minimum	EA	7.26	21.00	28.26
Average	"	11.00	31.50	42.50
Maximum	"	16.00	63.00	79.00
15890.30 **Flexible Ductwork**				
Flexible duct, 1.25" fiberglass				
5" dia.	LF	3.31	3.15	6.46
6" dia.	"	3.68	3.50	7.18
7" dia.	"	4.54	3.71	8.25
8" dia.	"	4.76	3.94	8.70
10" dia.	"	6.34	4.51	10.85
12" dia.	"	6.93	4.85	11.78
Flexible duct connector, 3" wide fabric	"	2.31	10.50	12.81
15910.10 **Dampers**				
Horizontal parallel aluminum backdraft damper				
12" x 12"	EA	58.00	15.75	73.75
16" x 16"	"	60.00	18.00	78.00
15940.10 **Diffusers**				
Ceiling diffusers, round, baked enamel finish				
6" dia.	EA	40.25	21.00	61.25
8" dia.	"	48.50	26.25	74.75
10" dia.	"	54.00	26.25	80.25
12" dia.	"	69.00	26.25	95.25
Rectangular				
6x6"	EA	43.00	21.00	64.00
9x9"	"	52.00	31.50	83.50
12x12"	"	76.00	31.50	108
15x15"	"	95.00	31.50	127
18x18"	"	120	31.50	152
15940.40 **Registers And Grilles**				
Lay in flush mounted, perforated face, return				
6x6/24x24	EA	51.00	25.25	76.25
8x8/24x24	"	51.00	25.25	76.25
9x9/24x24	"	55.00	25.25	80.25
10x10/24x24	"	60.00	25.25	85.25
12x12/24x24	"	60.00	25.25	85.25
Rectangular, ceiling return, single deflection				
10x10	EA	30.75	31.50	62.25
12x12	"	35.75	31.50	67.25
14x14	"	43.50	31.50	75.00

Air Distribution	UNIT	MAT.	INST.	TOTAL
15940.40 **Registers And Grilles** *(Cont.)*				
16x8	EA	35.75	31.50	67.25
16x16	"	35.75	31.50	67.25
Wall, return air register				
12x12	EA	51.00	15.75	66.75
16x16	"	75.00	15.75	90.75
18x18	"	89.00	15.75	105
Ceiling, return air grille				
6x6	EA	29.50	21.00	50.50
8x8	"	36.75	25.25	62.00
10x10	"	45.50	25.25	70.75
Ceiling, exhaust grille, aluminum egg crate				
6x6	EA	20.25	21.00	41.25
8x8	"	20.25	25.25	45.50
10x10	"	22.50	25.25	47.75
12x12	"	27.75	31.50	59.25

Basic Materials	UNIT	MAT.	INST.	TOTAL

16050.30   Bus Duct

	UNIT	MAT.	INST.	TOTAL
Bus duct, 100a, plug-in				
10', 600v	EA	320	200	520
With ground	"	430	310	740
Circuit breakers, with enclosure				
1 pole				
15a-60a	EA	320	73.00	393
70a-100a	"	370	92.00	462
2 pole				
15a-60a	EA	450	81.00	531
70a-100a	"	540	95.00	635
Circuit breaker, adapter cubicle				
225a	EA	4,920	110	5,030
400a	"	5,810	120	5,930
Fusible switches, 240v, 3 phase				
30a	EA	540	73.00	613
60a	"	660	92.00	752
100a	"	870	110	980
200a	"	1,530	150	1,680

16110.20   Conduit Specialties

	UNIT	MAT.	INST.	TOTAL
Rod beam clamp, 1/2"	EA	6.94	3.66	10.60
Hanger rod				
3/8"	LF	1.46	2.93	4.39
1/2"	"	3.65	3.66	7.31
All thread rod				
1/4"	LF	0.47	2.21	2.68
3/8"	"	0.53	2.93	3.46
1/2"	"	1.00	3.66	4.66
5/8"	"	1.76	5.86	7.62
Hanger channel, 1-1/2"				
No holes	EA	4.62	2.21	6.83
Holes	"	5.70	2.21	7.91
Channel strap				
1/2"	EA	1.44	3.66	5.10
3/4"	"	1.93	3.66	5.59
Conduit penetrations, roof and wall, 8" thick				
1/2"	EA		45.00	45.00
3/4"	"		45.00	45.00
1"	"		59.00	59.00
Threaded rod couplings				
1/4"	EA	1.68	3.66	5.34
3/8"	"	1.77	3.66	5.43
1/2"	"	2.00	3.66	5.66
5/8"	"	3.08	3.66	6.74
3/4"	"	3.37	3.66	7.03
Hex nuts, 1/4"	"	0.19	3.66	3.85
3/8"	"	0.29	3.66	3.95
1/2"	"	0.62	3.66	4.28
5/8"	"	1.35	3.66	5.01
3/4"	"	1.79	3.66	5.45
Square nuts				
1/4"	EA	0.16	3.66	3.82
3/8"	"	0.32	3.66	3.98

Basic Materials	UNIT	MAT.	INST.	TOTAL
16110.20	**Conduit Specialties** *(Cont.)*			
1/2"	EA	0.55	3.66	4.21
5/8"	"	0.73	3.66	4.39
3/4"	"	1.29	3.66	4.95
Flat washers, material only				
1/4"	EA			0.19
3/8"	"			0.26
1/2"	"			0.36
5/8"	"			0.73
3/4"	"			1.02
Lockwashers				
1/4"	EA			0.12
3/8"	"			0.20
1/2"	"			0.25
5/8"	"			0.43
3/4"	"			0.73
16110.21	**Aluminum Conduit**			
Aluminum conduit				
1/2"	LF	2.21	2.21	4.42
3/4"	"	2.84	2.93	5.77
1"	"	3.99	3.66	7.65
90 deg. elbow				
1/2"	EA	17.00	14.00	31.00
3/4"	"	23.25	18.25	41.50
1"	"	32.25	22.50	54.75
Coupling				
1/2"	EA	3.48	3.66	7.14
3/4"	"	5.28	4.34	9.62
1"	"	6.97	5.86	12.83
16110.22	**EMT Conduit**			
EMT conduit				
1/2"	LF	0.60	2.21	2.81
3/4"	"	1.09	2.93	4.02
1"	"	1.82	3.66	5.48
90 deg. elbow				
1/2"	EA	5.63	6.51	12.14
3/4"	"	6.19	7.32	13.51
1"	"	9.55	7.81	17.36
Connector, steel compression				
1/2"	EA	1.72	6.51	8.23
3/4"	"	3.30	6.51	9.81
1"	"	4.97	6.51	11.48
Coupling, steel, compression				
1/2"	EA	2.93	4.34	7.27
3/4"	"	3.99	4.34	8.33
1"	"	6.03	4.34	10.37
1 hole strap, steel				
1/2"	EA	0.19	2.93	3.12
3/4"	"	0.24	2.93	3.17
1"	"	0.37	2.93	3.30
Connector, steel set screw				
1/2"	EA	1.32	5.09	6.41

Basic Materials	UNIT	MAT.	INST.	TOTAL
16110.22 EMT Conduit *(Cont.)*				
3/4"	EA	2.11	5.09	7.20
1"	"	3.64	5.09	8.73
Insulated throat				
1/2"	EA	1.74	5.09	6.83
3/4"	"	2.83	5.09	7.92
1"	"	4.68	5.09	9.77
Connector, die cast set screw				
1/2"	EA	0.90	4.34	5.24
3/4"	"	1.54	4.34	5.88
1"	"	2.89	4.34	7.23
Insulated throat				
1/2"	EA	1.92	4.34	6.26
3/4"	"	3.10	4.34	7.44
1"	"	5.40	4.34	9.74
Coupling, steel set screw				
1/2"	EA	2.37	2.93	5.30
3/4"	"	3.57	2.93	6.50
1"	"	5.79	2.93	8.72
Diecast set screw				
1/2"	EA	0.83	2.93	3.76
3/4"	"	1.34	2.93	4.27
1"	"	2.22	2.93	5.15
1 hole malleable straps				
1/2"	EA	0.38	2.93	3.31
3/4"	"	0.52	2.93	3.45
1"	"	0.86	2.93	3.79
EMT to rigid compression coupling				
1/2"	EA	4.47	7.32	11.79
3/4"	"	6.39	7.32	13.71
1"	"	9.73	11.00	20.73
Set screw couplings				
1/2"	EA	1.17	7.32	8.49
3/4"	"	1.77	7.32	9.09
1"	"	2.95	10.75	13.70
Set screw offset connectors				
1/2"	EA	2.62	7.32	9.94
3/4"	"	3.53	7.32	10.85
1"	"	6.40	10.75	17.15
Compression offset connectors				
1/2"	EA	4.33	7.32	11.65
3/4"	"	5.48	7.32	12.80
1"	"	7.94	10.75	18.69
Type "LB" set screw condulets				
1/2"	EA	12.25	16.75	29.00
3/4"	"	15.00	21.75	36.75
1"	"	23.00	28.00	51.00
Type "T" set screw condulets				
1/2"	EA	15.50	21.75	37.25
3/4"	"	19.50	29.25	48.75
1"	"	28.00	32.50	60.50
Type "C" set screw condulets				
1/2"	EA	13.00	18.25	31.25
3/4"	"	16.25	21.75	38.00

Basic Materials	UNIT	MAT.	INST.	TOTAL
16110.22 EMT Conduit *(Cont.)*				
1"	EA	24.00	28.00	52.00
Type "LL" set screw condulets				
1/2"	EA	13.00	18.25	31.25
3/4"	"	15.75	21.75	37.50
1"	"	24.00	28.00	52.00
Type "LR" set screw condulets				
1/2"	EA	13.00	18.25	31.25
3/4"	"	15.75	21.75	37.50
1"	"	24.00	28.00	52.00
Type "LB" compression condulets				
1/2"	EA	29.00	21.75	50.75
3/4"	"	43.00	36.50	79.50
1"	"	55.00	36.50	91.50
Type "T" compression condulets				
1/2"	EA	39.00	29.25	68.25
3/4"	"	51.00	32.50	83.50
1"	"	80.00	45.00	125
Condulet covers				
1/2"	EA	1.79	9.01	10.80
3/4"	"	2.17	9.01	11.18
1"	"	2.97	9.01	11.98
Clamp type entrance caps				
1/2"	EA	9.61	18.25	27.86
3/4"	"	11.25	21.75	33.00
1"	"	13.25	29.25	42.50
Slip fitter type entrance caps				
1/2"	EA	7.01	18.25	25.26
3/4"	"	8.38	21.75	30.13
1"	"	10.00	29.25	39.25
16110.23 Flexible Conduit				
Flexible conduit, steel				
3/8"	LF	0.75	2.21	2.96
1/2	"	0.85	2.21	3.06
3/4"	"	1.17	2.93	4.10
1"	"	2.23	2.93	5.16
Flexible conduit, liquid tight				
3/8"	LF	2.10	2.21	4.31
1/2"	"	2.37	2.21	4.58
3/4"	"	3.23	2.93	6.16
1"	"	4.87	2.93	7.80
Connector, straight				
3/8"	EA	3.54	5.86	9.40
1/2"	"	3.79	5.86	9.65
3/4"	"	4.82	6.51	11.33
1"	"	8.60	7.32	15.92
Straight insulated throat connectors				
3/8"	EA	4.30	9.01	13.31
1/2"	"	4.30	9.01	13.31
3/4"	"	6.29	10.75	17.04
1"	"	9.72	10.75	20.47
90 deg connectors				
3/8"	EA	6.00	10.75	16.75

Basic Materials	UNIT	MAT.	INST.	TOTAL
16110.23 Flexible Conduit *(Cont.)*				
1/2"	EA	6.00	10.75	16.75
3/4"	"	9.64	12.50	22.14
1"	"	18.50	13.25	31.75
90 degree insulated throat connectors				
3/8"	EA	7.42	10.75	18.17
1/2"	"	7.42	10.75	18.17
3/4"	"	11.25	12.50	23.75
1"	"	21.25	13.00	34.25
Flexible aluminum conduit				
3/8"	LF	0.44	2.21	2.65
1/2"	"	0.52	2.21	2.73
3/4"	"	0.72	2.93	3.65
1"	"	1.36	2.93	4.29
Connector, straight				
3/8"	EA	1.35	7.32	8.67
1/2"	"	1.88	7.32	9.20
3/4"	"	2.05	7.81	9.86
1"	"	7.52	9.01	16.53
Straight insulated throat connectors				
3/8"	EA	1.32	6.51	7.83
1/2"	"	2.65	6.51	9.16
3/4"	"	2.81	6.51	9.32
1"	"	6.84	7.32	14.16
90 deg connectors				
3/8"	EA	2.16	10.75	12.91
1/2"	"	3.65	10.75	14.40
3/4"	"	5.73	10.75	16.48
1"	"	9.94	12.50	22.44
90 deg insulated throat connectors				
3/8"	EA	2.68	10.75	13.43
1/2"	"	4.05	10.75	14.80
3/4"	"	6.76	10.75	17.51
1"	"	11.00	12.50	23.50
16110.24 Galvanized Conduit				
Galvanized rigid steel conduit				
1/2"	LF	2.82	2.93	5.75
3/4"	"	3.13	3.66	6.79
1"	"	4.51	4.34	8.85
1-1/4"	"	6.24	5.86	12.10
1-1/2"	"	7.34	6.51	13.85
2"	"	9.34	7.32	16.66
90 degree ell				
1/2"	EA	7.51	18.25	25.76
3/4"	"	7.85	22.50	30.35
1"	"	12.00	28.00	40.00
1-1/4"	"	16.50	32.50	49.00
1-1/2"	"	20.50	36.50	57.00
2"	"	29.50	39.00	68.50
Couplings, with set screws				
1/2"	EA	3.76	3.66	7.42
3/4"	"	4.96	4.34	9.30
1"	"	7.90	5.86	13.76

Basic Materials	UNIT	MAT.	INST.	TOTAL
16110.24	**Galvanized Conduit** *(Cont.)*			
1-1/4"	EA	13.50	7.32	20.82
1-1/2"	"	17.50	9.01	26.51
2"	"	39.25	10.75	50.00
Split couplings				
1/2"	EA	3.20	14.00	17.20
3/4"	"	4.15	18.25	22.40
1"	"	5.84	20.25	26.09
1-1/4"	"	11.50	22.50	34.00
1-1/2"	"	14.75	28.00	42.75
2"	"	34.25	41.75	76.00
Erickson couplings				
1/2"	EA	4.80	32.50	37.30
3/4"	"	5.87	36.50	42.37
1"	"	10.25	45.00	55.25
1-1/4"	"	18.50	65.00	83.50
1-1/2"	"	24.00	73.00	97.00
2"	"	46.50	98.00	145
Seal fittings				
1/2"	EA	15.75	48.75	64.50
3/4"	"	17.50	59.00	76.50
1"	"	22.25	73.00	95.25
1-1/4"	"	23.00	84.00	107
1-1/2"	"	40.25	98.00	138
2"	"	44.00	120	164
Entrance fitting, (weather head), threaded				
1/2"	EA	7.34	32.50	39.84
3/4"	"	9.00	36.50	45.50
1"	"	11.50	41.75	53.25
1-1/4"	"	15.00	53.00	68.00
1-1/2"	"	26.50	59.00	85.50
2"	"	40.50	65.00	106
Locknuts				
1/2"	EA	0.18	3.66	3.84
3/4"	"	0.24	3.66	3.90
1"	"	0.37	3.66	4.03
1-1/4"	"	0.51	3.66	4.17
1-1/2"	"	0.86	4.34	5.20
2"	"	1.23	4.34	5.57
Plastic conduit bushings				
1/2"	EA	0.37	9.01	9.38
3/4"	"	0.57	10.75	11.32
1"	"	0.80	14.00	14.80
1-1/4"	"	1.05	16.25	17.30
1-1/2"	"	1.43	18.25	19.68
2"	"	3.24	22.50	25.74
Conduit bushings, steel				
1/2"	EA	0.51	9.01	9.52
3/4"	"	0.65	10.75	11.40
1"	"	1.00	14.00	15.00
1-1/4"	"	1.43	16.25	17.68
1-1/2"	"	2.04	18.25	20.29
2"	"	4.14	22.50	26.64
Pipe cap				

Basic Materials	UNIT	MAT.	INST.	TOTAL
16110.24	**Galvanized Conduit** *(Cont.)*			
1/2"	EA	0.48	3.66	4.14
3/4"	"	0.52	3.66	4.18
1"	"	0.84	3.66	4.50
1-1/4"	"	1.45	5.86	7.31
1-1/2"	"	2.25	5.86	8.11
2"	"	2.54	5.86	8.40
Threaded couplings				
1/2"	EA	1.81	3.66	5.47
3/4"	"	2.22	4.34	6.56
1"	"	3.30	5.86	9.16
1-1/4"	"	4.12	6.51	10.63
1-1/2"	"	5.06	7.32	12.38
2"	"	6.88	7.81	14.69
Threadless couplings				
1/2"	EA	7.37	7.32	14.69
3/4"	"	7.67	9.01	16.68
1"	"	9.96	10.75	20.71
1-1/4"	"	11.50	14.00	25.50
1-1/2"	"	13.75	18.25	32.00
2"	"	20.25	22.50	42.75
Threadless connectors				
1/2"	EA	3.50	7.32	10.82
3/4"	"	5.61	9.01	14.62
1"	"	8.86	10.75	19.61
1-1/4"	"	15.00	14.00	29.00
1-1/2"	"	23.00	18.25	41.25
2"	"	44.00	22.50	66.50
Setscrew connectors				
1/2"	EA	2.56	5.86	8.42
3/4"	"	3.55	6.51	10.06
1"	"	5.53	7.32	12.85
1-1/4"	"	9.82	9.01	18.83
1-1/2"	"	14.25	10.75	25.00
2"	"	28.25	14.00	42.25
Clamp type entrance caps				
1/2"	EA	8.58	22.50	31.08
3/4"	"	10.00	28.00	38.00
1"	"	14.00	32.50	46.50
1-1/4"	"	16.50	36.50	53.00
1-1/2"	"	29.75	45.00	74.75
2"	"	35.75	53.00	88.75
"LB" condulets				
1/2"	EA	10.25	22.50	32.75
3/4"	"	12.25	28.00	40.25
1"	"	18.50	32.50	51.00
1-1/4"	"	31.75	36.50	68.25
1-1/2"	"	41.50	45.00	86.50
2"	"	69.00	53.00	122
"T" condulets				
1/2"	EA	12.75	28.00	40.75
3/4"	"	15.25	32.50	47.75
1"	"	23.00	36.50	59.50
1-1/4"	"	33.75	41.75	75.50

Basic Materials	UNIT	MAT.	INST.	TOTAL
16110.24 **Galvanized Conduit** *(Cont.)*				
1-1/2"	EA	45.00	45.00	90.00
2"	"	70.00	53.00	123
"X" condulets				
1/2"	EA	18.75	32.50	51.25
3/4"	"	20.25	36.50	56.75
1"	"	33.50	41.75	75.25
1-1/4"	"	43.75	45.00	88.75
1-1/2"	"	56.00	48.75	105
2"	"	120	64.00	184
Blank steel condulet covers				
1/2"	EA	2.89	7.32	10.21
3/4"	"	3.59	7.32	10.91
1"	"	4.88	7.32	12.20
1-1/4"	"	5.98	9.01	14.99
1-1/2"	"	6.28	9.01	15.29
2"	"	10.50	9.01	19.51
Solid condulet gaskets				
1/2"	EA	2.39	3.66	6.05
3/4"	"	2.58	3.66	6.24
1"	"	2.99	3.66	6.65
1-1/4"	"	3.71	5.86	9.57
1-1/2"	"	3.91	5.86	9.77
2"	"	4.38	5.86	10.24
One-hole malleable straps				
1/2"	EA	0.38	2.93	3.31
3/4"	"	0.52	2.93	3.45
1"	"	0.75	2.93	3.68
1-1/4"	"	1.51	3.66	5.17
1-1/2"	"	1.74	3.66	5.40
2"	"	3.42	3.66	7.08
One-hole steel straps				
1/2"	EA	0.09	2.93	3.02
3/4"	"	0.14	2.93	3.07
1"	"	0.24	2.93	3.17
1-1/4"	"	0.34	3.66	4.00
1-1/2"	"	0.46	3.66	4.12
2"	"	0.60	3.66	4.26
Grounding locknuts				
1/2"	EA	2.29	5.86	8.15
3/4"	"	2.90	5.86	8.76
1"	"	4.19	5.86	10.05
1-1/4"	"	4.50	6.51	11.01
1-1/2"	"	4.69	6.51	11.20
2"	"	6.96	6.51	13.47
Insulated grounding metal bushings				
1/2"	EA	1.63	14.00	15.63
3/4"	"	2.40	16.25	18.65
1"	"	3.43	18.25	21.68
1-1/4"	"	5.48	22.50	27.98
1-1/2"	"	6.83	28.00	34.83
2"	"	9.89	32.50	42.39

16 ELECTRICAL

Basic Materials	UNIT	MAT.	INST.	TOTAL
16110.25 **Plastic Conduit**				
PVC conduit, schedule 40				
1/2"	LF	0.73	2.21	2.94
3/4"	"	0.91	2.21	3.12
1"	"	1.32	2.93	4.25
1-1/4"	"	1.82	2.93	4.75
1-1/2"	"	2.17	3.66	5.83
2"	"	2.78	3.66	6.44
Couplings				
1/2"	EA	0.45	3.66	4.11
3/4"	"	0.53	3.66	4.19
1"	"	0.84	3.66	4.50
1-1/4"	"	1.11	4.34	5.45
1-1/2"	"	1.54	4.34	5.88
2"	"	2.02	4.34	6.36
90 degree elbows				
1/2"	EA	1.74	7.32	9.06
3/4"	"	1.90	9.01	10.91
1"	"	3.01	9.01	12.02
1-1/4"	"	4.20	10.75	14.95
1-1/2"	"	5.69	14.00	19.69
2"	"	7.94	16.25	24.19
Terminal adapters				
1/2"	EA	0.66	7.32	7.98
3/4"	"	1.06	7.32	8.38
1"	"	1.33	7.32	8.65
1-1/4"	"	1.67	11.75	13.42
1-1/2"	"	2.13	11.75	13.88
2"	"	2.94	11.75	14.69
LB conduit body				
1/2"	EA	5.70	14.00	19.70
3/4"	"	7.35	14.00	21.35
1	"	8.10	14.00	22.10
1-1/4"	"	12.25	22.50	34.75
1-1/2"	"	14.75	22.50	37.25
2"	"	26.25	22.50	48.75
PVC cement				
1 pint	EA			15.75
1 quart	"			23.00
16110.27 **Plastic Coated Conduit**				
Rigid steel conduit, plastic coated				
1/2"	LF	7.06	3.66	10.72
3/4"	"	8.20	4.34	12.54
1"	"	10.50	5.86	16.36
1-1/4"	"	13.50	7.32	20.82
1-1/2"	"	16.25	9.01	25.26
2"	"	21.25	10.75	32.00
90 degree elbows				
1/2"	EA	27.75	22.50	50.25
3/4"	"	28.75	28.00	56.75
1"	"	33.00	32.50	65.50
1-1/4"	"	40.75	36.50	77.25
1-1/2"	"	50.00	45.00	95.00

156

Basic Materials	UNIT	MAT.	INST.	TOTAL
16110.27 **Plastic Coated Conduit** *(Cont.)*				
2"	EA	70.00	59.00	129
Couplings				
1/2"	EA	7.94	4.34	12.28
3/4"	"	8.30	5.86	14.16
1"	"	11.00	6.51	17.51
1-1/4"	"	12.75	7.81	20.56
1-1/2"	"	18.00	9.01	27.01
2"	"	22.50	10.75	33.25
1 hole conduit straps				
3/4"	EA	12.25	3.66	15.91
1"	"	12.75	3.66	16.41
1-1/4"	"	18.50	4.34	22.84
1-1/2"	"	19.50	4.34	23.84
2"	"	28.50	4.34	32.84
16110.28 **Steel Conduit**				
Intermediate metal conduit (IMC)				
1/2"	LF	2.02	2.21	4.23
3/4"	"	2.48	2.93	5.41
1"	"	3.76	3.66	7.42
1-1/4"	"	4.81	4.34	9.15
1-1/2"	"	6.02	5.86	11.88
2"	"	7.86	6.51	14.37
90 degree ell				
1/2"	EA	15.50	18.25	33.75
3/4"	"	16.25	22.50	38.75
1"	"	24.75	28.00	52.75
1-1/4"	"	34.50	32.50	67.00
1-1/2"	"	42.50	36.50	79.00
2"	"	61.00	41.75	103
Couplings				
1/2"	EA	3.79	3.66	7.45
3/4"	"	4.66	4.34	9.00
1"	"	6.90	5.86	12.76
1-1/4"	"	8.64	6.51	15.15
1-1/2"	"	11.00	7.32	18.32
2"	"	14.50	7.81	22.31
16110.35 **Surface Mounted Raceway**				
Single Raceway				
3/4" x 17/32" Conduit	LF	1.83	2.93	4.76
Mounting Strap	EA	0.49	3.90	4.39
Connector	"	0.66	3.90	4.56
Elbow				
45 degree	EA	8.38	3.66	12.04
90 degree	"	2.67	3.66	6.33
internal	"	3.36	3.66	7.02
external	"	3.10	3.66	6.76
Switch	"	21.75	29.25	51.00
Utility Box	"	14.50	29.25	43.75
Receptacle	"	25.75	29.25	55.00
3/4" x 21/32" Conduit	LF	2.09	2.93	5.02
Mounting Strap	EA	0.77	3.90	4.67

Basic Materials	UNIT	MAT.	INST.	TOTAL
16110.35 **Surface Mounted Raceway** *(Cont.)*				
Connector	EA	0.79	3.90	4.69
Elbow				
45 degree	EA	10.25	3.66	13.91
90 degree	"	2.85	3.66	6.51
internal	"	3.87	3.66	7.53
external	"	3.87	3.66	7.53
Switch	"	21.75	29.25	51.00
Utility Box	"	14.50	29.25	43.75
Receptacle	"	25.75	29.25	55.00
16120.41 **Aluminum Conductors**				
Type XHHW, stranded aluminum, 600v				
#8	LF	0.36	0.36	0.72
#6	"	0.38	0.44	0.82
#4	"	0.47	0.58	1.05
#2	"	0.65	0.65	1.30
1/0	"	1.02	0.80	1.82
2/0	"	1.33	0.88	2.21
3/0	"	1.65	1.02	2.67
4/0	"	1.85	1.09	2.94
Type S.E.U. cable				
#8/3	LF	1.52	1.83	3.35
#6/3	"	1.52	2.05	3.57
#4/3	"	1.96	2.54	4.50
#2/3	"	2.61	2.79	5.40
#1/3	"	3.55	2.93	6.48
1/0-3	"	3.99	3.08	7.07
2/0-3	"	4.58	3.25	7.83
3/0-3	"	6.39	3.78	10.17
4/0-3	"	6.42	4.18	10.60
Type S.E.R. cable with ground				
#8/3	LF	1.84	2.05	3.89
#6/3	"	2.09	2.54	4.63
16120.43 **Copper Conductors**				
Copper conductors, type THW, solid				
#14	LF	0.12	0.29	0.41
#12	"	0.18	0.36	0.54
#10	"	0.28	0.44	0.72
THHN-THWN, solid				
#14	LF	0.12	0.29	0.41
#12	"	0.18	0.36	0.54
#10	"	0.28	0.44	0.72
Stranded				
#14	LF	0.12	0.29	0.41
#12	"	0.18	0.36	0.54
#10	"	0.28	0.44	0.72
#8	"	0.49	0.58	1.07
#6	"	0.77	0.65	1.42
#4	"	1.22	0.73	1.95
#2	"	1.70	0.88	2.58
#1	"	2.15	1.02	3.17
1/0	"	2.65	1.17	3.82

Basic Materials	UNIT	MAT.	INST.	TOTAL
16120.43 **Copper Conductors** *(Cont.)*				
2/0	LF	3.27	1.46	4.73
3/0	"	4.11	1.83	5.94
4/0	"	5.14	2.05	7.19
Bare stranded wire				
#8	LF	0.42	0.58	1.00
#6	"	0.71	0.73	1.44
#4	"	1.11	0.73	1.84
#2	"	1.77	0.80	2.57
#1	"	2.22	1.02	3.24
Type "BX" solid armored cable				
#14/2	LF	0.81	1.83	2.64
#14/3	"	1.28	2.05	3.33
#14/4	"	1.80	2.25	4.05
#12/2	"	0.83	2.05	2.88
#12/3	"	1.34	2.25	3.59
#12/4	"	1.85	2.54	4.39
#10/2	"	1.55	2.25	3.80
#10/3	"	2.22	2.54	4.76
#10/4	"	3.45	2.93	6.38
Steel type, metal clad cable, solid, with ground				
#14/2	LF	0.67	1.31	1.98
#14/3	"	1.03	1.46	2.49
#14/4	"	1.38	1.67	3.05
#12/2	"	0.69	1.46	2.15
#12/3	"	1.14	1.83	2.97
#12/4	"	1.54	2.21	3.75
#10/2	"	1.43	1.67	3.10
#10/3	"	1.99	2.05	4.04
#10/4	"	3.09	2.44	5.53
16120.47 **Sheathed Cable**				
Non-metallic sheathed cable				
Type NM cable with ground				
#14/2	LF	0.28	1.09	1.37
#12/2	"	0.44	1.17	1.61
#10/2	"	0.69	1.30	1.99
#8/2	"	1.13	1.46	2.59
#6/2	"	1.78	1.83	3.61
#14/3	"	0.39	1.89	2.28
#12/3	"	0.62	1.95	2.57
#10/3	"	0.99	1.98	2.97
#8/3	"	1.66	2.02	3.68
#6/3	"	2.68	2.05	4.73
#4/3	"	5.55	2.34	7.89
#2/3	"	8.34	2.54	10.88
Type U.F. cable with ground				
#14/2	LF	0.33	1.17	1.50
#12/2	"	0.49	1.39	1.88
#14/3	"	0.46	1.46	1.92
#12/3	"	0.70	1.60	2.30
Type S.F.U. cable, 3 conductor				
#8	LF	1.41	2.05	3.46
#6	"	2.47	2.25	4.72

Basic Materials	UNIT	MAT.	INST.	TOTAL
16120.47 **Sheathed Cable** *(Cont.)*				
Type SER cable, 4 conductor				
#6	LF	3.54	2.66	6.20
#4	"	4.96	2.85	7.81
Flexible cord, type STO cord				
#18/2	LF	0.60	0.29	0.89
#18/3	"	0.70	0.36	1.06
#18/4	"	0.97	0.43	1.40
#16/2	"	0.69	0.29	0.98
#16/3	"	0.58	0.32	0.90
#16/4	"	0.81	0.36	1.17
#14/2	"	1.08	0.36	1.44
#14/3	"	0.99	0.45	1.44
#14/4	"	1.22	0.51	1.73
#12/2	"	1.37	0.43	1.80
#12/3	"	1.03	0.48	1.51
#12/4	"	1.49	0.58	2.07
#10/2	"	1.70	0.51	2.21
#10/3	"	1.63	0.58	2.21
#10/4	"	2.53	0.65	3.18
16130.40 **Boxes**				
Round cast box, type SEH				
1/2"	EA	20.75	25.50	46.25
3/4"	"	20.75	30.75	51.50
SEHC				
1/2"	EA	24.50	25.50	50.00
3/4"	"	24.50	30.75	55.25
SEHL				
1/2"	EA	25.25	25.50	50.75
3/4"	"	24.50	32.50	57.00
SEHT				
1/2"	EA	27.25	30.75	58.00
3/4"	"	27.25	36.50	63.75
SEHX				
1/2"	EA	29.25	36.50	65.75
3/4"	"	29.25	45.00	74.25
Blank cover	"	4.98	10.75	15.73
1/2", hub cover	"	4.76	10.75	15.51
Cover with gasket	"	5.21	13.00	18.21
Rectangle, type FS boxes				
1/2"	EA	10.75	25.50	36.25
3/4"	"	11.25	29.25	40.50
1"	"	12.25	36.50	48.75
FSA				
1/2"	EA	19.00	25.50	44.50
3/4"	"	17.75	29.25	47.00
FSC				
1/2"	EA	11.75	25.50	37.25
3/4"	"	13.00	30.75	43.75
1"	"	16.00	36.50	52.50
FSL				
1/2"	EA	19.00	25.50	44.50
3/4"	"	18.75	29.25	48.00

Basic Materials	UNIT	MAT.	INST.	TOTAL
16130.40 **Boxes** *(Cont.)*				
FSR				
1/2"	EA	19.75	25.50	45.25
3/4"	"	20.00	29.25	49.25
FSS				
1/2"	EA	11.75	25.50	37.25
3/4"	"	13.00	29.25	42.25
FSLA				
1/2"	EA	8.06	25.50	33.56
3/4"	"	9.13	29.25	38.38
FSCA				
1/2"	EA	23.75	25.50	49.25
3/4"	"	23.00	29.25	52.25
FSCC				
1/2"	EA	14.50	29.25	43.75
3/4"	"	21.50	36.50	58.00
FSCT				
1/2"	EA	14.50	29.25	43.75
3/4"	"	18.00	36.50	54.50
1"	"	14.75	41.75	56.50
FST				
1/2"	EA	21.25	36.50	57.75
3/4"	"	21.25	41.75	63.00
FSX				
1/2"	EA	24.25	45.00	69.25
3/4"	"	22.25	53.00	75.25
FSCD boxes				
1/2"	EA	20.25	45.00	65.25
3/4"	"	21.25	53.00	74.25
Rectangle, type FS, 2 gang boxes				
1/2"	EA	22.50	25.50	48.00
3/4"	"	23.25	29.25	52.50
1"	"	24.50	36.50	61.00
FSC, 2 gang boxes				
1/2"	EA	24.00	25.50	49.50
3/4"	"	26.50	29.25	55.75
1"	"	32.25	36.50	68.75
FSS, 2 gang boxes				
3/4"	EA	25.00	29.25	54.25
FS, tandem boxes				
1/2"	EA	25.00	29.25	54.25
3/4"	"	25.75	32.50	58.25
FSC, tandem boxes				
1/2"	EA	33.75	29.25	63.00
3/4"	"	36.00	32.50	68.50
FS, three gang boxes				
3/4"	EA	36.75	32.50	69.25
1"	"	40.50	36.50	77.00
FSS, three gang boxes, 3/4"	"	47.25	36.50	83.75
Weatherproof cast aluminum boxes, 1 gang, 3 outlets				
1/2"	EA	6.74	29.25	35.99
3/4"	"	7.30	36.50	43.80
2 gang, 3 outlets				
1/2"	EA	12.75	36.50	49.25

Basic Materials	UNIT	MAT.	INST.	TOTAL
16130.40 **Boxes** *(Cont.)*				
3/4"	EA	13.75	39.00	52.75
1 gang, 4 outlets				
1/2"	EA	11.75	45.00	56.75
3/4"	"	13.00	53.00	66.00
2 gang, 4 outlets				
1/2"	EA	12.25	45.00	57.25
3/4"	"	13.75	53.00	66.75
1 gang, 5 outlets				
1/2"	EA	9.73	53.00	62.73
3/4"	"	11.50	59.00	70.50
2 gang, 5 outlets				
1/2"	EA	17.50	53.00	70.50
3/4"	"	21.25	59.00	80.25
2 gang, 6 outlets				
1/2"	EA	20.00	62.00	82.00
3/4"	"	21.25	66.00	87.25
2 gang, 7 outlets				
1/2"	EA	21.25	73.00	94.25
3/4"	"	26.25	80.00	106
Weatherproof and type FS box covers, blank, 1 gang	"	3.06	10.75	13.81
Tumbler switch, 1 gang	"	6.30	10.75	17.05
1 gang, single recept	"	3.97	10.75	14.72
Duplex recept	"	5.06	10.75	15.81
Despard	"	5.08	10.75	15.83
Red pilot light	"	24.00	10.75	34.75
SW and				
Single recept	EA	10.75	14.75	25.50
Duplex recept	"	8.72	14.75	23.47
2 gang				
Blank	EA	3.19	13.25	16.44
Tumbler switch	"	4.19	13.25	17.44
Single recept	"	4.19	13.25	17.44
Duplex recept	"	4.19	13.25	17.44
3 gang				
Blank	EA	7.30	14.75	22.05
Tumbler switch	"	9.06	14.75	23.81
4 gang				
Tumbler switch	EA	11.50	18.25	29.75
Box covers				
Surface	EA	16.00	14.75	30.75
Sealing	"	17.50	14.75	32.25
Dome	"	24.25	14.75	39.00
1/2" nipple	"	31.00	14.75	45.75
3/4" nipple	"	31.75	14.75	46.50
16130.60 **Pull And Junction Boxes**				
4"				
Octagon box	EA	3.96	8.37	12.33
Box extension	"	6.67	4.34	11.01
Plaster ring	"	3.66	4.34	8.00
Cover blank	"	1.61	4.34	5.95
Square box	"	5.70	8.37	14.07
Box extension	"	5.59	4.34	9.93

Basic Materials	UNIT	MAT.	INST.	TOTAL
16130.60 **Pull And Junction Boxes** *(Cont.)*				
Plaster ring	EA	3.06	4.34	7.40
Cover blank	"	1.57	4.34	5.91
Switch and device boxes				
2 gang	EA	17.25	8.37	25.62
3 gang	"	30.25	8.37	38.62
4 gang	"	40.50	11.75	52.25
Device covers				
2 gang	EA	13.75	4.34	18.09
3 gang	"	14.25	4.34	18.59
4 gang	"	19.25	4.34	23.59
Handy box	"	4.25	8.37	12.62
Extension	"	4.00	4.34	8.34
Switch cover	"	2.12	4.34	6.46
Switch box with knockout	"	6.38	10.75	17.13
Weatherproof cover, spring type	"	11.75	5.86	17.61
Cover plate, dryer receptacle 1 gang plastic	"	1.81	7.32	9.13
For 4" receptacle, 2 gang	"	3.23	7.32	10.55
Duplex receptacle cover plate, plastic	"	0.79	4.34	5.13
4", vertical bracket box, 1-1/2" with				
RMX clamps	EA	8.22	10.75	18.97
BX clamps	"	8.82	10.75	19.57
4", octagon device cover				
1 switch	EA	4.82	4.34	9.16
1 duplex recept	"	4.82	4.34	9.16
4", octagon swivel hanger box, 1/2" hub	"	12.75	4.34	17.09
3/4" hub	"	14.75	4.34	19.09
4" octagon adjustable bar hangers				
18-1/2"	EA	5.99	3.66	9.65
26-1/2"	"	6.54	3.66	10.20
With clip				
18-1/2"	EA	4.43	3.66	8.09
26-1/2"	"	4.97	3.66	8.63
4", square face bracket boxes, 1-1/2"				
RMX	EA	9.81	10.75	20.56
BX	"	10.75	10.75	21.50
4" square to round plaster rings	"	3.26	4.34	7.60
2 gang device plaster rings	"	3.37	4.34	7.71
Surface covers				
1 gang switch	EA	2.94	4.34	7.28
2 gang switch	"	3.01	4.34	7.35
1 single recept	"	4.43	4.34	8.77
1 20a twist lock recept	"	5.54	4.34	9.88
1 30a twist lock recept	"	7.10	4.34	11.44
1 duplex recept	"	2.74	4.34	7.08
2 duplex recept	"	2.74	4.34	7.08
Switch and duplex recept	"	4.58	4.34	8.92
4" plastic round boxes, ground straps				
Box only	EA	2.09	10.75	12.84
Box w/clamps	"	2.43	14.75	17.18
Box w/16" bar	"	5.16	16.75	21.91
Box w/24" bar	"	5.15	18.25	23.40
4" plastic round box covers				
Blank cover	EA	1.36	4.34	5.70

Basic Materials	UNIT	MAT.	INST.	TOTAL

16130.60   Pull And Junction Boxes (Cont.)

	UNIT	MAT.	INST.	TOTAL
Plaster ring	EA	2.22	4.34	6.56
4" plastic square boxes				
Box only	EA	1.61	10.75	12.36
Box w/clamps	"	1.99	14.75	16.74
Box w/hanger	"	2.46	18.25	20.71
Box w/nails and clamp	"	3.52	18.25	21.77
4" plastic square box covers				
Blank cover	EA	1.32	4.34	5.66
1 gang ring	"	1.61	4.34	5.95
2 gang ring	"	2.26	4.34	6.60
Round ring	"	1.80	4.34	6.14

16130.80   Receptacles

	UNIT	MAT.	INST.	TOTAL
Contractor grade duplex receptacles, 15a 120v				
Duplex	EA	1.60	14.75	16.35
125 volt, 20a, duplex, standard grade	"	12.00	14.75	26.75
Ground fault interrupter type	"	38.75	21.75	60.50
250 volt, 20a, 2 pole, single, ground type	"	20.00	14.75	34.75
120/208v, 4 pole, single receptacle, twist lock				
20a	EA	23.75	25.50	49.25
50a	"	45.25	25.50	70.75
125/250v, 3 pole, flush receptacle				
30a	EA	24.00	21.75	45.75
50a	"	29.75	21.75	51.50
60a	"	77.00	25.50	103
Dryer receptacle, 250v, 30a/50a, 3 wire	"	18.00	21.75	39.75
Clock receptacle, 2 pole, grounding type	"	12.00	14.75	26.75
125v, 20a single recept. grounding type				
Standard grade	EA	13.00	14.75	27.75
125/250v, 3 pole, 3 wire surface recepts				
30a	EA	20.25	21.75	42.00
50a	"	22.50	21.75	44.25
60a	"	49.50	25.50	75.00
Cord set, 3 wire, 6' cord				
30a	EA	18.25	21.75	40.00
50a	"	25.50	21.75	47.25
125/250v, 3 pole, 3 wire cap				
30a	EA	18.00	29.25	47.25
50a	"	32.75	29.25	62.00
60a	"	42.00	32.50	74.50

16199.10   Utility Poles & Fittings

	UNIT	MAT.	INST.	TOTAL
Wood pole, creosoted				
25'	EA	450	170	620
30'	"	540	220	760
Treated, wood preservative, 6"x6"				
8'	EA	100	36.50	137
10'	"	150	59.00	209
12'	"	160	65.00	225
14'	"	200	98.00	298
16'	"	250	120	370
18'	"	280	150	430
20'	"	360	150	510

Basic Materials	UNIT	MAT.	INST.	TOTAL
16199.10 Utility Poles & Fittings *(Cont.)*				
Aluminum, brushed, no base				
8'	EA	650	150	800
10'	"	750	200	950
15'	"	840	200	1,040
20'	"	1,020	230	1,250
25'	"	1,360	280	1,640
Steel, no base				
10'	EA	760	180	940
15'	"	830	220	1,050
20'	"	1,110	280	1,390
25'	"	1,260	330	1,590
Concrete, no base				
13'	EA	980	400	1,380
16'	"	1,360	530	1,890
18'	"	1,640	640	2,280
25'	"	2,010	730	2,740
16350.10 Circuit Breakers				
Load center circuit breakers, 240v				
1 pole, 10-60a	EA	21.25	18.25	39.50
2 pole				
10-60a	EA	43.25	29.25	72.50
70-100a	"	130	48.75	179
110-150a	"	280	53.00	333
Load center, G.F.I. breakers, 240v				
1 pole, 15-30a	EA	160	21.75	182
Tandem breakers, 240v				
1 pole, 15-30a	EA	35.00	29.25	64.25
2 pole, 15-30a	"	64.00	39.00	103
16365.10 Fuses				
Fuse, one-time, 250v				
30a	EA	2.77	3.66	6.43
60a	"	4.68	3.66	8.34
100a	"	19.50	3.66	23.16
16395.10 Grounding				
Ground rods, copper clad, 1/2" x				
6'	EA	17.00	48.75	65.75
8'	"	23.50	53.00	76.50
5/8" x				
6'	EA	22.75	53.00	75.75
8'	"	29.25	73.00	102
Ground rod clamp				
5/8"	EA	6.94	9.01	15.95
Ground rod couplings				
1/2"	EA	12.75	7.32	20.07
5/8"	"	17.75	7.32	25.07
Ground rod, driving stud				
1/2"	EA	10.25	7.32	17.57
5/8"	"	12.00	7.32	19.32
Ground rod clamps, #8-2 to				
1" pipe	EA	11.00	14.75	25.75

Basic Materials	UNIT	MAT.	INST.	TOTAL
16395.10 Grounding *(Cont.)*				
2" pipe	EA	13.75	18.25	32.00

Service And Distribution	UNIT	MAT.	INST.	TOTAL
16430.20 Metering				
Outdoor wp meter sockets, 1 gang, 240v, 1 phase				
Includes sealing ring, 100a	EA	47.25	110	157
150a	"	56.00	130	186
200a	"	71.00	150	221
Die cast hubs, 1-1/4"	"	6.52	23.50	30.02
1-1/2"	"	7.48	23.50	30.98
2"	"	9.05	23.50	32.55
16470.10 Panelboards				
Indoor load center, 1 phase 240v main lug only				
30a - 2 spaces	EA	32.25	150	182
100a - 8 spaces	"	100	180	280
150a - 16 spaces	"	270	220	490
200a - 24 spaces	"	550	250	800
200a - 42 spaces	"	570	290	860
Main circuit breaker				
100a - 8 spaces	EA	330	180	510
100a - 16 spaces	"	350	200	550
150a - 16 spaces	"	580	220	800
150a - 24 spaces	"	690	230	920
200a - 24 spaces	"	640	250	890
200a - 42 spaces	"	910	270	1,180
120/208v, flush, 3 ph., 4 wire, main only				
100a				
12 circuits	EA	1,020	370	1,390
20 circuits	"	1,410	460	1,870
30 circuits	"	2,090	510	2,600
225a				
30 circuits	EA	2,130	570	2,700
42 circuits	"	2,690	700	3,390
16490.10 Switches				
Photo electric switches				
1000 watt				
105-135v	EA	37.00	53.00	90.00
Dimmer switch and switch plate				
600w	EA	34.00	22.50	56.50
Time clocks with skip, 40a, 120v				
SPST	EA	100	55.00	155
Contractor grade wall switch 15a, 120v				
Single pole	EA	1.79	11.75	13.54

16 ELECTRICAL

Service And Distribution	UNIT	MAT.	INST.	TOTAL
16490.10	**Switches** *(Cont.)*			
Three way	EA	3.26	14.75	18.01
Four way	"	11.00	19.50	30.50
Specification grade toggle switches, 20a, 120-277v				
Single pole	EA	3.93	14.75	18.68
Double pole	"	9.43	21.75	31.18
3 way	"	10.25	18.25	28.50
4 way	"	31.00	21.75	52.75
Combination switch and pilot light, single pole	"	13.50	21.75	35.25
3 way	"	16.75	25.50	42.25
Combination switch and receptacle, single pole	"	19.50	21.75	41.25
3 way	"	24.00	21.75	45.75
Switch plates, plastic ivory				
1 gang	EA	0.43	5.86	6.29
2 gang	"	1.01	7.32	8.33
3 gang	"	1.58	8.74	10.32
4 gang	"	4.04	10.75	14.79
5 gang	"	4.23	11.75	15.98
6 gang	"	4.99	13.25	18.24
Stainless steel				
1 gang	EA	3.64	5.86	9.50
2 gang	"	5.06	7.32	12.38
3 gang	"	7.76	9.01	16.77
4 gang	"	13.25	10.75	24.00
5 gang	"	15.50	11.75	27.25
6 gang	"	19.50	13.25	32.75
Brass				
1 gang	EA	6.79	5.86	12.65
2 gang	"	14.50	7.32	21.82
3 gang	"	22.50	9.01	31.51
4 gang	"	26.00	10.75	36.75
5 gang	"	32.25	11.75	44.00
6 gang	"	38.75	13.25	52.00

Lighting	UNIT	MAT.	INST.	TOTAL
16510.05	**Interior Lighting**			
Recessed fluorescent fixtures, 2'x2'				
2 lamp	EA	72.00	53.00	125
4 lamp	"	98.00	53.00	151
Surface mounted incandescent fixtures				
40w	EA	110	48.75	159
75w	"	110	48.75	159
100w	"	120	48.75	169
150w	"	160	48.75	209
Pendant				
40w	EA	90.00	59.00	149

Lighting	UNIT	MAT.	INST.	TOTAL
16510.05		**Interior Lighting** *(Cont.)*		
75w	EA	99.00	59.00	158
100w	"	110	59.00	169
150w	"	130	59.00	189
Contractor grade recessed down lights				
100 watt housing only	EA	74.00	73.00	147
150 watt housing only	"	110	73.00	183
100 watt trim	"	61.00	36.50	97.50
150 watt trim	"	95.00	36.50	132
Recessed incandescent fixtures				
40w	EA	150	110	260
75w	"	160	110	270
100w	"	180	110	290
150w	"	190	110	300
Light track single circuit				
2'	EA	44.25	36.50	80.75
4'	"	52.00	36.50	88.50
8'	"	72.00	73.00	145
12'	"	100	110	210
Fittings and accessories				
Dead end	EA	17.50	10.75	28.25
Starter kit	"	23.50	18.25	41.75
Conduit feed	"	22.75	10.75	33.50
Straight connector	"	20.25	10.75	31.00
Center feed	"	32.25	10.75	43.00
L-connector	"	22.75	10.75	33.50
T-connector	"	30.50	10.75	41.25
X-connector	"	36.75	14.75	51.50
Cord and plug	"	36.75	7.32	44.07
Rigid corner	"	48.50	10.75	59.25
Flex connector	"	38.00	10.75	48.75
2 way connector	"	110	14.75	125
Spacer clip	"	1.64	3.66	5.30
Grid box	"	9.17	10.75	19.92
T-bar clip	"	2.45	3.66	6.11
Utility hook	"	7.10	10.75	17.85
Fixtures, square				
R-20	EA	45.00	10.75	55.75
R-30	"	70.00	10.75	80.75
40w flood	"	110	10.75	121
40w spot	"	110	10.75	121
100w flood	"	130	10.75	141
100w spot	"	100	10.75	111
Mini spot	"	42.75	10.75	53.50
Mini flood	"	99.00	10.75	110
Quartz, 500w	"	250	10.75	261
R-20 sphere	"	75.00	10.75	85.75
R-30 sphere	"	39.75	10.75	50.50
R-20 cylinder	"	53.00	10.75	63.75
R-30 cylinder	"	62.00	10.75	72.75
R-40 cylinder	"	63.00	10.75	73.75
R-30 wall wash	"	99.00	10.75	110
R-40 wall wash	"	120	10.75	131

Lighting	UNIT	MAT.	INST.	TOTAL
16510.10 **Industrial Lighting**				
Surface mounted fluorescent, wrap around lens				
1 lamp	EA	90.00	59.00	149
2 lamps	"	130	65.00	195
Wall mounted fluorescent				
2-20w lamps	EA	93.00	36.50	130
2-30w lamps	"	110	36.50	147
2-40w lamps	"	110	48.75	159
Strip fluorescent				
4'				
1 lamp	EA	45.50	48.75	94.25
2 lamps	"	55.00	48.75	104
8'				
1 lamp	EA	66.00	53.00	119
2 lamps	"	99.00	65.00	164
Compact fluorescent				
2-7w	EA	160	73.00	233
2-13w	"	180	98.00	278
16670.10 **Lightning Protection**				
Lightning protection				
Copper point, nickel plated, 12'				
1/2" dia.	EA	44.00	73.00	117
5/8" dia.	"	49.50	73.00	123
16750.20 **Signaling Systems**				
Contractor grade doorbell chime kit				
Chime	EA	36.75	73.00	110
Doorbutton	"	5.11	23.50	28.61
Transformer	"	16.25	36.50	52.75

Resistance Heating	UNIT	MAT.	INST.	TOTAL
16850.10 **Electric Heating**				
Baseboard heater				
2', 375w	EA	41.75	73.00	115
3', 500w	"	49.50	73.00	123
4', 750w	"	55.00	84.00	139
5', 935w	"	78.00	98.00	176
6', 1125w	"	92.00	120	212
7', 1310w	"	100	130	230
8', 1500w	"	120	150	270
9', 1680w	"	130	160	290
10', 1875w	"	180	170	350
Unit heater, wall mounted				
750w	EA	160	120	280
1500w	"	220	120	340

Resistance Heating	UNIT	MAT.	INST.	TOTAL
16850.10 — Electric Heating *(Cont.)*				
Thermostat				
Integral	EA	37.50	36.50	74.00
Line voltage	"	38.50	36.50	75.00
Electric heater connection	"	1.65	18.25	19.90
Fittings				
Inside corner	EA	24.25	29.25	53.50
Outside corner	"	26.50	29.25	55.75
Receptacle section	"	27.50	29.25	56.75
Blank section	"	34.00	29.25	63.25
Radiant ceiling heater panels				
500w	EA	390	73.00	463
750w	"	430	73.00	503
Unit heater thermostat	"	60.00	39.00	99.00
Mounting bracket	"	62.00	53.00	115
Relay	"	79.00	45.00	124
16910.40 — Control Cable				
Control cable, 600v, #14 THWN, PVC jacket				
2 wire	LF	0.38	0.58	0.96
4 wire	"	0.65	0.73	1.38

Solar	UNIT	MAT.	INST.	TOTAL
16990.10 — Solar Electrical Systems				
Photovoltaic, Full Grid-Tie System				
Panel array, inverter, mounts, racks, conduit, etc., 3,000				
Min.	EA			21,780
Ave.	"			29,040
Max.	"			36,300
4,000 Watt				
Min.	EA			29,040
Ave.	"			38,720
Max.	"			48,400
5,000 Watt				
Min.	EA			36,300
Ave.	"			48,400
Max.	"			60,500
Photovoltaic Components				
Polycristalline Rigid Panel, 200 watt				
Min.	EA	480	21.00	501
Ave.	"	540	21.00	561
Max.	"	280	21.00	301
215 Watt				
Min.	EA	520	21.00	541
Ave.	"	580	21.00	601
Max.	"	630	21.00	651

Solar	UNIT	MAT.	INST.	TOTAL
16990.10 **Solar Electrical Systems** *(Cont.)*				
230 Watt				
Min.	EA	570	21.00	591
Ave.	"	630	21.00	651
Max.	"	690	21.00	711
245 Watt				
Min.	EA	600	21.00	621
Ave.	"	670	21.00	691
Max.	"	730	21.00	751
Anodized aluminum rail, 8'				
Min.	EA	41.25	13.00	54.25
Ave.	"	43.50	13.00	56.50
Max.	"	46.00	13.00	59.00
10'				
Min.	EA	51.00	14.75	65.75
Ave.	"	54.00	14.75	68.75
Max.	"	58.00	14.75	72.75
12'				
Min.	EA	63.00	16.75	79.75
Ave.	"	65.00	16.75	81.75
Max.	"	68.00	16.75	84.75
Panel mounts, aluminum, mount tile, flush				
Min.	EA	34.00	39.00	73.00
Ave.	"	36.25	39.00	75.25
Max.	"	38.75	39.00	77.75
Standard				
Min.	EA	26.50	39.00	65.50
Ave.	"	29.00	39.00	68.00
Max.	"	31.50	39.00	70.50
Rail clamp, mid-clamp				
Min.	EA			8.47
Ave.	"			9.68
Max.	"			11.00
End-clamp				
Min.	EA			11.00
Ave.	"			12.25
Max.	"			13.75
Anodized aluminum, rail splice kit				
Min.	EA			17.00
Ave.	"			19.25
Max.	"			21.75
Power Distribution Panel				
Min.	EA	150	150	300
Ave.	"	170	150	320
Max.	"	200	150	350
Inverters				
Light capacity, micro inverter, 190 Watt				
Min.	EA	220	19.50	240
Ave.	"	250	19.50	270
Max.	"	270	19.50	290
Medium capacity, inverter, 1000 Watt				
Min.	EA	2,180	290	2,470
Ave.	"	2,420	290	2,710
Max.	"	2,660	290	2,950

Solar	UNIT	MAT.	INST.	TOTAL
16990.40 **Solar Electrical Systems** *(Cont.)*				
2500 Watt				
Min.	EA	3,630	290	3,920
Ave.	"	3,870	290	4,160
Max.	"	4,110	290	4,400
5000 Watt				
Min.	EA	5,810	290	6,100
Ave.	"	6,050	290	6,340
Max.	"	6,290	290	6,580
Circuits				
Combiner Box, 12 circuit				
Min.	EA	660	200	860
Ave.	"	790	200	990
Max.	"	880	200	1,080
28 circuit				
Min.	EA	1,070	200	1,270
Ave.	"	1,200	200	1,400
Max.	"	1,340	200	1,540
Solar Circuit Breaker, 15A, 150 VDC				
Min.	EA			13.25
Ave.	"			15.75
Max.	"			18.25
Wires and Conductors				
Cable, bare copper, 10 AWG, 500' coils				
Min.	EA			190
Ave.	"			200
Max.	"			210
Multi-Contact branch connector, MC4				
Min.	EA	34.00	19.50	53.50
Ave.	"	36.25	19.50	55.75
Max.	"	38.75	19.50	58.25

Man-Hour Tables

The man-hour productivities used to develop the labor costs are listed in the following section of this book. These productivities represent typical installation labor for thousands of construction items. The data takes into account all activities involved in normal construction under commonly experienced working conditions. As with the Costbook pages, these items are listed according to the 16 divisions. In order to best use the information in this book, please review this sample page and read the "Features in this Book" section.

Division

Broadscope Category

Mediumscope Category (First 5 Digits)

Detailed Descriptions
Complete descriptions of items may include information listed above a particular line. Review of the whole category is recommended for a complete description.

Unit of Measurement
Each item is defined in terms of the common estimating unit. Quantities listed are defined as man-hour per unit.

Man-Hours
Man-hour quantities represent typical installation times and take into account all activities involved in normal construction under commonly experienced working conditions.

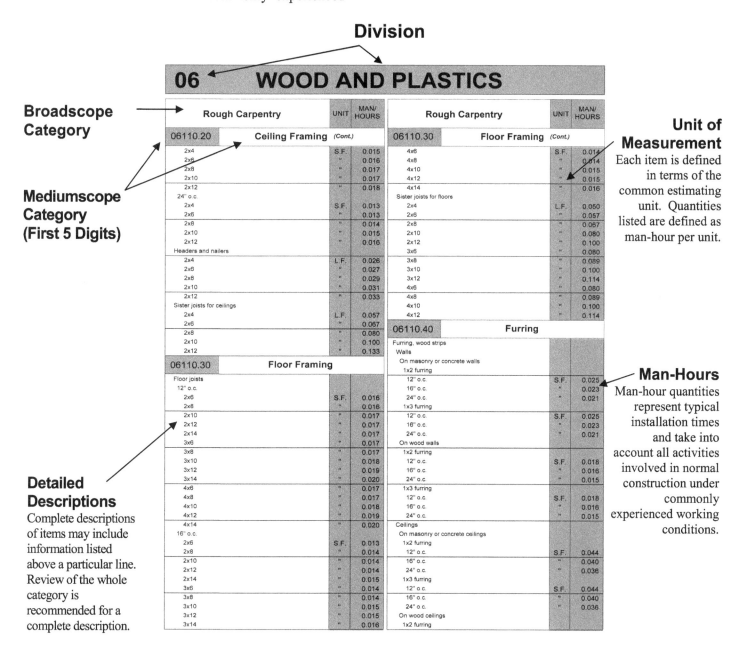

06 WOOD AND PLASTICS

Rough Carpentry	UNIT	MAN/HOURS
06110.20 Ceiling Framing *(Cont.)*		
2x4	S.F.	0.015
2x6	"	0.016
2x8	"	0.017
2x10	"	0.017
2x12	"	0.018
24" o.c.		
2x4	S.F.	0.013
2x6	"	0.013
2x8	"	0.014
2x10	"	0.015
2x12	"	0.016
Headers and nailers		
2x4	L.F.	0.026
2x6	"	0.027
2x8	"	0.029
2x10	"	0.031
2x12	"	0.033
Sister joists for ceilings		
2x4	L.F.	0.057
2x6	"	0.067
2x8	"	0.080
2x10	"	0.100
2x12	"	0.133
06110.30 Floor Framing		
Floor joists		
12" o.c.		
2x6	S.F.	0.016
2x8	"	0.016
2x10	"	0.017
2x12	"	0.017
2x14	"	0.017
3x6	"	0.017
3x8	"	0.017
3x10	"	0.018
3x12	"	0.019
3x14	"	0.020
4x6	"	0.017
4x8	"	0.017
4x10	"	0.018
4x12	"	0.019
4x14	"	0.020
16" o.c.		
2x6	S.F.	0.013
2x8	"	0.014
2x10	"	0.014
2x12	"	0.014
2x14	"	0.015
3x6	"	0.014
3x8	"	0.014
3x10	"	0.015
3x12	"	0.015
3x14	"	0.016

Rough Carpentry	UNIT	MAN/HOURS
06110.30 Floor Framing *(Cont.)*		
4x6	S.F.	0.014
4x8	"	0.014
4x10	"	0.015
4x12	"	0.015
4x14	"	0.016
Sister joists for floors		
2x4	L.F.	0.050
2x6	"	0.057
2x8	"	0.067
2x10	"	0.080
2x12	"	0.100
3x6	"	0.080
3x8	"	0.089
3x10	"	0.100
3x12	"	0.114
4x6	"	0.080
4x8	"	0.089
4x10	"	0.100
4x12	"	0.114
06110.40 Furring		
Furring, wood strips		
Walls		
On masonry or concrete walls		
1x2 furring		
12" o.c.	S.F.	0.025
16" o.c.	"	0.023
24" o.c.	"	0.021
1x3 furring		
12" o.c.	S.F.	0.025
16" o.c.	"	0.023
24" o.c.	"	0.021
On wood walls		
1x2 furring		
12" o.c.	S.F.	0.018
16" o.c.	"	0.016
24" o.c.	"	0.015
1x3 furring		
12" o.c.	S.F.	0.018
16" o.c.	"	0.016
24" o.c.	"	0.015
Ceilings		
On masonry or concrete ceilings		
1x2 furring		
12" o.c.	S.F.	0.044
16" o.c.	"	0.040
24" o.c.	"	0.036
1x3 furring		
12" o.c.	S.F.	0.044
16" o.c.	"	0.040
24" o.c.	"	0.036
On wood ceilings		
1x2 furring		

Site Remediation	UNIT	MAN/ HOURS
02115.66 **Septic Tank Removal**		
Remove septic tank		
1000 gals	EA	2.000
2000 gals	"	2.400

Site Preparation	UNIT	MAN/ HOURS
02210.10 **Soil Boring**		
Borings, uncased, stable earth		
2-1/2" dia.	LF	0.300
4" dia.	"	0.343
Cased, including samples		
2-1/2" dia.	LF	0.400
4" dia.	"	0.686
Drilling in rock		
No sampling	LF	0.632
With casing and sampling	"	0.800
Test pits		
Light soil	EA	4.000
Heavy soil	"	6.000
02210.20 **Rough Grading**		
Site grading, cut & fill, sandy clay, 200' haul, 75 hp dozer	CY	0.032
Spread topsoil by equipment on site	"	0.036
Site grading (cut and fill to 6") less than 1 acre		
75 hp dozer	CY	0.053
1.5 cy backhoe/loader	"	0.080

Demolition	UNIT	MAN/ HOURS
02220.10 **Complete Building Demolition**		
Wood frame	CF	0.003
02220.15 **Selective Building Demolition**		
Partition removal		
Concrete block partitions		
8" thick	SF	0.053
Brick masonry partitions		
4" thick	SF	0.040
8" thick	"	0.050

Demolition	UNIT	MAN/ HOURS
02220.15 **Selective Building Demolition** *(Cont.)*		
Stud partitions		
Metal or wood, with drywall both sides	SF	0.040
Door and frame removal		
Wood in framed wall		
2'6"x6'8"	EA	0.571
3'x6'8"	"	0.667
Ceiling removal		
Acoustical tile ceiling		
Adhesive fastened	SF	0.008
Furred and glued	"	0.007
Suspended grid	"	0.005
Drywall ceiling		
Furred and nailed	SF	0.009
Nailed to framing	"	0.008
Window removal		
Metal windows, trim included		
2'x3'	EA	0.800
3'x4'	"	1.000
4'x8'	"	2.000
Wood windows, trim included		
2'x3'	EA	0.444
3'x4'	"	0.533
6'x8'	"	0.800
Concrete block walls, not including toothing		
4" thick	SF	0.044
6" thick	"	0.047
8" thick	"	0.050
Rubbish handling		
Load in dumpster or truck		
Minimum	CF	0.018
Maximum	"	0.027
Rubbish hauling		
Hand loaded on trucks, 2 mile trip	CY	0.320
Machine loaded on trucks, 2 mile trip	"	0.240

Selective Site Demolition	UNIT	MAN/ HOURS
02225.13 **Core Drilling**		
Concrete		
6" thick		
3" dia.	EA	0.571
8" thick		
3" dia.	EA	0.800

02 SITE CONSTRUCTION

Selective Site Demolition		UNIT	MAN/HOURS
02225.15	**Curb & Gutter Demolition**		
Removal, plain concrete curb		LF	0.060
Plain concrete curb and 2' gutter		"	0.083
02225.20	**Fence Demolition**		
Remove fencing			
Chain link, 8' high			
For disposal		LF	0.040
For reuse		"	0.100
Wood			
4' high		SF	0.027
Masonry			
8" thick			
4' high		SF	0.080
6' high		"	0.100
02225.25	**Guardrail Demolition**		
Remove standard guardrail			
Steel		LF	0.080
Wood		"	0.062
02225.30	**Hydrant Demolition**		
Remove and reset fire hydrant		EA	12.000
02225.40	**Pavement And Sidewalk**		
Concrete pavement, 6" thick			
No reinforcement		SY	0.160
With wire mesh		"	0.240
With rebars		"	0.300
Sidewalk, 4" thick, with disposal		"	0.080
02225.42	**Drainage Piping Demolition**		
Remove drainage pipe, not including excavation			
12" dia.		LF	0.100
18" dia.		"	0.126
02225.43	**Gas Piping Demolition**		
Remove welded steel pipe, not including excavation			
4" dia.		LF	0.150
5" dia.		"	0.240
02225.45	**Sanitary Piping Demolition**		
Remove sewer pipe, not including excavation			
4" dia.		LF	0.096
02225.48	**Water Piping Demolition**		
Remove water pipe, not including excavation			
4" dia.		LF	0.109

Selective Site Demolition		UNIT	MAN/HOURS
02225.50	**Saw Cutting Pavement**		
Pavement, bituminous			
2" thick		LF	0.016
3" thick		"	0.020
Concrete pavement, with wire mesh			
4" thick		LF	0.031
5" thick		"	0.033
Plain concrete, unreinforced			
4" thick		LF	0.027
5" thick		"	0.031

Site Clearing		UNIT	MAN/HOURS
02230.10	**Clear Wooded Areas**		
Clear wooded area			
Light density		ACRE	60.000
Medium density		"	80.000
Heavy density		"	96.000
02230.50	**Tree Cutting & Clearing**		
Cut trees and clear out stumps			
9" to 12" dia.		EA	4.800
To 24" dia.		"	6.000
24" dia. and up		"	8.000
Loading and trucking			
For machine load, per load, round trip			
1 mile		EA	0.960
3 mile		"	1.091
5 mile		"	1.200
10 mile		"	1.600
20 mile		"	2.400
Hand loaded, round trip			
1 mile		EA	2.000
3 mile		"	2.286
5 mile		"	2.667
10 mile		"	3.200
20 mile		"	4.000

02 SITE CONSTRUCTION

Earthwork, Excavation & Fill	UNIT	MAN/HOURS
02315.10 — **Base Course**		
Base course, crushed stone		
3" thick	SY	0.004
4" thick	"	0.004
6" thick	"	0.005
Base course, bank run gravel		
4" deep	SY	0.004
6" deep	"	0.005
Prepare and roll sub base		
Minimum	SY	0.004
Average	"	0.005
Maximum	"	0.007
02315.20 — **Borrow**		
Borrow fill, F.O.B. at pit		
Sand, haul to site, round trip		
10 mile	CY	0.080
20 mile	"	0.133
30 mile	"	0.200
Place borrow fill and compact		
Less than 1 in 4 slope	CY	0.040
Greater than 1 in 4 slope	"	0.053
02315.30 — **Bulk Excavation**		
Excavation, by small dozer		
Large areas	CY	0.016
Small areas	"	0.027
Trim banks	"	0.040
Hydraulic excavator		
1 cy capacity		
Light material	CY	0.040
Medium material	"	0.048
Wet material	"	0.060
Blasted rock	"	0.069
1-1/2 cy capacity		
Light material	CY	0.010
Medium material	"	0.013
Wet material	"	0.016
Wheel mounted front-end loader		
7/8 cy capacity		
Light material	CY	0.020
Medium material	"	0.023
Wet material	"	0.027
Blasted rock	"	0.032
1-1/2 cy capacity		
Light material	CY	0.011
Medium material	"	0.012
Wet material	"	0.013
Blasted rock	"	0.015
2-1/2 cy capacity		
Light material	CY	0.009
Medium material	"	0.010
Wet material	"	0.011
Blasted rock	"	0.011

Earthwork, Excavation & Fill	UNIT	MAN/HOURS
02315.30 — **Bulk Excavation** (Cont.)		
Track mounted front-end loader		
1-1/2 cy capacity		
Light material	CY	0.013
Medium material	"	0.015
Wet material	"	0.016
Blasted rock	"	0.018
2-3/4 cy capacity		
Light material	CY	0.008
Medium material	"	0.009
Wet material	"	0.010
Blasted rock	"	0.011
02315.40 — **Building Excavation**		
Structural excavation, unclassified earth		
3/8 cy backhoe	CY	0.107
3/4 cy backhoe	"	0.080
1 cy backhoe	"	0.067
Foundation backfill and compaction by machine	"	0.160
02315.45 — **Hand Excavation**		
Excavation		
To 2' deep		
Normal soil	CY	0.889
Sand and gravel	"	0.800
Medium clay	"	1.000
Heavy clay	"	1.143
Loose rock	"	1.333
To 6' deep		
Normal soil	CY	1.143
Sand and gravel	"	1.000
Medium clay	"	1.333
Heavy clay	"	1.600
Loose rock	"	2.000
Backfilling foundation without compaction, 6" lifts	"	0.500
Compaction of backfill around structures or in trench		
By hand with air tamper	CY	0.571
By hand with vibrating plate tamper	"	0.533
1 ton roller	"	0.400
Miscellaneous hand labor		
Trim slopes, sides of excavation	SF	0.001
Trim bottom of excavation	"	0.002
Excavation around obstructions and services	CY	2.667
02315.50 — **Roadway Excavation**		
Roadway excavation		
1/4 mile haul	CY	0.016
2 mile haul	"	0.027
5 mile haul	"	0.040
Spread base course	"	0.020
Roll and compact	"	0.027

02 SITE CONSTRUCTION

Earthwork, Excavation & Fill	UNIT	MAN/ HOURS
02315.60 — **Trenching**		
Trenching and continuous footing excavation		
By gradall		
1 cy capacity		
Light soil	CY	0.023
Medium soil	"	0.025
Heavy/wet soil	"	0.027
Loose rock	"	0.029
Blasted rock	"	0.031
By hydraulic excavator		
1/2 cy capacity		
Light soil	CY	0.027
Medium soil	"	0.029
Heavy/wet soil	"	0.032
Loose rock	"	0.036
Blasted rock	"	0.040
1 cy capacity		
Light soil	CY	0.019
Medium soil	"	0.020
Heavy/wet soil	"	0.021
Loose rock	"	0.023
Blasted rock	"	0.025
1-1/2 cy capacity		
Light soil	CY	0.017
Medium soil	"	0.018
Heavy/wet soil	"	0.019
Loose rock	"	0.020
Blasted rock	"	0.021
2 cy capacity		
Light soil	CY	0.016
Medium soil	"	0.017
Heavy/wet soil	"	0.018
Loose rock	"	0.019
Blasted rock	"	0.020
Hand excavation		
Bulk, wheeled 100'		
Normal soil	CY	0.889
Sand or gravel	"	0.800
Medium clay	"	1.143
Heavy clay	"	1.600
Loose rock	"	2.000
Trenches, up to 2' deep		
Normal soil	CY	1.000
Sand or gravel	"	0.889
Medium clay	"	1.333
Heavy clay	"	2.000
Loose rock	"	2.667
Trenches, to 6' deep		
Normal soil	CY	1.143
Sand or gravel	"	1.000
Medium clay	"	1.600
Heavy clay	"	2.667
Loose rock	"	4.000
Backfill trenches		

Earthwork, Excavation & Fill	UNIT	MAN/ HOURS
02315.60 — **Trenching** (Cont.)		
With compaction		
By hand	CY	0.667
By 60 hp tracked dozer	"	0.020
02315.70 — **Utility Excavation**		
Trencher, sandy clay, 8" wide trench		
18" deep	LF	0.018
24" deep	"	0.020
36" deep	"	0.023
Trench backfill, 95% compaction		
Tamp by hand	CY	0.500
Vibratory compaction	"	0.400
Trench backfilling, with borrow sand, place & compact	"	0.400
02315.80 — **Hauling Material**		
Haul material by 10 cy dump truck, round trip distance		
1 mile	CY	0.044
2 mile	"	0.053
5 mile	"	0.073
10 mile	"	0.080
20 mile	"	0.089
30 mile	"	0.107

Soil Stabilization & Treatment	UNIT	MAN/ HOURS
02340.05 — **Soil Stabilization**		
Straw bale secured with rebar	LF	0.027
Filter barrier, 18" high filter fabric	"	0.080
Sediment fence, 36" fabric with 6" mesh	"	0.100
Soil stabilization with tar paper, burlap, straw and stakes	SF	0.001
02360.20 — **Soil Treatment**		
Soil treatment, termite control pretreatment		
Under slabs	SF	0.004
By walls	"	0.005
02370.40 — **Riprap**		
Riprap		
Crushed stone blanket, max size 2-1/2"	TON	0.533
Stone, quarry run, 300 lb. stones	"	0.492
400 lb. stones	"	0.457
500 lb. stones	"	0.427
750 lb. stones	"	0.400
Dry concrete riprap in bags 3" thick, 80 lb. per bag	BAG	0.027

Piles And Caissons		UNIT	MAN/HOURS
02455.60	**Steel Piles**		
H-section piles			
8x8			
36 lb/ft			
30' long		LF	0.080
40' long		"	0.064
Tapered friction piles, fluted casing, up to 50'			
With 4000 psi concrete no reinforcing			
12" dia.		LF	0.048
14" dia.		"	0.049
02455.65	**Steel Pipe Piles**		
Concrete filled, 3000# concrete, up to 40'			
8" dia.		LF	0.069
10" dia.		"	0.071
12" dia.		"	0.074
Pipe piles, non-filled			
8" dia.		LF	0.053
10" dia.		"	0.055
12" dia.		"	0.056
Splice			
8" dia.		EA	1.600
10" dia.		"	1.600
12" dia.		"	2.000
Standard point			
8" dia.		EA	1.600
10" dia.		"	1.600
12" dia.		"	2.000
Heavy duty point			
8" dia.		EA	2.000
10" dia.		"	2.000
12" dia.		"	2.667
02455.80	**Wood And Timber Piles**		
Treated wood piles, 12" butt, 8" tip			
25' long		LF	0.096
30' long		"	0.080
35' long		"	0.069
40' long		"	0.060
02465.50	**Prestressed Piling**		
Prestressed concrete piling, less than 60' long			
10" sq.		LF	0.040
12" sq.		"	0.042
Straight cylinder, less than 60' long			
12" dia.		LF	0.044
14" dia.		"	0.045

Utility Services		UNIT	MAN/HOURS
02510.10	**Wells**		
Domestic water, drilled and cased			
4" dia.		LF	0.480
6" dia.		"	0.533
02510.40	**Ductile Iron Pipe**		
Ductile iron pipe, cement lined, slip-on joints			
4"		LF	0.067
6"		"	0.071
8"		"	0.075
Mechanical joint pipe			
4"		LF	0.092
6"		"	0.100
8"		"	0.109
Fittings, mechanical joint			
90 degree elbow			
4"		EA	0.533
6"		"	0.615
8"		"	0.800
45 degree elbow			
4"		EA	0.533
6"		"	0.615
8"		"	0.800
02510.60	**Plastic Pipe**		
PVC, class 150 pipe			
4" dia.		LF	0.060
6" dia.		"	0.065
8" dia.		"	0.069
Schedule 40 pipe			
1-1/2" dia.		LF	0.047
2" dia.		"	0.050
2-1/2" dia.		"	0.053
3" dia.		"	0.057
4" dia.		"	0.067
6" dia.		"	0.080
90 degree elbows			
1"		EA	0.133
1-1/2"		"	0.133
2"		"	0.145
2-1/2"		"	0.160
3"		"	0.178
4"		"	0.200
6"		"	0.267
45 degree elbows			
1"		EA	0.133
1-1/2"		"	0.133
2"		"	0.145
2-1/2"		"	0.160
3"		"	0.178
4"		"	0.200
6"		"	0.267
Tees			
1"		EA	0.160

Utility Services		UNIT	MAN/ HOURS
02510.60	**Plastic Pipe** *(Cont.)*		
1-1/2"		EA	0.160
2"		"	0.178
2-1/2"		"	0.200
3"		"	0.229
4"		"	0.267
6"		"	0.320
Couplings			
1"		EA	0.133
1-1/2"		"	0.133
2"		"	0.145
2-1/2"		"	0.160
3"		"	0.178
4"		"	0.200
6"		"	0.267
Drainage pipe			
PVC schedule 80			
1" dia.		LF	0.047
1-1/2" dia.		"	0.047
ABS, 2" dia.		"	0.050
2-1/2" dia.		"	0.053
3" dia.		"	0.057
4" dia.		"	0.067
6" dia.		"	0.080
8" dia.		"	0.063
10" dia.		"	0.075
12" dia.		"	0.080
90 degree elbows			
1"		EA	0.133
1-1/2"		"	0.133
2"		"	0.145
2-1/2"		"	0.160
3"		"	0.178
4"		"	0.200
6"		"	0.267
45 degree elbows			
1"		EA	0.133
1-1/2"		"	0.133
2"		"	0.145
2-1/2"		"	0.160
3"		"	0.178
4"		"	0.200
6"		"	0.267

Sanitary Sewer		UNIT	MAN/ HOURS
02530.20	**Vitrified Clay Pipe**		
Vitrified clay pipe, extra strength			
6" dia.		LF	0.109
8" dia.		"	0.114
10" dia.		"	0.120
02530.30	**Manholes**		
Precast sections, 48" dia.			
Base section		EA	2.000
1'0" riser		"	1.600
1'4" riser		"	1.714
2'8" riser		"	1.846
4'0" riser		"	2.000
2'8" cone top		"	2.400
Precast manholes, 48" dia.			
4' deep		EA	4.800
6' deep		"	6.000
7' deep		"	6.857
8' deep		"	8.000
10' deep		"	9.600
Cast-in-place, 48" dia., with frame and cover			
5' deep		EA	12.000
6' deep		"	13.714
8' deep		"	16.000
10' deep		"	19.200
Brick manholes, 48" dia. with cover, 8" thick			
4' deep		EA	8.000
6' deep		"	8.889
8' deep		"	10.000
10' deep		"	11.429
Frames and covers, 24" diameter			
300 lb		EA	0.800
400 lb		"	0.889
Steps for manholes			
7" x 9"		EA	0.160
8" x 9"		"	0.178
02530.40	**Sanitary Sewers**		
Clay			
6" pipe		LF	0.080
PVC			
4" pipe		LF	0.060
6" pipe		"	0.063
02540.10	**Drainage Fields**		
Perforated PVC pipe, for drain field			
4" pipe		LF	0.053
6" pipe		"	0.057

02   SITE CONSTRUCTION

Sanitary Sewer	UNIT	MAN/HOURS
02540.50 **Septic Tanks**		
Septic tank, precast concrete		
1000 gals	EA	4.000
2000 gals	"	6.000
Leaching pit, precast concrete, 72" diameter		
3' deep	EA	3.000
6' deep	"	3.429
8' deep	"	4.000
02630.70 **Underdrain**		
Drain tile, clay		
6" pipe	LF	0.053
8" pipe	"	0.056
Porous concrete, standard strength		
6" pipe	LF	0.053
8" pipe	"	0.056
Corrugated metal pipe, perforated type		
6" pipe	LF	0.060
8" pipe	"	0.063
Perforated clay pipe		
6" pipe	LF	0.069
8" pipe	"	0.071
Drain tile, concrete		
6" pipe	LF	0.053
8" pipe	"	0.056
Perforated rigid PVC underdrain pipe		
4" pipe	LF	0.040
6" pipe	"	0.048
8" pipe	"	0.053
Underslab drainage, crushed stone		
3" thick	SF	0.008
4" thick	"	0.009
6" thick	"	0.010
Plastic filter fabric for drain lines	"	0.008

Paving	UNIT	MAN/HOURS
02740.20 **Asphalt Surfaces**		
Asphalt wearing surface, flexible pavement		
1" thick	SY	0.016
1-1/2" thick	"	0.019
Binder course		
1-1/2" thick	SY	0.018
2" thick	"	0.022
Bituminous sidewalk, no base		
2" thick	SY	0.028
3" thick	"	0.030

Rigid Pavement	UNIT	MAN/HOURS
02750.10 **Concrete Paving**		
Concrete paving, reinforced, 5000 psi concrete		
6" thick	SY	0.150
7" thick	"	0.160
8" thick	"	0.171

Site Improvements	UNIT	MAN/HOURS
02820.10 **Chain Link Fence**		
Chain link fence, 9 ga., galvanized, with posts 10' o.c.		
4' high	LF	0.057
5' high	"	0.073
6' high	"	0.100
Corner or gate post, 3" post		
4' high	EA	0.267
5' high	"	0.296
6' high	"	0.348
Gate with gate posts, galvanized, 3' wide		
4' high	EA	2.000
5' high	"	2.667
6' high	"	2.667
Fabric, galvanized chain link, 2" mesh, 9 ga.		
4' high	LF	0.027
5' high	"	0.032
6' high	"	0.040
Line post, no rail fitting, galvanized, 2-1/2" dia.		
4' high	EA	0.229
5' high	"	0.250
6' high	"	0.267
Vinyl coated, 9 ga., with posts 10' o.c.		
4' high	LF	0.057
5' high	"	0.073
6' high	"	0.100
Gate, with posts, 3' wide		
4' high	EA	2.000
5' high	"	2.667
6' high	"	2.667
Fabric, vinyl, chain link, 2" mesh, 9 ga.		
4' high	LF	0.027
5' high	"	0.032
6' high	"	0.040
Swing gates, galvanized, 4' high		
Single gate		

Site Improvements	UNIT	MAN/HOURS
02820.10 **Chain Link Fence** *(Cont.)*		
3' wide	EA	2.000
4' wide	"	2.000
6' high		
Single gate		
3' wide	EA	2.667
4' wide	"	2.667
02880.70 **Recreational Courts**		
Walls, galvanized steel		
8' high	LF	0.160
10' high	"	0.178
12' high	"	0.211
Vinyl coated		
8' high	LF	0.160
10' high	"	0.178
12' high	"	0.211
Gates, galvanized steel		
Single, 3' transom		
3'x7'	EA	4.000
4'x7'	"	4.571
5'x7'	"	5.333
6'x7'	"	6.400
Vinyl coated		
Single, 3' transom		
3'x7'	EA	4.000
4'x7'	"	4.571
5'x7'	"	5.333
6'x7'	"	6.400

Planting	UNIT	MAN/HOURS
02910.10 **Topsoil**		
Spread topsoil, with equipment		
Minimum	CY	0.080
Maximum	"	0.100
By hand		
Minimum	CY	0.800
Maximum	"	1.000
Area prep. seeding (grade, rake and clean)		
Square yard	SY	0.006
By acre	ACRE	32.000
Remove topsoil and stockpile on site		
4" deep	CY	0.067
6" deep	"	0.062
Spreading topsoil from stock pile		
By loader	CY	0.073

Planting	UNIT	MAN/HOURS
02910.10 **Topsoil** *(Cont.)*		
By hand	CY	0.800
Top dress by hand	SY	0.008
Place imported top soil		
By loader		
4" deep	SY	0.008
6" deep	"	0.009
By hand		
4" deep	SY	0.089
6" deep	"	0.100
Plant bed preparation, 18" deep		
With backhoe/loader	SY	0.020
By hand	"	0.133
02920.10 **Fertilizing**		
Fertilizing (23#/1000 sf)		
By square yard	SY	0.002
By acre	ACRE	10.000
Liming (70#/1000 sf)		
By square yard	SY	0.003
By acre	ACRE	13.333
02920.30 **Seeding**		
Mechanical seeding, 175 lb/acre		
By square yard	SY	0.002
By acre	ACRE	8.000
450 lb/acre		
By square yard	SY	0.002
By acre	ACRE	10.000
Seeding by hand, 10 lb per 100 s.y.		
By square yard	SY	0.003
By acre	ACRE	13.333
Reseed disturbed areas	SF	0.004
02930.10 **Plants**		
Euonymus coloratus, 18" (Purple Wintercreeper)	EA	0.133
Hedera Helix, 2-1/4" pot (English ivy)	"	0.133
Liriope muscari, 2" clumps	"	0.080
Santolina, 12"	"	0.080
Vinca major or minor, 3" pot	"	0.080
Cortaderia argentia, 2 gallon (Pampas Grass)	"	0.080
Ophiopogan japonicus, 1 quart (4" pot)	"	0.080
Ajuga reptans, 2-3/4" pot (carpet bugle)	"	0.080
Pachysandra terminalis, 2-3/4" pot (Japanese Spurge)	"	0.080
02930.30 **Shrubs**		
Juniperus conferia litoralis, 18"-24" (Shore Juniper)	EA	0.320
Horizontalis plumosa, 18"-24" (Andorra Juniper)	"	0.320
Sabina tamar-iscfolia-tamarix juniper, 18"-24"	"	0.320
Chin San Jose, 18"-24" (San Jose Juniper)	"	0.320
Sargenti, 18"-24" (Sargent's Juniper)	"	0.320
Nandina domestica, 18"-24" (Heavenly Bamboo)	"	0.320
Raphiolepis Indica Springtime, 18"-24"	"	0.320
Osmanthus Heterophyllus Gulftide, 18"-24"	"	0.320

Planting	UNIT	MAN/ HOURS
02930.30 **Shrubs** *(Cont.)*		
Ilex Cornuta Burfordi Nana, 18"-24"	EA	0.320
Glabra, 18"-24" (Inkberry Holly)	"	0.320
Azalea, Indica types, 18"-24"	"	0.320
Kurume types, 18"-24"	"	0.320
Berberis Julianae, 18"-24" (Wintergreen Barberry)	"	0.320
Pieris Japonica Japanese, 18"-24"	"	0.320
Ilex Cornuta Rotunda, 18"-24"	"	0.320
Juniperus Horiz. Plumosa, 24"-30"	"	0.400
Rhodopendrow Hybrids, 24"-30"	"	0.400
Aucuba Japonica Varigata, 24"-30"	"	0.400
Ilex Crenata Willow Leaf, 24"-30"	"	0.400
Cleyera Japonica, 30"-36"	"	0.500
Pittosporum Tobira, 30"-36"	"	0.500
Prumus Laurocerasus, 30"-36"	"	0.500
Ilex Cornuta Burfordi, 30"-36" (Burford Holly)	"	0.500
Abelia Grandiflora, 24"-36" (Yew Podocarpus)	"	0.400
Podocarpos Macrophylla, 24"-36"	"	0.400
Pyracantha Coccinea Lalandi, 3'-4' (Firethorn)	"	0.500
Photinia Frazieri, 3'-4' (Red Photinia)	"	0.500
Forsythia Suspensa, 3'-4' (Weeping Forsythia)	"	0.500
Camellia Japonica, 3'-4' (Common Camellia)	"	0.500
Juniperus Chin Torulosa, 3'-4' (Hollywood Juniper)	"	0.500
Cupressocyparis Leylandi, 3'-4'	"	0.500
Ilex Opaca Fosteri, 5'-6' (Foster's Holly)	"	0.667
Opaca, 5'-6' (American Holly)	"	0.667
Nyrica Cerifera, 4'-5' (Southern Wax Myrtles)	"	0.571
Ligustrum Japonicum, 4'-5' (Japanese Privet)	"	0.571
02930.60 **Trees**		
Cornus Florida, 5'-6' (White flowering Dogwood)	EA	0.667
Prunus Serrulata Kwanzan, 6'-8' (Kwanzan Cherry)	"	0.800
Caroliniana, 6'-8' (Carolina Cherry Laurel)	"	0.800
Cercis Canadensis, 6'-8' (Eastern Redbud)	"	0.800
Koelreuteria Paniculata, 8'-10' (Goldenrain Tree)	"	1.000
Acer Platanoides, 1-3/4"-2" (11'-13')	"	1.333
Rubrum, 1-3/4"-2" (11'-13') (Red Maple)	"	1.333
Saccharum, 1-3/4"-2" (Sugar Maple)	"	1.333
Fraxinus Pennsylvanica, 1-3/4"-2"	"	1.333
Celtis Occidentalis, 1-3/4"-2"	"	1.333
Glenditsia Triacantos Inermis, 2"	"	1.333
Prunus Cerasifera 'Thundercloud', 6'-8'	"	0.800
Yeodensis, 6'-8' (Yoshino Cherry)	"	0.800
Lagerstroemia Indica, 8'-10' (Crapemyrtle)	"	1.000
Crataegus Phaenopyrum, 8'-10'	"	1.000
Quercus Borealis, 1-3/4"-2" (Northern Red Oak)	"	1.333
Quercus Acutissima, 1-3/4"-2" (8'-10')	"	1.333
Saliz Babylonica, 1-3/4"-2" (Weeping Willow)	"	1.333
Tilia Cordata Greenspire, 1-3/4"-2" (10'-12')	"	1.333
Malus, 2"-2-1/2" (8'-10') (Flowering Crabapple)	"	1.333
Platanus Occidentalis, (12'-14')	"	1.600
Pyrus Calleryana Bradford, 2"-2-1/2"	"	1.333
Quercus Palustris, 2"-2-1/2" (12'-14') (Pin Oak)	"	1.333
Phellos, 2-1/2"-3" (Willow Oak)	"	1.600

Planting	UNIT	MAN/ HOURS
02930.60 **Trees** *(Cont.)*		
Nigra, 2"-2-1/2" (Water Oak)	EA	1.333
Magnolia Soulangeana, 4'-5' (Saucer Magnolia)	"	0.667
Grandiflora, 6'-8' (Southern Magnolia)	"	0.800
Cedrus Deodara, 10'-12' (Deodare Cedar)	"	1.333
Gingko Biloba, 10'-12' (2"-2-1/2")	"	1.333
Pinus Thunbergi, 5'-6' (Japanese Black Pine)	"	0.667
Strobus, 6'-8' (White Pine)	"	0.800
Taeda, 6'-8' (Loblolly Pine)	"	0.800
Quercus Virginiana, 2"-2-1/2" (Live Oak)	"	1.600
02935.10 **Shrub & Tree Maintenance**		
Moving shrubs on site		
3' high	EA	0.800
4' high	"	0.889
Moving trees on site		
6' high	EA	0.533
8' high	"	0.600
10' high	"	0.800
Palm trees		
10' high	EA	0.800
40' high	"	4.800
02935.30 **Weed Control**		
Weed control, bromicil, 15 lb./acre, wettable powder	ACRE	4.000
Vegetation control, by application of plant killer	SY	0.003
Weed killer, lawns and fields	"	0.002
02945.20 **Landscape Accessories**		
Steel edging, 3/16" x 4"	LF	0.010
Landscaping stepping stones, 15"x15", white	EA	0.040
Wood chip mulch	CY	0.533
2" thick	SY	0.016
4" thick	"	0.023
6" thick	"	0.029
Gravel mulch, 3/4" stone	CY	0.800
White marble chips, 1" deep	SF	0.008
Peat moss		
2" thick	SY	0.018
4" thick	"	0.027
6" thick	"	0.033
Landscaping timbers, treated lumber		
4" x 4"	LF	0.027
6" x 6"	"	0.029
8" x 8"	"	0.033

Formwork		UNIT	MAN/ HOURS
03110.05	**Beam Formwork**		
Beam forms, job built			
Beam bottoms			
1 use		SF	0.133
4 uses		"	0.118
5 uses		"	0.114
Beam sides			
1 use		SF	0.089
5 uses		"	0.073
03110.15	**Column Formwork**		
Column, square forms, job built			
8" x 8" columns			
1 use		SF	0.160
5 uses		"	0.138
12" x 12" columns			
1 use		SF	0.145
5 uses		"	0.127
Round fiber forms, 1 use			
10" dia.		LF	0.160
12" dia.		"	0.163
03110.18	**Curb Formwork**		
Curb forms			
Straight, 6" high			
1 use		LF	0.080
5 uses		"	0.067
Curved, 6" high			
1 use		LF	0.100
5 uses		"	0.082
03110.25	**Equipment Pad Formwork**		
Equipment pad, job built			
1 use		SF	0.100
3 uses		"	0.089
5 uses		"	0.080
03110.35	**Footing Formwork**		
Wall footings, job built, continuous			
1 use		SF	0.080
3 uses		"	0.073
5 uses		"	0.067
03110.50	**Grade Beam Formwork**		
Grade beams, job built			
1 use		SF	0.080
3 uses		"	0.073
5 uses		"	0.067

Formwork		UNIT	MAN/ HOURS
03110.53	**Pile Cap Formwork**		
Pile cap forms, job built			
Square			
1 use		SF	0.100
5 uses		"	0.080
03110.55	**Slab / Mat Formwork**		
Mat foundations, job built			
1 use		SF	0.100
3 uses		"	0.089
5 uses		"	0.080
Edge forms			
6" high			
1 use		LF	0.073
3 uses		"	0.067
5 uses		"	0.062
03110.65	**Wall Formwork**		
Wall forms, exterior, job built			
Up to 8' high wall			
1 use		SF	0.080
3 uses		"	0.073
5 uses		"	0.067
Retaining wall forms			
1 use		SF	0.089
3 uses		"	0.080
5 uses		"	0.073
Column pier and pilaster			
1 use		SF	0.160
5 uses		"	0.114
03110.90	**Miscellaneous Formwork**		
Keyway forms (5 uses)			
2 x 4		LF	0.040
2 x 6		"	0.044
Bulkheads			
Walls, with keyways			
3 piece		LF	0.080
Ground slab, with keyway			
2 piece		LF	0.057
3 piece		"	0.062
Chamfer strips			
Wood			
1/2" wide		LF	0.018
3/4" wide		"	0.018
1" wide		"	0.018
PVC			
1/2" wide		LF	0.018
3/4" wide		"	0.018
1" wide		"	0.018

03 CONCRETE

Reinforcement		UNIT	MAN/HOURS
03210.05	**Beam Reinforcing**		
Beam-girders			
#3 - #4		TON	20.000
#5 - #6		"	16.000
03210.15	**Column Reinforcing**		
Columns			
#3 - #4		TON	22.857
#5 - #6		"	17.778
03210.20	**Elevated Slab Reinforcing**		
Elevated slab			
#3 - #4		TON	10.000
#5 - #6		"	8.889
03210.25	**Equip. Pad Reinforcing**		
Equipment pad			
#3 - #4		TON	16.000
#5 - #6		"	14.545
03210.35	**Footing Reinforcing**		
Footings			
#3 - #4		TON	13.333
#5 - #6		"	11.429
#7 - #8		"	10.000
Straight dowels, 24" long			
3/4" dia. (#6)		EA	0.080
5/8" dia. (#5)		"	0.067
1/2" dia. (#4)		"	0.057
03210.45	**Foundation Reinforcing**		
Foundations			
#3 - #4		TON	13.333
#5 - #6		"	11.429
#7 - #8		"	10.000
03210.50	**Grade Beam Reinforcing**		
Grade beams			
#3 - #4		TON	12.308
#5 - #6		"	10.667
#7 - #8		"	9.412
03210.53	**Pile Cap Reinforcing**		
Pile caps			
#3 - #4		TON	20.000
#5 - #6		"	17.778
#7 - #8		"	16.000

Reinforcement		UNIT	MAN/HOURS
03210.55	**Slab / Mat Reinforcing**		
Bars, slabs			
#3 - #4		TON	13.333
#5 - #6		"	11.429
Wire mesh, slabs			
Galvanized			
4x4			
W1.4xW1.4		SF	0.005
W2.0xW2.0		"	0.006
W2.9xW2.9		"	0.006
W4.0xW4.0		"	0.007
6x6			
W1.4xW1.4		SF	0.004
W2.0xW2.0		"	0.004
W2.9xW2.9		"	0.005
W4.0xW4.0		"	0.005
03210.65	**Wall Reinforcing**		
Walls			
#3 - #4		TON	11.429
#5 - #6		"	10.000
Masonry wall (horizontal)			
#3 - #4		TON	32.000
#5 - #6		"	26.667
Galvanized			
#3 - #4		TON	32.000
#5 - #6		"	26.667
Masonry wall (vertical)			
#3 - #4		TON	40.000
#5 - #6		"	32.000
Galvanized			
#3 - #4		TON	40.000
#5 - #6		"	32.000

Accessories		UNIT	MAN/HOURS
03250.40	**Concrete Accessories**		
Expansion joint, poured			
Asphalt			
1/2" x 1"		LF	0.016
1" x 2"		"	0.017
Expansion joint, premolded, in slabs			
Asphalt			
1/2" x 6"		LF	0.020
1" x 12"		"	0.027
Cork			
1/2" x 6"		LF	0.020

03 CONCRETE

Accessories	UNIT	MAN/ HOURS
03250.40 Concrete Accessories *(Cont.)*		
1" x 12"	LF	0.027
Neoprene sponge		
1/2" x 6"	LF	0.020
1" x 12"	"	0.027
Polyethylene foam		
1/2" x 6"	LF	0.020
1" x 12"	"	0.027
Polyurethane foam		
1/2" x 6"	LF	0.020
1" x 12"	"	0.027
Polyvinyl chloride foam		
1/2" x 6"	LF	0.020
1" x 12"	"	0.027
Rubber, gray sponge		
1/2" x 6"	LF	0.020
1" x 12"	"	0.027
Asphalt felt control joints or bond breaker, screed joints		
4" slab	LF	0.016
6" slab	"	0.018
8" slab	"	0.020
Waterstops		
Polyvinyl chloride		
Ribbed		
3/16" thick x		
4" wide	LF	0.040
6" wide	"	0.044
1/2" thick x		
9" wide	LF	0.050
Ribbed with center bulb		
3/16" thick x 9" wide	LF	0.050
3/8" thick x 9" wide	"	0.050
Dumbbell type, 3/8" thick x 6" wide	"	0.044
Plain, 3/8" thick x 9" wide	"	0.050
Center bulb, 3/8" thick x 9" wide	"	0.050
Rubber		
Vapor barrier		
4 mil polyethylene	SF	0.003
6 mil polyethylene	"	0.003
Gravel porous fill, under floor slabs, 3/4" stone	CY	1.333
Reinforcing accessories		
Beam bolsters		
1-1/2" high, plain	LF	0.008
Galvanized	"	0.008
3" high		
Plain	LF	0.010
Galvanized	"	0.010
Slab bolsters		
1" high		
Plain	LF	0.004
Galvanized	"	0.004
2" high		
Plain	LF	0.004
Galvanized	"	0.004

Accessories	UNIT	MAN/ HOURS
03250.40 Concrete Accessories *(Cont.)*		
Chairs, high chairs		
3" high		
Plain	EA	0.020
Galvanized	"	0.020
8" high		
Plain	EA	0.023
Galvanized	"	0.023
Continuous, high chair		
3" high		
Plain	LF	0.005
Galvanized	"	0.005

Cast-in-place Concrete	UNIT	MAN/ HOURS
03350.10 Concrete Finishes		
Floor finishes		
Broom	SF	0.011
Screed	"	0.010
Darby	"	0.010
Steel float	"	0.013
Wall finishes		
Burlap rub, with cement paste	SF	0.013
03360.10 Pneumatic Concrete		
Pneumatic applied concrete (gunite)		
2" thick	SF	0.030
3" thick	"	0.040
4" thick	"	0.048
Finish surface		
Minimum	SF	0.040
Maximum	"	0.080
03370.10 Curing Concrete		
Sprayed membrane		
Slabs	SF	0.002
Walls	"	0.002
Curing paper		
Slabs	SF	0.002
Walls	"	0.002
Burlap		
7.5 oz.	SF	0.003
12 oz.	"	0.003

Placing Concrete	UNIT	MAN/ HOURS
03380.05 Beam Concrete		
Beams and girders		
2500# or 3000# concrete		
By crane	CY	0.960
By pump	"	0.873
By hand buggy	"	0.800
3500# or 4000# concrete		
By crane	CY	0.960
By pump	"	0.873
By hand buggy	"	0.800
03380.15 Column Concrete		
Columns		
2500# or 3000# concrete		
By crane	CY	0.873
By pump	"	0.800
3500# or 4000# concrete		
By crane	CY	0.873
By pump	"	0.800
03380.20 Elevated Slab Concrete		
Elevated slab		
2500# or 3000# concrete		
By crane	CY	0.480
By pump	"	0.369
By hand buggy	"	0.800
03380.25 Equipment Pad Concrete		
Equipment pad		
2500# or 3000# concrete		
By chute	CY	0.267
By pump	"	0.686
By crane	"	0.800
3500# or 4000# concrete		
By chute	CY	0.267
By pump	"	0.686
03380.35 Footing Concrete		
Continuous footing		
2500# or 3000# concrete		
By chute	CY	0.267
By pump	"	0.600
By crane	"	0.686
Spread footing		
2500# or 3000# concrete		
By chute	CY	0.267
By pump	"	0.640
By crane	"	0.738

Placing Concrete	UNIT	MAN/ HOURS
03380.50 Grade Beam Concrete		
Grade beam		
2500# or 3000# concrete		
By chute	CY	0.267
By crane	"	0.686
By pump	"	0.600
By hand buggy	"	0.800
3500# or 4000# concrete		
By chute	CY	0.267
By crane	"	0.686
By pump	"	0.600
By hand buggy	"	0.800
03380.53 Pile Cap Concrete		
Pile cap		
2500# or 3000 concrete		
By chute	CY	0.267
By crane	"	0.800
By pump	"	0.686
By hand buggy	"	0.800
03380.55 Slab / Mat Concrete		
Slab on grade		
2500# or 3000# concrete		
By chute	CY	0.200
By crane	"	0.400
By pump	"	0.343
By hand buggy	"	0.533
03380.58 Sidewalks		
Walks, cast in place with wire mesh, base not incl.		
4" thick	SF	0.027
5" thick	"	0.032
6" thick	"	0.040
03380.65 Wall Concrete		
Walls		
2500# or 3000# concrete		
To 4'		
By chute	CY	0.229
By crane	"	0.800
By pump	"	0.738
To 8'		
By crane	CY	0.873
By pump	"	0.800
Filled block (CMU)		
3000# concrete, by pump		
4" wide	SF	0.034
6" wide	"	0.040
8" wide	"	0.048

Placing Concrete	UNIT	MAN/HOURS
03400.90 **Precast Specialties**		
Precast concrete, coping, 4' to 8' long		
12" wide	LF	0.060
10" wide	"	0.069
Splash block, 30"x12"x4"	EA	0.400
Stair unit, per riser	"	0.400
Sun screen and trellis, 8' long, 12" high		
4" thick blades	EA	0.300

Grout	UNIT	MAN/HOURS
03600.10 **Grouting**		
Grouting for bases		
Non-metallic grout		
1" deep	SF	0.160
2" deep	"	0.178
Portland cement grout (1 cement to 3 sand)		
1/2" joint thickness		
6" wide joints	LF	0.027
8" wide joints	"	0.032
1" joint thickness		
4" wide joints	LF	0.025
6" wide joints	"	0.028

Mortar, Grout And Accessories	UNIT	MAN/ HOURS
04100.10 **Masonry Grout**		
Grout, non shrink, non-metallic, trowelable	CF	0.016
Grout door frame, hollow metal		
Single	EA	0.600
Double	"	0.632
Grout-filled concrete block (CMU)		
4" wide	SF	0.020
6" wide	"	0.022
8" wide	"	0.024
12" wide	"	0.025
Grout-filled individual CMU cells		
4" wide	LF	0.012
6" wide	"	0.012
8" wide	"	0.012
10" wide	"	0.014
12" wide	"	0.014
Bond beams or lintels, 8" deep		
6" thick	LF	0.022
8" thick	"	0.024
10" thick	"	0.027
12" thick	"	0.030
Cavity walls		
2" thick	SF	0.032
3" thick	"	0.032
4" thick	"	0.034
6" thick	"	0.040
04150.10 **Masonry Accessories**		
Foundation vents	EA	0.320
Bar reinforcing		
Horizontal		
#3 - #4	LB	0.032
#5 - #6	"	0.027
Vertical		
#3 - #4	LB	0.040
#5 - #6	"	0.032
Horizontal joint reinforcing		
Truss type		
4" wide, 6" wall	LF	0.003
6" wide, 8" wall	"	0.003
8" wide, 10" wall	"	0.003
10" wide, 12" wall	"	0.004
12" wide, 14" wall	"	0.004
Ladder type		
4" wide, 6" wall	LF	0.003
6" wide, 8" wall	"	0.003
8" wide, 10" wall	"	0.003
10" wide, 12" wall	"	0.003
Rectangular wall ties		
3/16" dia., galvanized		
2" x 6"	EA	0.013
2" x 8"	"	0.013
2" x 10"	"	0.013
2" x 12"	"	0.013

Mortar, Grout And Accessories	UNIT	MAN/ HOURS
04150.10 **Masonry Accessories** *(Cont.)*		
4" x 6"	EA	0.016
4" x 8"	"	0.016
4" x 10"	"	0.016
4" x 12"	"	0.016
1/4" dia., galvanized		
2" x 6"	EA	0.013
2" x 8"	"	0.013
2" x 10"	"	0.013
2" x 12"	"	0.013
4" x 6"	"	0.016
4" x 8"	"	0.016
4" x 10"	"	0.016
4" x 12"	"	0.016
"Z" type wall ties, galvanized		
6" long		
1/8" dia.	EA	0.013
3/16" dia.	"	0.013
1/4" dia.	"	0.013
8" long		
1/8" dia.	EA	0.013
3/16" dia.	"	0.013
1/4" dia.	"	0.013
10" long		
1/8" dia.	EA	0.013
3/16" dia.	"	0.013
1/4" dia.	"	0.013
Dovetail anchor slots		
Galvanized steel, filled		
24 ga.	LF	0.020
20 ga.	"	0.020
16 oz. copper, foam filled	"	0.020
Dovetail anchors		
16 ga.		
3-1/2" long	EA	0.013
5-1/2" long	"	0.013
12 ga.		
3-1/2" long	EA	0.013
5-1/2" long	"	0.013
Dovetail, triangular galvanized ties, 12 ga.		
3" x 3"	EA	0.013
5" x 5"	"	0.013
7" x 7"	"	0.013
7" x 9"	"	0.013
Brick anchors		
Corrugated, 3-1/2" long		
16 ga.	EA	0.013
12 ga.	"	0.013
Non-corrugated, 3-1/2" long		
16 ga.	EA	0.013
12 ga.	"	0.013
Cavity wall anchors, corrugated, galvanized		
5" long		
16 ga.	EA	0.013

04 MASONRY

## Mortar, Grout And Accessories

	UNIT	MAN/HOURS

04150.10 — Masonry Accessories *(Cont.)*

	UNIT	MAN/HOURS
12 ga.	EA	0.013
7" long		
28 ga.	EA	0.013
24 ga.	"	0.013
22 ga.	"	0.013
16 ga.	"	0.013
Mesh ties, 16 ga., 3" wide		
8" long	EA	0.013
12" long	"	0.013
20" long	"	0.013
24" long	"	0.013

04150.20 — Masonry Control Joints

	UNIT	MAN/HOURS
Control joint, cross shaped PVC	LF	0.020
Closed cell joint filler		
1/2"	LF	0.020
3/4"	"	0.020
Rubber, for		
4" wall	LF	0.020
PVC, for		
4" wall	LF	0.020

04150.50 — Masonry Flashing

	UNIT	MAN/HOURS
Through-wall flashing		
5 oz. coated copper	SF	0.067
0.030" elastomeric	"	0.053

Unit Masonry

	UNIT	MAN/HOURS

04210.10 — Brick Masonry

	UNIT	MAN/HOURS
Standard size brick, running bond		
Face brick, red (6.4/sf)		
Veneer	SF	0.133
Cavity wall	"	0.114
9" solid wall	"	0.229
Common brick (6.4/sf)		
Select common for veneers	SF	0.133
Back-up		
4" thick	SF	0.100
8" thick	"	0.160
Glazed brick (7.4/sf)		
Veneer	SF	0.145
Buff or gray face brick (6.4/sf)		
Veneer	SF	0.133
Cavity wall	"	0.114
Jumbo or oversize brick (3/sf)		

Unit Masonry

04210.10 — Brick Masonry *(Cont.)*

	UNIT	MAN/HOURS
4" veneer	SF	0.080
4" back-up	"	0.067
8" back-up	"	0.114
Norman brick, red face, (4.5/sf)		
4" veneer	SF	0.100
Cavity wall	"	0.089
Chimney, standard brick, including flue		
16" x 16"	LF	0.800
16" x 20"	"	0.800
16" x 24"	"	0.800
20" x 20"	"	1.000
20" x 24"	"	1.000
20" x 32"	"	1.143
Window sill, face brick on edge	"	0.200

04210.60 — Pavers, Masonry

	UNIT	MAN/HOURS
Brick walk laid on sand, sand joints		
Laid flat, (4.5 per sf)	SF	0.089
Laid on edge, (7.2 per sf)	"	0.133
Precast concrete patio blocks		
2" thick		
Natural	SF	0.027
Colors	"	0.027
Exposed aggregates, local aggregate		
Natural	SF	0.027
Colors	"	0.027
Granite or limestone aggregate	"	0.027
White tumblestone aggregate	"	0.027
Stone pavers, set in mortar		
Bluestone		
1" thick		
Irregular	SF	0.200
Snapped rectangular	"	0.160
1-1/2" thick, random rectangular	"	0.200
2" thick, random rectangular	"	0.229
Slate		
Natural cleft		
Irregular, 3/4" thick	SF	0.229
Random rectangular		
1-1/4" thick	SF	0.200
1-1/2" thick	"	0.222
Granite blocks		
3" thick, 3" to 6" wide		
4" to 12" long	SF	0.267
6" to 15" long	"	0.229
Crushed stone, white marble, 3" thick	"	0.016

04220.10 — Concrete Masonry Units

	UNIT	MAN/HOURS
Hollow, load bearing		
4"	SF	0.059
6"	"	0.062
8"	"	0.067
10"	"	0.073

Unit Masonry	UNIT	MAN/HOURS
04220.10 **Concrete Masonry Units** *(Cont.)*		
12"	SF	0.080
Solid, load bearing		
4"	SF	0.059
6"	"	0.062
8"	"	0.067
10"	"	0.073
12"	"	0.080
Back-up block, 8" x 16"		
2"	SF	0.046
4"	"	0.047
6"	"	0.050
8"	"	0.053
10"	"	0.057
12"	"	0.062
Foundation wall, 8" x 16"		
6"	SF	0.057
8"	"	0.062
10"	"	0.067
12"	"	0.073
Solid		
6"	SF	0.062
8"	"	0.067
10"	"	0.073
12"	"	0.080
Exterior, styrofoam inserts, std weight, 8" x 16"		
6"	SF	0.062
8"	"	0.067
10"	"	0.073
12"	"	0.080
Lightweight		
6"	SF	0.062
8"	"	0.067
10"	"	0.073
12"	"	0.080
Acoustical slotted block		
4"	SF	0.073
6"	"	0.073
8"	"	0.080
Filled cavities		
4"	SF	0.089
6"	"	0.094
8"	"	0.100
Hollow, split face		
4"	SF	0.059
6"	"	0.062
8"	"	0.067
10"	"	0.073
12"	"	0.080
Split rib profile		
4"	SF	0.073
6"	"	0.073
8"	"	0.080
10"	"	0.080

Unit Masonry	UNIT	MAN/HOURS
04220.10 **Concrete Masonry Units** *(Cont.)*		
12"	SF	0.080
Solar screen concrete block		
4" thick		
6" x 6"	SF	0.178
8" x 8"	"	0.160
12" x 12"	"	0.123
8" thick		
8" x 16"	SF	0.114
Vertical reinforcing		
4' o.c., add 5% to labor		
2'8" o.c., add 15% to labor		
Interior partitions, add 10% to labor		
04220.90 **Bond Beams & Lintels**		
Bond beam, no grout or reinforcement		
8" x 16" x		
4" thick	LF	0.062
6" thick	"	0.064
8" thick	"	0.067
10" thick	"	0.070
12" thick	"	0.073
Beam lintel, no grout or reinforcement		
8" x 16" x		
10" thick	LF	0.080
12" thick	"	0.089
Precast masonry lintel		
6 lf, 8" high x		
4" thick	LF	0.133
6" thick	"	0.133
8" thick	"	0.145
10" thick	"	0.145
10 lf, 8" high x		
4" thick	LF	0.080
6" thick	"	0.080
8" thick	"	0.089
10" thick	"	0.089
Steel angles and plates		
Minimum	LB	0.011
Maximum	"	0.020
Various size angle lintels		
1/4" stock		
3" x 3"	LF	0.050
3" x 3-1/2"	"	0.050
3/8" stock		
3" x 4"	LF	0.050
3-1/2" x 4"	"	0.050
4" x 4"	"	0.050
5" x 3-1/2"	"	0.050
6" x 3-1/2"	"	0.050
1/2" stock		
6" x 4"	LF	0.050

Unit Masonry		UNIT	MAN/ HOURS
04270.10	**Glass Block**		
Glass block, 4" thick			
6" x 6"		SF	0.267
8" x 8"		"	0.200
12" x 12"		"	0.160
04295.10	**Parging / Masonry Plaster**		
Parging			
1/2" thick		SF	0.053
3/4" thick		"	0.067
1" thick		"	0.080

Stone		UNIT	MAN/ HOURS
04400.10	**Stone**		
Rubble stone			
Walls set in mortar			
8" thick		SF	0.200
12" thick		"	0.320
18" thick		"	0.400
24" thick		"	0.533
Dry set wall			
8" thick		SF	0.133
12" thick		"	0.200
18" thick		"	0.267
24" thick		"	0.320
Thresholds, 7/8" thick, 3' long, 4" to 6" wide			
Plain		EA	0.667
Beveled		"	0.667
Window sill			
6" wide, 2" thick		LF	0.320
Stools			
5" wide, 7/8" thick		LF	0.320
Granite veneer facing panels, polished			
7/8" thick			
Black		SF	0.320
Gray		"	0.320
Slate, panels			
1" thick		SF	0.320
Sills or stools			
1" thick			
6" wide		LF	0.320
10" wide		"	0.348

Stone		UNIT	MAN/ HOURS
04520.10	**Restoration And Cleaning**		
Masonry cleaning			
Washing brick			
Smooth surface		SF	0.013
Rough surface		"	0.018
Steam clean masonry			
Smooth face			
Minimum		SF	0.010
Maximum		"	0.015
Rough face			
Minimum		SF	0.013
Maximum		"	0.020

Refractories		UNIT	MAN/ HOURS
04550.10	**Flue Liners**		
Flue liners			
Rectangular			
8" x 12"		LF	0.133
12" x 12"		"	0.145
12" x 18"		"	0.160
16" x 16"		"	0.178
18" x 18"		"	0.190
20" x 20"		"	0.200
24" x 24"		"	0.229
Round			
18" dia.		LF	0.190
24" dia.		"	0.229

Metal Fastening	UNIT	MAN/ HOURS
05050.10 **Structural Welding**		
Welding		
Single pass		
1/8"	LF	0.040
3/16"	"	0.053
1/4"	"	0.067
05050.95 **Metal Lintels**		
Lintels, steel		
Plain	LB	0.020
Galvanized	"	0.020
05120.10 **Beams, Girders, Columns, Trusses**		
Beams and girders, A-36		
Welded	TON	4.800
Bolted	"	4.364
Columns		
Pipe		
6" dia.	LB	0.005
Structural tube		
6" square		
Light sections	TON	9.600

Cold Formed Framing	UNIT	MAN/ HOURS
05410.10 **Metal Framing**		
Furring channel, galvanized		
Beams and columns, 3/4"		
12" o.c.	SF	0.080
16" o.c.	"	0.073
Walls, 3/4"		
12" o.c.	SF	0.040
16" o.c.	"	0.033
24" o.c.	"	0.027
1-1/2"		
12" o.c.	SF	0.040
16" o.c.	"	0.033
24" o.c.	"	0.027
Stud, load bearing		
16" o.c.		
16 ga.		
2-1/2"	SF	0.036
3-5/8"	"	0.036
4"	"	0.036
6"	"	0.040
18 ga.		
2-1/2"	SF	0.036

Cold Formed Framing	UNIT	MAN/ HOURS
05410.10 **Metal Framing** *(Cont.)*		
3-5/8"	SF	0.036
4"	"	0.036
6"	"	0.040
8"	"	0.040
20 ga.		
2-1/2"	SF	0.036
3-5/8"	"	0.036
4"	"	0.036
6"	"	0.040
8"	"	0.040
24" o.c.		
16 ga.		
2-1/2"	SF	0.031
3-5/8"	"	0.031
4"	"	0.031
6"	"	0.033
8"	"	0.033
18 ga.		
2-1/2"	SF	0.031
3-5/8"	"	0.031
4"	"	0.031
6"	"	0.033
8"	"	0.033
20 ga.		
2-1/2"	SF	0.031
3-5/8"	"	0.031
4"	"	0.031
6"	"	0.033
8"	"	0.033

Metal Fabrications	UNIT	MAN/ HOURS
05520.10 **Railings**		
Railing, pipe		
1-1/4" diameter, welded steel		
2-rail		
Primed	LF	0.160
Galvanized	"	0.160
3-rail		
Primed	LF	0.200
Galvanized	"	0.200
Wall mounted, single rail, welded steel		
Primed	LF	0.123
Galvanized	"	0.123

Metal Fabrications		UNIT	MAN/ HOURS
05520.10	**Railings** *(Cont.)*		
Wall mounted, single rail, welded steel			
Primed		LF	0.123
Galvanized		"	0.123
Wall mounted, single rail, welded steel			
Primed		LF	0.133
Galvanized		"	0.133

Misc. Fabrications		UNIT	MAN/ HOURS
05700.10	**Ornamental Metal**		
Railings, square bars, 6" o.c., shaped top rails			
Steel		LF	0.400
Aluminum		"	0.400
Bronze		"	0.533
Stainless steel		"	0.533
Laminated metal or wood handrails			
2-1/2" round or oval shape		LF	0.400

Fasteners And Adhesives	UNIT	MAN/HOURS
06050.10 **Accessories**		
Column/post base, cast aluminum		
4" x 4"	EA	0.200
6" x 6"	"	0.200
Bridging, metal, per pair		
12" o.c.	EA	0.080
16" o.c.	"	0.073
Anchors		
Bolts, threaded two ends, with nuts and washers		
1/2" dia.		
4" long	EA	0.050
7-1/2" long	"	0.050
3/4" dia.		
7-1/2" long	EA	0.050
15" long	"	0.050
Framing anchors		
10 gauge	EA	0.067
Bolts, carriage		
1/4 x 4	EA	0.080
5/16 x 6	"	0.084
3/8 x 6	"	0.084
1/2 x 6	"	0.084
Joist and beam hangers		
18 ga.		
2 x 4	EA	0.080
2 x 6	"	0.080
2 x 8	"	0.080
2 x 10	"	0.089
2 x 12	"	0.100
16 ga.		
3 x 6	EA	0.089
3 x 8	"	0.089
3 x 10	"	0.094
3 x 12	"	0.107
3 x 14	"	0.114
4 x 6	"	0.089
4 x 8	"	0.089
4 x 10	"	0.094
4 x 12	"	0.107
4 x 14	"	0.114
Rafter anchors, 18 ga., 1-1/2" wide		
5-1/4" long	EA	0.067
10-3/4" long	"	0.067
Shear plates		
2-5/8" dia.	EA	0.062
4" dia.	"	0.067
Sill anchors		
Embedded in concrete	EA	0.080
Split rings		
2-1/2" dia.	EA	0.089
4" dia.	"	0.100
Strap ties, 14 ga., 1-3/8" wide		
12" long	EA	0.067

Fasteners And Adhesives	UNIT	MAN/HOURS
06050.10 **Accessories** *(Cont.)*		
18" long	EA	0.073
24" long	"	0.080
36" long	"	0.089
Toothed rings		
2-5/8" dia.	EA	0.133
4" dia.	"	0.160

Rough Carpentry	UNIT	MAN/HOURS
06110.10 **Blocking**		
Steel construction		
Walls		
2x4	LF	0.053
2x6	"	0.062
2x8	"	0.067
2x10	"	0.073
2x12	"	0.080
Ceilings		
2x4	LF	0.062
2x6	"	0.073
2x8	"	0.080
2x10	"	0.089
2x12	"	0.100
Wood construction		
Walls		
2x4	LF	0.044
2x6	"	0.050
2x8	"	0.053
2x10	"	0.057
2x12	"	0.062
Ceilings		
2x4	LF	0.050
2x6	"	0.057
2x8	"	0.062
2x10	"	0.067
2x12	"	0.073
06110.20 **Ceiling Framing**		
Ceiling joists		
12" o.c.		
2x4	SF	0.019
2x6	"	0.020
2x8	"	0.021
2x10	"	0.022
2x12	"	0.024
16" o.c.		

Rough Carpentry	UNIT	MAN/ HOURS	Rough Carpentry	UNIT	MAN/ HOURS
06110.20 Ceiling Framing *(Cont.)*			**06110.30** Floor Framing *(Cont.)*		
2x4	SF	0.015	4x6	SF	0.014
2x6	"	0.016	4x8	"	0.014
2x8	"	0.017	4x10	"	0.015
2x10	"	0.017	4x12	"	0.015
2x12	"	0.018	4x14	"	0.016
24" o.c.			Sister joists for floors		
2x4	SF	0.013	2x4	LF	0.050
2x6	"	0.013	2x6	"	0.057
2x8	"	0.014	2x8	"	0.067
2x10	"	0.015	2x10	"	0.080
2x12	"	0.016	2x12	"	0.100
Headers and nailers			3x6	"	0.080
2x4	LF	0.026	3x8	"	0.089
2x6	"	0.027	3x10	"	0.100
2x8	"	0.029	3x12	"	0.114
2x10	"	0.031	4x6	"	0.080
2x12	"	0.033	4x8	"	0.089
Sister joists for ceilings			4x10	"	0.100
2x4	LF	0.057	4x12	"	0.114
2x6	"	0.067	**06110.40** Furring		
2x8	"	0.080	Furring, wood strips		
2x10	"	0.100	Walls		
2x12	"	0.133	On masonry or concrete walls		
06110.30 Floor Framing			1x2 furring		
Floor joists			12" o.c.	SF	0.025
12" o.c.			16" o.c.	"	0.023
2x6	SF	0.016	24" o.c.	"	0.021
2x8	"	0.016	1x3 furring		
2x10	"	0.017	12" o.c.	SF	0.025
2x12	"	0.017	16" o.c.	"	0.023
2x14	"	0.017	24" o.c.	"	0.021
3x6	"	0.017	On wood walls		
3x8	"	0.017	1x2 furring		
3x10	"	0.018	12" o.c.	SF	0.018
3x12	"	0.019	16" o.c.	"	0.016
3x14	"	0.020	24" o.c.	"	0.015
4x6	"	0.017	1x3 furring		
4x8	"	0.017	12" o.c.	SF	0.018
4x10	"	0.018	16" o.c.	"	0.016
4x12	"	0.019	24" o.c.	"	0.015
4x14	"	0.020	Ceilings		
16" o.c.			On masonry or concrete ceilings		
2x6	SF	0.013	1x2 furring		
2x8	"	0.014	12" o.c.	SF	0.044
2x10	"	0.014	16" o.c.	"	0.040
2x12	"	0.014	24" o.c.	"	0.036
2x14	"	0.015	1x3 furring		
3x6	"	0.014	12" o.c.	SF	0.044
3x8	"	0.014	16" o.c.	"	0.040
3x10	"	0.015	24" o.c.	"	0.036
3x12	"	0.015	On wood ceilings		
3x14	"	0.016	1x2 furring		

Rough Carpentry	UNIT	MAN/HOURS
06110.40 **Furring** *(Cont.)*		
12" o.c.	SF	0.030
16" o.c.	"	0.027
24" o.c.	"	0.024
1x3		
12" o.c.	SF	0.030
16" o.c.	"	0.027
24" o.c.	"	0.024
06110.50 **Roof Framing**		
Roof framing		
Rafters, gable end		
0-2 pitch (flat to 2-in-12)		
12" o.c.		
2x4	SF	0.017
2x6	"	0.017
2x8	"	0.018
2x10	"	0.019
2x12	"	0.020
16" o.c.		
2x6	SF	0.014
2x8	"	0.015
2x10	"	0.015
2x12	"	0.016
24" o.c.		
2x6	SF	0.012
2x8	"	0.013
2x10	"	0.013
2x12	"	0.013
4-6 pitch (4-in-12 to 6-in-12)		
12" o.c.		
2x4	SF	0.017
2x6	"	0.018
2x8	"	0.019
2x10	"	0.020
2x12	"	0.021
16" o.c.		
2x6	SF	0.015
2x8	"	0.015
2x10	"	0.016
2x12	"	0.017
24" o.c.		
2x6	SF	0.013
2x8	"	0.013
2x10	"	0.014
2x12	"	0.015
8-12 pitch (8-in-12 to 12-in-12)		
12" o.c.		
2x4	SF	0.018
2x6	"	0.019
2x8	"	0.020
2x10	"	0.021
2x12	"	0.022
16" o.c.		

Rough Carpentry	UNIT	MAN/HOURS
06110.50 **Roof Framing** *(Cont.)*		
2x6	SF	0.015
2x8	"	0.016
2x10	"	0.017
2x12	"	0.017
24" o.c.		
2x6	SF	0.013
2x8	"	0.013
2x10	"	0.014
2x12	"	0.014
Ridge boards		
2x6	LF	0.040
2x8	"	0.044
2x10	"	0.050
2x12	"	0.057
Hip rafters		
2x6	LF	0.029
2x8	"	0.030
2x10	"	0.031
2x12	"	0.032
Jack rafters		
4-6 pitch (4-in-12 to 6-in-12)		
16" o.c.		
2x6	SF	0.024
2x8	"	0.024
2x10	"	0.026
2x12	"	0.027
24" o.c.		
2x6	SF	0.018
2x8	"	0.019
2x10	"	0.020
2x12	"	0.020
8-12 pitch (8-in-12 to 12-in-12)		
16" o.c.		
2x6	SF	0.025
2x8	"	0.026
2x10	"	0.027
2x12	"	0.028
24" o.c.		
2x6	SF	0.019
2x8	"	0.020
2x10	"	0.020
2x12	"	0.021
Sister rafters		
2x4	LF	0.057
2x6	"	0.067
2x8	"	0.080
2x10	"	0.100
2x12	"	0.133
Fascia boards		
2x4	LF	0.040
2x6	"	0.040
2x8	"	0.044
2x10	"	0.044

Rough Carpentry		UNIT	MAN/HOURS
06110.50	**Roof Framing** *(Cont.)*		
2x12		LF	0.050
Cant strips			
Fiber			
3x3		LF	0.023
4x4		"	0.024
Wood			
3x3		LF	0.024
06110.60	**Sleepers**		
Sleepers, over concrete			
12" o.c.			
1x2		SF	0.018
1x3		"	0.019
2x4		"	0.022
2x6		"	0.024
16" o.c.			
1x2		SF	0.016
1x3		"	0.016
2x4		"	0.019
2x6		"	0.020
06110.65	**Soffits**		
Soffit framing			
2x3		LF	0.057
2x4		"	0.062
2x6		"	0.067
2x8		"	0.073
06110.70	**Wall Framing**		
Framing wall, studs			
12" o.c.			
2x3		SF	0.015
2x4		"	0.015
2x6		"	0.016
2x8		"	0.017
16" o.c.			
2x3		SF	0.013
2x4		"	0.013
2x6		"	0.013
2x8		"	0.014
24" o.c.			
2x3		SF	0.011
2x4		"	0.011
2x6		"	0.011
2x8		"	0.012
Plates, top or bottom			
2x3		LF	0.024
2x4		"	0.025
2x6		"	0.027
2x8		"	0.029
Headers, door or window			
2x6			
Single			

Rough Carpentry		UNIT	MAN/HOURS
06110.70	**Wall Framing** *(Cont.)*		
3' long		EA	0.400
6' long		"	0.500
Double			
3' long		EA	0.444
6' long		"	0.571
2x8			
Single			
4' long		EA	0.500
8' long		"	0.615
Double			
4' long		EA	0.571
8' long		"	0.727
2x10			
Single			
5' long		EA	0.615
10' long		"	0.800
Double			
5' long		EA	0.667
10' long		"	0.800
2x12			
Single			
6' long		EA	0.615
12' long		"	0.800
Double			
6' long		EA	0.727
12' long		"	0.889
06115.10	**Floor Sheathing**		
Sub-flooring, plywood, CDX			
1/2" thick		SF	0.010
5/8" thick		"	0.011
3/4" thick		"	0.013
Structural plywood			
1/2" thick		SF	0.010
5/8" thick		"	0.011
3/4" thick		"	0.012
Board type subflooring			
1x6			
Minimum		SF	0.018
Maximum		"	0.020
1x8			
Minimum		SF	0.017
Maximum		"	0.019
1x10			
Minimum		SF	0.016
Maximum		"	0.018
Underlayment			
Hardboard, 1/4" tempered		SF	0.010
Plywood, CDX			
3/8" thick		SF	0.010
1/2" thick		"	0.011
5/8" thick		"	0.011
3/4" thick		"	0.012

Rough Carpentry	UNIT	MAN/HOURS
06115.20　　Roof Sheathing		
Sheathing		
Plywood, CDX		
3/8" thick	SF	0.010
1/2" thick	"	0.011
5/8" thick	"	0.011
3/4" thick	"	0.012
Structural plywood		
3/8" thick	SF	0.010
1/2" thick	"	0.011
5/8" thick	"	0.011
3/4" thick	"	0.012
06115.30　　Wall Sheathing		
Sheathing		
Plywood, CDX		
3/8" thick	SF	0.012
1/2" thick	"	0.012
5/8" thick	"	0.013
3/4" thick	"	0.015
Waferboard		
3/8" thick	SF	0.012
1/2" thick	"	0.012
5/8" thick	"	0.013
3/4" thick	"	0.015
Structural plywood		
3/8" thick	SF	0.012
1/2" thick	"	0.012
5/8" thick	"	0.013
3/4" thick	"	0.015
Gypsum, 1/2" thick	"	0.012
Asphalt impregnated fiberboard, 1/2" thick	"	0.012
06125.10　　Wood Decking		
Decking, T&G solid		
Cedar		
3" thick	SF	0.020
4" thick	"	0.021
Fir		
3" thick	SF	0.020
4" thick	"	0.021
Southern yellow pine		
3" thick	SF	0.023
4" thick	"	0.025
White pine		
3" thick	SF	0.020
4" thick	"	0.021
06130.10　　Heavy Timber		
Mill framed structures		
Beams to 20' long		
Douglas fir		
6x8	LF	0.080
6x10	"	0.083

Rough Carpentry	UNIT	MAN/HOURS
06130.10　　Heavy Timber *(Cont.)*		
6x12	LF	0.089
6x14	"	0.092
6x16	"	0.096
8x10	"	0.083
8x12	"	0.089
8x14	"	0.092
8x16	"	0.096
Southern yellow pine		
6x8	LF	0.080
6x10	"	0.083
6x12	"	0.089
6x14	"	0.092
6x16	"	0.096
8x10	"	0.083
8x12	"	0.089
8x14	"	0.092
8x16	"	0.096
Columns to 12' high		
Douglas fir		
6x6	LF	0.120
8x8	"	0.120
10x10	"	0.133
12x12	"	0.133
Southern yellow pine		
6x6	LF	0.120
8x8	"	0.120
10x10	"	0.133
12x12	"	0.133
Posts, treated		
4x4	LF	0.032
6x6	"	0.040
06190.20　　Wood Trusses		
Truss, fink, 2x4 members		
3-in-12 slope		
24' span	EA	0.686
26' span	"	0.686
28' span	"	0.727
30' span	"	0.727
34' span	"	0.774
38' span	"	0.774
5-in-12 slope		
24' span	EA	0.706
28' span	"	0.727
30' span	"	0.750
32' span	"	0.750
40' span	"	0.800
Gable, 2x4 members		
5-in-12 slope		
24' span	EA	0.706
26' span	"	0.706
28' span	"	0.727
30' span	"	0.750

Rough Carpentry	UNIT	MAN/HOURS
06190.20 **Wood Trusses** *(Cont.)*		
32' span	EA	0.750
36' span	"	0.774
40' span	"	0.800
King post type, 2x4 members		
4-in-12 slope		
16' span	EA	0.649
18' span	"	0.667
24' span	"	0.706
26' span	"	0.706
30' span	"	0.750
34' span	"	0.750
38' span	"	0.774
42' span	"	0.828

Finish Carpentry	UNIT	MAN/HOURS
06200.10 **Finish Carpentry**		
Mouldings and trim		
Apron, flat		
9/16 x 2	LF	0.040
9/16 x 3-1/2	"	0.042
Base		
Colonial		
7/16 x 2-1/4	LF	0.040
7/16 x 3	"	0.040
7/16 x 3-1/4	"	0.040
9/16 x 3	"	0.042
9/16 x 3-1/4	"	0.042
11/16 x 2-1/4	"	0.044
Ranch		
7/16 x 2-1/4	LF	0.040
7/16 x 3-1/4	"	0.040
9/16 x 2-1/4	"	0.042
9/16 x 3	"	0.042
9/16 x 3-1/4	"	0.042
Casing		
11/16 x 2-1/2	LF	0.036
11/16 x 3-1/2	"	0.038
Chair rail		
9/16 x 2-1/2	LF	0.040
9/16 x 3-1/2	"	0.040
Closet pole		
1-1/8" dia.	LF	0.053
1-5/8" dia.	"	0.053
Cove		
9/16 x 1-3/4	LF	0.040

Finish Carpentry	UNIT	MAN/HOURS
06200.10 **Finish Carpentry** *(Cont.)*		
11/16 x 2-3/4	LF	0.040
Crown		
9/16 x 1-5/8	LF	0.053
9/16 x 2-5/8	"	0.062
11/16 x 3-5/8	"	0.067
11/16 x 4-1/4	"	0.073
11/16 x 5-1/4	"	0.080
Drip cap		
1-1/16 x 1-5/8	LF	0.040
Glass bead		
3/8 x 3/8	LF	0.050
1/2 x 9/16	"	0.050
5/8 x 5/8	"	0.050
3/4 x 3/4	"	0.050
Half round		
1/2	LF	0.032
5/8	"	0.032
3/4	"	0.032
Lattice		
1/4 x 7/8	LF	0.032
1/4 x 1-1/8	"	0.032
1/4 x 1-3/8	"	0.032
1/4 x 1-3/4	"	0.032
1/4 x 2	"	0.032
Ogee molding		
5/8 x 3/4	LF	0.040
11/16 x 1-1/8	"	0.040
11/16 x 1-3/8	"	0.040
Parting bead		
3/8 x 7/8	LF	0.050
Quarter round		
1/4 x 1/4	LF	0.032
3/8 x 3/8	"	0.032
1/2 x 1/2	"	0.032
11/16 x 11/16	"	0.035
3/4 x 3/4	"	0.035
1-1/16 x 1-1/16	"	0.036
Railings, balusters		
1-1/8 x 1-1/8	LF	0.080
1-1/2 x 1-1/2	"	0.073
Screen moldings		
1/4 x 3/4	LF	0.067
5/8 x 5/16	"	0.067
Shoe		
7/16 x 11/16	LF	0.032
Sash beads		
1/2 x 3/4	LF	0.067
1/2 x 7/8	"	0.067
1/2 x 1-1/8	"	0.073
5/8 x 7/8	"	0.073
Stop		
5/8 x 1-5/8		
Colonial	LF	0.050

06   WOOD AND PLASTICS

Finish Carpentry	UNIT	MAN/HOURS
06200.10 Finish Carpentry *(Cont.)*		
Ranch	LF	0.050
Stools		
11/16 x 2-1/4	LF	0.089
11/16 x 2-1/2	"	0.089
11/16 x 5-1/4	"	0.100
Exterior trim, casing, select pine, 1x3	"	0.040
Douglas fir		
1x3	LF	0.040
1x4	"	0.040
1x6	"	0.044
1x8	"	0.050
Cornices, white pine, #2 or better		
1x2	LF	0.040
1x4	"	0.040
1x6	"	0.044
1x8	"	0.047
1x10	"	0.050
1x12	"	0.053
Shelving, pine		
1x8	LF	0.062
1x10	"	0.064
1x12	"	0.067
Plywood shelf, 3/4", with edge band, 12" wide	"	0.080
Adjustable shelf, and rod, 12" wide		
3' to 4' long	EA	0.200
5' to 8' long	"	0.267
Prefinished wood shelves with brackets and supports		
8" wide		
3' long	EA	0.200
4' long	"	0.200
6' long	"	0.200
10" wide		
3' long	EA	0.200
4' long	"	0.200
6' long	"	0.200
06220.10 Millwork		
Countertop, laminated plastic		
25" x 7/8" thick		
Minimum	LF	0.200
Average	"	0.267
Maximum	"	0.320
25" x 1-1/4" thick		
Minimum	LF	0.267
Average	"	0.320
Maximum	"	0.400
Add for cutouts	EA	0.500
Backsplash, 4" high, 7/8" thick	LF	0.160
Plywood, sanded, A-C		
1/4" thick	SF	0.027
3/8" thick	"	0.029
1/2" thick	"	0.031
A-D		

Finish Carpentry	UNIT	MAN/HOURS
06220.10 Millwork *(Cont.)*		
1/4" thick	SF	0.027
3/8" thick	"	0.029
1/2" thick	"	0.031
Base cabinet, 34-1/2" high, 24" deep, hardwood		
Minimum	LF	0.320
Average	"	0.400
Maximum	"	0.533
Wall cabinets		
Minimum	LF	0.267
Average	"	0.320
Maximum	"	0.400
Oil borne		
Water borne		

Architectural Woodwork	UNIT	MAN/HOURS
06420.10 Panel Work		
Hardboard, tempered, 1/4" thick		
Natural faced	SF	0.020
Plastic faced	"	0.023
Pegboard, natural	"	0.020
Plastic faced	"	0.023
Untempered, 1/4" thick		
Natural faced	SF	0.020
Plastic faced	"	0.023
Pegboard, natural	"	0.020
Plastic faced	"	0.023
Plywood unfinished, 1/4" thick		
Birch		
Natural	SF	0.027
Select	"	0.027
Knotty pine	"	0.027
Cedar (closet lining)		
Standard boards T&G	SF	0.027
Particle board	"	0.027
Plywood, prefinished, 1/4" thick, premium grade		
Birch veneer	SF	0.032
Cherry veneer	"	0.032
Chestnut veneer	"	0.032
Lauan veneer	"	0.032
Mahogany veneer	"	0.032
Oak veneer (red)	"	0.032
Pecan veneer	"	0.032
Rosewood veneer	"	0.032
Teak veneer	"	0.032
Walnut veneer	"	0.032

Architectural Woodwork		UNIT	MAN/ HOURS
06430.10	**Stairwork**		
Risers, 1x8, 42" wide			
White oak		EA	0.400
Pine		"	0.400
Treads, 1-1/16" x 9-1/2" x 42"			
White oak		EA	0.500
06440.10	**Columns**		
Column, hollow, round wood			
12" diameter			
10' high		EA	0.800
12' high		"	0.857
14' high		"	0.960
16' high		"	1.200
24" diameter			
16' high		EA	1.200
18' high		"	1.263
20' high		"	1.263
22' high		"	1.333
24' high		"	1.333

Moisture Protection	UNIT	MAN/HOURS
07100.10 Waterproofing		
Membrane waterproofing, elastomeric		
Butyl		
1/32" thick	SF	0.032
1/16" thick	"	0.033
Neoprene		
1/32" thick	SF	0.032
1/16" thick	"	0.033
Plastic vapor barrier (polyethylene)		
4 mil	SF	0.003
6 mil	"	0.003
10 mil	"	0.004
Bituminous membrane, asphalt felt, 15 lb.		
One ply	SF	0.020
Two ply	"	0.024
Three ply	"	0.029
Bentonite waterproofing, panels		
3/16" thick	SF	0.020
1/4" thick	"	0.020
07150.10 Dampproofing		
Silicone dampproofing, sprayed on		
Concrete surface		
1 coat	SF	0.004
2 coats	"	0.006
Concrete block		
1 coat	SF	0.005
2 coats	"	0.007
Brick		
1 coat	SF	0.006
2 coats	"	0.008
07160.10 Bituminous Dampproofing		
Building paper, asphalt felt		
15 lb	SF	0.032
30 lb	"	0.033
Asphalt, troweled, cold, primer plus		
1 coat	SF	0.027
2 coats	"	0.040
3 coats	"	0.050
Fibrous asphalt, hot troweled, primer plus		
1 coat	SF	0.032
2 coats	"	0.044
3 coats	"	0.057
Asphaltic paint dampproofing, per coat		
Brush on	SF	0.011
Spray on	"	0.009
07190.10 Vapor Barriers		
Vapor barrier, polyethylene		
2 mil	SF	0.004
6 mil	"	0.004
8 mil	"	0.004
10 mil	"	0.004

Insulation	UNIT	MAN/HOURS
07210.10 Batt Insulation		
Ceiling, fiberglass, unfaced		
3-1/2" thick, R11	SF	0.009
6" thick, R19	"	0.011
9" thick, R30	"	0.012
Suspended ceiling, unfaced		
3-1/2" thick, R11	SF	0.009
6" thick, R19	"	0.010
9" thick, R30	"	0.011
Crawl space, unfaced		
3-1/2" thick, R11	SF	0.012
6" thick, R19	"	0.013
9" thick, R30	"	0.015
Wall, fiberglass		
Paper backed		
2" thick, R7	SF	0.008
3" thick, R8	"	0.009
4" thick, R11	"	0.009
6" thick, R19	"	0.010
Foil backed, 1 side		
2" thick, R7	SF	0.008
3" thick, R11	"	0.009
4" thick, R14	"	0.009
6" thick, R21	"	0.010
Foil backed, 2 sides		
2" thick, R7	SF	0.009
3" thick, R11	"	0.010
4" thick, R14	"	0.011
6" thick, R21	"	0.011
Unfaced		
2" thick, R7	SF	0.008
3" thick, R9	"	0.009
4" thick, R11	"	0.009
6" thick, R19	"	0.010
Mineral wool batts		
Paper backed		
2" thick, R6	SF	0.008
4" thick, R12	"	0.009
6" thick, R19	"	0.010
Fasteners, self adhering, attached to ceiling deck		
2-1/2" long	EA	0.013
4-1/2" long	"	0.015
Capped, self-locking washers	"	0.008
07210.20 Board Insulation		
Perlite board, roof		
1.00" thick, R2.78	SF	0.007
1.50" thick, R4.17	"	0.007
Rigid urethane		
1" thick, R6.67	SF	0.007
1.50" thick, R11.11	"	0.007
Polystyrene		
1.0" thick, R4.17	SF	0.007
1.5" thick, R6.26	"	0.007

Insulation	UNIT	MAN/HOURS
07210.60 Loose Fill Insulation		
Blown-in type		
Fiberglass		
5" thick, R11	SF	0.007
6" thick, R13	"	0.008
9" thick, R19	"	0.011
Rockwool, attic application		
6" thick, R13	SF	0.008
8" thick, R19	"	0.010
10" thick, R22	"	0.012
12" thick, R26	"	0.013
15" thick, R30	"	0.016
Poured type		
Fiberglass		
1" thick, R4	SF	0.005
2" thick, R8	"	0.006
3" thick, R12	"	0.007
4" thick, R16	"	0.008
Mineral wool		
1" thick, R3	SF	0.005
2" thick, R6	"	0.006
3" thick, R9	"	0.007
4" thiok, R12	"	0.008
Vermiculite or perlite		
2" thick, R4.8	SF	0.006
3" thick, R7.2	"	0.007
4" thick, R9.6	"	0.008
Masonry, poured vermiculite or perlite		
4" block	SF	0.004
6" block	"	0.005
8" block	"	0.006
10" block	"	0.006
12" block	"	0.007
07210.70 Sprayed Insulation		
Foam, sprayed on		
Polystyrene		
1" thick, R4	SF	0.008
2" thick, R8	"	0.011
Urethane		
1" thick, R4	SF	0.008
2" thick, R8	"	0.011

Shingles And Tiles	UNIT	MAN/HOURS
07310.10 Asphalt Shingles		
Standard asphalt shingles, strip shingles		
210 lb/square	SQ	0.800
235 lb/square	"	0.889
240 lb/square	"	1.000
260 lb/square	"	1.143
300 lb/square	"	1.333
385 lb/square	"	1.600
Roll roofing, mineral surface		
90 lb	SQ	0.571
110 lb	"	0.667
140 lb	"	0.800
07310.50 Metal Shingles		
Aluminum, .020" thick		
Plain	SQ	1.600
Colors	"	1.600
Steel, galvanized		
26 ga.		
Plain	SQ	1.600
Colors	"	1.600
24 ga.		
Plain	SQ	1.600
Colors	"	1.600
Porcelain enamel, 22 ga.		
Minimum	SQ	2.000
Average	"	2.000
Maximum	"	2.000
07310.60 Slate Shingles		
Slate shingles		
Pennsylvania		
Ribbon	SQ	4.000
Clear	"	4.000
Vermont		
Black	SQ	4.000
Gray	"	4.000
Green	"	4.000
Red	"	4.000
07310.70 Wood Shingles		
Wood shingles, on roofs		
White cedar, #1 shingles		
4" exposure	SQ	2.667
5" exposure	"	2.000
#2 shingles		
4" exposure	SQ	2.667
5" exposure	"	2.000
Resquared and rebutted		
4" exposure	SQ	2.667
5" exposure	"	2.000
On walls		
White cedar, #1 shingles		
4" exposure	SQ	4.000

Shingles And Tiles		UNIT	MAN/ HOURS
07310.70	**Wood Shingles** *(Cont.)*		
5" exposure		SQ	3.200
6" exposure		"	2.667
#2 shingles			
4" exposure		SQ	4.000
5" exposure		"	3.200
6" exposure		"	2.667
07310.80	**Wood Shakes**		
Shakes, hand split, 24" red cedar, on roofs			
5" exposure		SQ	4.000
7" exposure		"	3.200
9" exposure		"	2.667
On walls			
6" exposure		SQ	4.000
8" exposure		"	3.200
10" exposure		"	2.667
07460.10	**Metal Siding Panels**		
Aluminum siding panels			
Corrugated			
Plain finish			
.024"		SF	0.032
.032"		"	0.032
Painted finish			
.024"		SF	0.032
.032"		"	0.032
Steel siding panels			
Corrugated			
22 ga.		SF	0.053
24 ga.		"	0.053
Ribbed, sheets, galvanized			
22 ga.		SF	0.032
24 ga.		"	0.032
Primed			
24 ga.		SF	0.032
26 ga.		"	0.032
07460.50	**Plastic Siding**		
Horizontal vinyl siding, solid			
8" wide			
Standard		SF	0.031
Insulated		"	0.031
10" wide			
Standard		SF	0.029
Insulated		"	0.029
Vinyl moldings for doors and windows		LF	0.032
07460.60	**Plywood Siding**		
Rough sawn cedar, 3/8" thick		SF	0.027
Fir, 3/8" thick		"	0.027
Texture 1-11, 5/8" thick			
Cedar		SF	0.029
Fir		"	0.029

Shingles And Tiles		UNIT	MAN/ HOURS
07460.60	**Plywood Siding** *(Cont.)*		
Redwood		SF	0.029
Southern Yellow Pine		"	0.029
07460.80	**Wood Siding**		
Beveled siding, cedar			
A grade			
1/2 x 8		SF	0.032
3/4 x 10		"	0.027
Clear			
1/2 x 6		SF	0.040
1/2 x 8		"	0.032
3/4 x 10		"	0.027
B grade			
1/2 x 6		SF	0.040
1/2 x 8		"	0.032
3/4 x 10		"	0.027
Board and batten			
Cedar			
1x6		SF	0.040
1x8		"	0.032
1x10		"	0.029
1x12		"	0.026
Pine			
1x6		SF	0.040
1x8		"	0.032
1x10		"	0.029
1x12		"	0.026
Redwood			
1x6		SF	0.040
1x8		"	0.032
1x10		"	0.029
1x12		"	0.026
Tongue and groove			
Cedar			
1x4		SF	0.044
1x6		"	0.042
1x8		"	0.040
1x10		"	0.038
Pine			
1x4		SF	0.044
1x6		"	0.042
1x8		"	0.040
1x10		"	0.038
Redwood			
1x4		SF	0.044
1x6		"	0.042
1x8		"	0.040
1x10		"	0.038

Membrane Roofing	UNIT	MAN/ HOURS
07510.10 **Built-up Asphalt Roofing**		
Built-up roofing, asphalt felt, including gravel		
2 ply	SQ	2.000
3 ply	"	2.667
4 ply	"	3.200
Cant strip, 4" x 4"		
Treated wood	LF	0.023
Foamglass	"	0.020
New gravel for built-up roofing, 400 lb/sq	SQ	1.600
07530.10 **Single-ply Roofing**		
Elastic sheet roofing		
Neoprene, 1/16" thick	SF	0.010
PVC		
45 mil	SF	0.010
Flashing		
Pipe flashing, 90 mil thick		
1" pipe	EA	0.200
Neoprene flashing, 60 mil thick strip		
6" wide	LF	0.067
12" wide	"	0.100

Flashing And Sheet Metal	UNIT	MAN/ HOURS
07610.10 **Metal Roofing**		
Sheet metal roofing, copper, 16 oz, batten seam	SQ	5.333
Standing seam	"	5.000
Aluminum roofing, natural finish		
Corrugated, on steel frame		
.0175" thick	SQ	2.286
.0215" thick	"	2.286
.024" thick	"	2.286
.032" thick	"	2.286
V-beam, on steel frame		
.032" thick	SQ	2.286
.040" thick	"	2.286
.050" thick	"	2.286
Ridge cap		
.019" thick	LF	0.027
Corrugated galvanized steel roofing, on steel frame		
28 ga.	SQ	2.286
26 ga.	"	2.286
24 ga.	"	2.286
22 ga.	"	2.286
26 ga., factory insulated with 1" polystyrene	"	3.200
Ridge roll		
10" wide	LF	0.027

Flashing And Sheet Metal	UNIT	MAN/ HOURS
07610.10 **Metal Roofing** *(Cont.)*		
20" wide	LF	0.032
07620.10 **Flashing And Trim**		
Counter flashing		
Aluminum, .032"	SF	0.080
Stainless steel, .015"	"	0.080
Copper		
16 oz.	SF	0.080
20 oz.	"	0.080
24 oz.	"	0.080
32 oz.	"	0.080
Valley flashing		
Aluminum, .032"	SF	0.050
Stainless steel, .015	"	0.050
Copper		
16 oz.	SF	0.050
20 oz.	"	0.067
24 oz.	"	0.050
32 oz.	"	0.050
Base flashing		
Aluminum, .040"	SF	0.067
Stainless steel, .018"	"	0.067
Copper		
16 oz.	SF	0.067
20 oz.	"	0.050
24 oz.	"	0.067
32 oz.	"	0.067
Flashing and trim, aluminum		
.019" thick	SF	0.057
.032" thick	"	0.057
.040" thick	"	0.062
Neoprene sheet flashing, .060" thick	"	0.050
Copper, paper backed		
2 oz.	SF	0.080
5 oz.	"	0.080
07620.20 **Gutters And Downspouts**		
Copper gutter and downspout		
Downspouts, 16 oz. copper		
Round		
3" dia.	LF	0.053
4" dia.	"	0.053
Rectangular, corrugated		
2" x 3"	LF	0.050
3" x 4"	"	0.050
Rectangular, flat surface		
2" x 3"	LF	0.053
3" x 4"	"	0.053
Lead-coated copper downspouts		
Round		
3" dia.	LF	0.050
4" dia.	"	0.057
Rectangular, corrugated		

Flashing And Sheet Metal	UNIT	MAN/ HOURS
07620.20 **Gutters And Downspouts** *(Cont.)*		
2" x 3"	LF	0.053
3" x 4"	"	0.053
Rectangular, plain		
2" x 3"	LF	0.053
3" x 4"	"	0.053
Gutters, 16 oz. copper		
Half round		
4" wide	LF	0.080
5" wide	"	0.089
Type K		
4" wide	LF	0.080
5" wide	"	0.089
Lead-coated copper gutters		
Half round		
4" wide	LF	0.080
6" wide	"	0.089
Type K		
4" wide	LF	0.080
5" wide	"	0.089
Aluminum gutter and downspout		
Downspouts		
2" x 3"	LF	0.053
3" x 4"	"	0.057
4" x 5"	"	0.062
Round		
3" dia.	LF	0.053
4" dia.	"	0.057
Gutters, stock units		
4" wide	LF	0.084
5" wide	"	0.089
Galvanized steel gutter and downspout		
Downspouts, round corrugated		
3" dia.	LF	0.053
4" dia.	"	0.053
5" dia.	"	0.057
6" dia.	"	0.057
Rectangular		
2" x 3"	LF	0.053
3" x 4"	"	0.050
4" x 4"	"	0.050
Gutters, stock units		
5" wide		
Plain	LF	0.089
Painted	"	0.089
6" wide		
Plain	LF	0.094
Painted	"	0.094

Skylights	UNIT	MAN/ HOURS
07810.10 **Plastic Skylights**		
Single thickness, not including mounting curb		
2' x 4'	EA	1.000
4' x 4'	"	1.333
5' x 5'	"	2.000
6' x 8'	"	2.667
Double thickness, not including mounting curb		
2' x 4'	EA	1.000
4' x 4'	"	1.333
5' x 5'	"	2.000
6' x 8'	"	2.667

Joint Sealants	UNIT	MAN/ HOURS
07920.10 **Caulking**		
Caulk exterior, two component		
1/4 x 1/2	LF	0.040
3/8 x 1/2	"	0.044
1/2 x 1/2	"	0.050
Caulk interior, single component		
1/4 x 1/2	LF	0.038
3/8 x 1/2	"	0.042
1/2 x 1/2	"	0.047

08 DOORS AND WINDOWS

Metal Doors & Transoms	UNIT	MAN/ HOURS
08110.10 **Metal Doors**		
Flush hollow metal, std. duty, 20 ga., 1-3/8" thick		
2-6 x 6-8	EA	0.889
2-8 x 6-8	"	0.889
3-0 x 6-8	"	0.889
1-3/4" thick		
2-6 x 6-8	EA	0.889
2-8 x 6-8	"	0.889
3-0 x 6-8	"	0.889
2-6 x 7-0	"	0.889
2-8 x 7-0	"	0.889
3-0 x 7-0	"	0.889
Heavy duty, 20 ga., unrated, 1-3/4"		
2-8 x 6-8	EA	0.889
3-0 x 6-8	"	0.889
2-8 x 7-0	"	0.889
3-0 x 7-0	"	0.889
3-4 x 7-0	"	0.889
18 ga., 1-3/4", unrated door		
2-0 x 7-0	EA	0.889
2-4 x 7-0	"	0.889
2-6 x 7-0	"	0.889
2-8 x 7-0	"	0.889
3-0 x 7-0	"	0.889
3-4 x 7-0	"	0.889
2", unrated door		
2-0 x 7-0	EA	1.000
2-4 x 7-0	"	1.000
2-6 x 7-0	"	1.000
2-8 x 7-0	"	1.000
3-0 x 7-0	"	1.000
3-4 x 7-0	"	1.000
08110.40 **Metal Door Frames**		
Hollow metal, stock, 18 ga., 4-3/4" x 1-3/4"		
2-0 x 7-0	EA	1.000
2-4 x 7-0	"	1.000
2-6 x 7-0	"	1.000
2-8 x 7-0	"	1.000
3-0 x 7-0	"	1.000
4-0 x 7-0	"	1.333
5-0 x 7-0	"	1.333
6-0 x 7-0	"	1.333
16 ga., 6-3/4" x 1-3/4"		
2-0 x 7-0	EA	1.000
2-4 x 7-0	"	1.000
2-6 x 7-0	"	1.000
2-8 x 7-0	"	1.000
3-0 x 7-0	"	1.000
4-0 x 7-0	"	1.333
6-0 x 7-0	"	1.333

Wood And Plastic	UNIT	MAN/ HOURS
08210.10 **Wood Doors**		
Solid core, 1-3/8" thick		
Birch faced		
2-4 x 7-0	EA	1.000
2-8 x 7-0	"	1.000
3-0 x 7-0	"	1.000
3-4 x 7-0	"	1.000
2-4 x 6-8	"	1.000
2-6 x 6-8	"	1.000
2-8 x 6-8	"	1.000
3-0 x 6-8	"	1.000
Lauan faced		
2-4 x 6-8	EA	1.000
2-8 x 6-8	"	1.000
3-0 x 6-8	"	1.000
3-4 x 6-8	"	1.000
Tempered hardboard faced		
2-4 x 7-0	EA	1.000
2-8 x 7-0	"	1.000
3-0 x 7-0	"	1.000
3-4 x 7-0	"	1.000
Hollow core, 1-3/8" thick		
Birch faced		
2-4 x 7-0	EA	1.000
2-8 x 7-0	"	1.000
3-0 x 7-0	"	1.000
3-4 x 7-0	"	1.000
Lauan faced		
2-4 x 6-8	EA	1.000
2-6 x 6-8	"	1.000
2-8 x 6-8	"	1.000
3-0 x 6-8	"	1.000
3-4 x 6-8	"	1.000
Tempered hardboard faced		
2-4 x 7-0	EA	1.000
2-6 x 7-0	"	1.000
2-8 x 7-0	"	1.000
3-0 x 7-0	"	1.000
3-4 x 7-0	"	1.000
Solid core, 1-3/4" thick		
Birch faced		
2-4 x 7-0	EA	1.000
2-6 x 7-0	"	1.000
2-8 x 7-0	"	1.000
3-0 x 7-0	"	1.000
3-4 x 7-0	"	1.000
Lauan faced		
2-4 x 7-0	EA	1.000
2-6 x 7-0	"	1.000
2-8 x 7-0	"	1.000
3-4 x 7-0	"	1.000
3-0 x 7-0	"	1.000
Tempered hardboard faced		
2-4 x 7-0	EA	1.000

Wood And Plastic	UNIT	MAN/HOURS
08210.10 **Wood Doors** *(Cont.)*		
2-6 x 7-0	EA	1.000
2-8 x 7-0	"	1.000
3-0 x 7-0	"	1.000
3-4 x 7-0	"	1.000
Hollow core, 1-3/4" thick		
Birch faced		
2-4 x 7-0	EA	1.000
2-6 x 7-0	"	1.000
2-8 x 7-0	"	1.000
3-0 x 7-0	"	1.000
3-4 x 7-0	"	1.000
Lauan faced		
2-4 x 6-8	EA	1.000
2-6 x 6-8	"	1.000
2-8 x 6-8	"	1.000
3-0 x 6-8	"	1.000
3-4 x 6-8	"	1.000
Tempered hardboard		
2-4 x 7-0	EA	1.000
2-6 x 7-0	"	1.000
2-8 x 7-0	"	1.000
3-0 x 7-0	"	1.000
3-4 x 7-0	"	1.000
Add-on, louver	"	0.800
Glass	"	0.800
Exterior doors, 3-0 x 7-0 x 2-1/2", solid core		
Carved		
One face	EA	2.000
Two faces	"	2.000
Closet doors, 1-3/4" thick		
Bi-fold or bi-passing, includes frame and trim		
Paneled		
4-0 x 6-8	EA	1.333
6-0 x 6-8	"	1.333
Louvered		
4-0 x 6-8	EA	1.333
6-0 x 6-8	"	1.333
Flush		
4-0 x 6-8	EA	1.333
6-0 x 6-8	"	1.333
Primed		
4-0 x 6-8	EA	1.333
6-0 x 6-8	"	1.333
08210.90 **Wood Frames**		
Frame, interior, pine		
2-6 x 6-8	EA	1.143
2-8 x 6-8	"	1.143
3-0 x 6-8	"	1.143
5-0 x 6-8	"	1.143
6-0 x 6-8	"	1.143
2-6 x 7-0	"	1.143
2-8 x 7-0	"	1.143

Wood And Plastic	UNIT	MAN/HOURS
08210.90 **Wood Frames** *(Cont.)*		
3-0 x 7-0	EA	1.143
5-0 x 7-0	"	1.600
6-0 x 7-0	"	1.600
Exterior, custom, with threshold, including trim		
Walnut		
3-0 x 7-0	EA	2.000
6-0 x 7-0	"	2.000
Oak		
3-0 x 7-0	EA	2.000
6-0 x 7-0	"	2.000
Pine		
2-4 x 7-0	EA	1.600
2-6 x 7-0	"	1.600
2-8 x 7-0	"	1.600
3-0 x 7-0	"	1.600
3-4 x 7-0	"	1.600
6-0 x 7-0	"	2.667
08300.10 **Special Doors**		
Sliding glass doors		
Tempered plate glass, 1/4" thick		
6' wide		
Economy grade	EA	2.667
Premium grade	"	2.667
12' wide		
Economy grade	EA	4.000
Premium grade	"	4.000
Insulating glass, 5/8" thick		
6' wide		
Economy grade	EA	2.667
Premium grade	"	2.667
12' wide		
Economy grade	EA	4.000
Premium grade	"	4.000
1" thick		
6' wide		
Economy grade	EA	2.667
Premium grade	"	2.667
12' wide		
Economy grade	EA	4.000
Premium grade	"	4.000
Residential storm door		
Minimum	EA	1.333
Average	"	1.333
Maximum	"	2.000
08520.10 **Aluminum Windows**		
Jalousie		
3-0 x 4-0	EA	1.000
3-0 x 5-0	"	1.000
Fixed window		
6 sf to 8 sf	SF	0.114
12 sf to 16 sf	"	0.089

Wood And Plastic	UNIT	MAN/ HOURS
08520.10 **Aluminum Windows** *(Cont.)*		
Projecting window		
6 sf to 8 sf	SF	0.200
12 sf to 16 sf	"	0.133
Horizontal sliding		
6 sf to 8 sf	SF	0.100
12 sf to 16 sf	"	0.080
Double hung		
6 sf to 8 sf	SF	0.160
10 sf to 12 sf	"	0.133
Storm window, 0.5 cfm, up to		
60 u.i. (united inches)	EA	0.400
70 u.i.	"	0.400
80 u.i.	"	0.400
90 u.i.	"	0.444
100 u.i.	"	0.444
2.0 cfm, up to		
60 u.i.	EA	0.400
70 u.i.	"	0.400
80 u.i.	"	0.400
90 u.i.	"	0.444
100 u.i.	"	0.444

Wood And Plastic	UNIT	MAN/ HOURS
08600.10 **Wood Windows**		
Double hung		
24" x 36"		
Minimum	EA	0.800
Average	"	1.000
Maximum	"	1.333
24" x 48"		
Minimum	EA	0.800
Average	"	1.000
Maximum	"	1.333
30" x 48"		
Minimum	EA	0.889
Average	"	1.143
Maximum	"	1.600
30" x 60"		
Minimum	EA	0.889
Average	"	1.143
Maximum	"	1.600
Casement		
1 leaf, 22" x 38" high		
Minimum	EA	0.800
Average	"	1.000

Wood And Plastic	UNIT	MAN/ HOURS
08600.10 **Wood Windows** *(Cont.)*		
Maximum	EA	1.333
2 leaf, 50" x 50" high		
Minimum	EA	1.000
Average	"	1.333
Maximum	"	2.000
3 leaf, 71" x 62" high		
Minimum	EA	1.000
Average	"	1.333
Maximum	"	2.000
4 leaf, 95" x 75" high		
Minimum	EA	1.143
Average	"	1.600
Maximum	"	2.667
5 leaf, 119" x 75" high		
Minimum	EA	1.143
Average	"	1.600
Maximum	"	2.667
Picture window, fixed glass, 54" x 54" high		
Minimum	EA	1.000
Average	"	1.143
Maximum	"	1.333
68" x 55" high		
Minimum	EA	1.000
Average	"	1.143
Maximum	"	1.333
Sliding, 40" x 31" high		
Minimum	EA	0.800
Average	"	1.000
Maximum	"	1.333
52" x 39" high		
Minimum	EA	1.000
Average	"	1.143
Maximum	"	1.333
64" x 72" high		
Minimum	EA	1.000
Average	"	1.333
Maximum	"	1.600
Awning windows		
34" x 21" high		
Minimum	EA	0.800
Average	"	1.000
Maximum	"	1.333
40" x 21" high		
Minimum	EA	0.889
Average	"	1.143
Maximum	"	1.600
48" x 27" high		
Minimum	EA	0.889
Average	"	1.143
Maximum	"	1.600
60" x 36" high		
Minimum	EA	1.000

Wood And Plastic	UNIT	MAN/ HOURS
08600.10 **Wood Windows** *(Cont.)*		
Average	EA	1.333
Maximum	"	1.600
Window frame, milled		
Minimum	LF	0.160
Average	"	0.200
Maximum	"	0.267

Hardware	UNIT	MAN/ HOURS
08710.20 **Locksets**		
Latchset, heavy duty		
Cylindrical	EA	0.500
Mortise	"	0.800
Lockset, heavy duty		
Cylindrical	EA	0.500
Mortise	"	0.800
Lockset		
Privacy (bath or bedroom)	EA	0.667
Entry lock	"	0.667
08710.30 **Closers**		
Door closers		
Standard	EA	1.000
Heavy duty	"	1.000
08710.40 **Door Trim**		
Panic device		
Mortise	EA	2.000
Vertical rod	"	2.000
Labeled, rim type	"	2.000
Mortise	"	2.000
Vertical rod	"	2.000
08710.60 **Weatherstripping**		
Weatherstrip, head and jamb, metal strip, neoprene bulb		
Standard duty	LF	0.044
Heavy duty	"	0.050
Spring type		
Metal doors	EA	2.000
Wood doors	"	2.667
Sponge type with adhesive backing	"	0.800
Thresholds		
Bronze	LF	0.200
Aluminum		

Hardware	UNIT	MAN/ HOURS
08710.60 **Weatherstripping** *(Cont.)*		
Plain	LF	0.200
Vinyl insert	"	0.200
Aluminum with grit	"	0.200
Steel		
Plain	LF	0.200
Interlocking	"	0.667

Glazing	UNIT	MAN/ HOURS
08810.10 **Glass Glazing**		
Sheet glass, 1/8" thick	SF	0.044
Plate glass, bronze or grey, 1/4" thick	"	0.073
Clear	"	0.073
Polished	"	0.073
Plexiglass		
1/8" thick	SF	0.073
1/4" thick	"	0.044
Float glass, clear		
3/16" thick	SF	0.067
1/4" thick	"	0.073
3/8" thick	"	0.100
Tinted glass, polished plate, twin ground		
3/16" thick	SF	0.067
1/4" thick	"	0.073
3/8" thick	"	0.100
Insulating glass, two lites, clear float glass		
1/2" thick	SF	0.133
5/8" thick	"	0.160
3/4" thick	"	0.200
7/8" thick	"	0.229
1" thick	"	0.267
Glass seal edge		
3/8" thick	SF	0.133
Tinted glass		
1/2" thick	SF	0.133
1" thick	"	0.267
Tempered, clear		
1" thick	SF	0.267
Plate mirror glass		
1/4" thick		
15 sf	SF	0.080
Over 15 sf	"	0.073

Glazed Curtain Walls	UNIT	MAN/ HOURS
08910.10 **Glazed Curtain Walls**		
Curtain wall, aluminum system, framing sections		
2" x 3"		
Jamb	LF	0.067
Horizontal	"	0.067
Mullion	"	0.067
2" x 4"		
Jamb	LF	0.100
Horizontal	"	0.100
Mullion	"	0.100
3" x 5-1/2"		
Jamb	LF	0.100
Horizontal	"	0.100
Mullion	"	0.100
4" corner mullion	"	0.133
Coping sections		
1/8" x 8"	LF	0.133
1/8" x 9"	"	0.133
1/8" x 12-1/2"	"	0.160
Sill section		
1/8" x 6"	LF	0.080
1/8" x 7"	"	0.080
1/8" x 8-1/2"	"	0.080
Column covers, aluminum		
1/8" x 26"	LF	0.200
1/8" x 34"	"	0.211
1/8" x 38"	"	0.211
Doors		
Aluminum framed, standard hardware		
Narrow stile		
2-6 x 7-0	EA	4.000
3-0 x 7-0	"	4.000
3-6 x 7-0	"	4.000

Support Systems	UNIT	MAN/ HOURS
09110.10 **Metal Studs**		
Studs, non load bearing, galvanized		
2-1/2", 20 ga.		
12" o.c.	SF	0.017
16" o.c.	"	0.013
25 ga.		
12" o.c.	SF	0.017
16" o.c.	"	0.013
24" o.c.	"	0.011
3-5/8", 20 ga.		
12" o.c.	SF	0.020
16" o.c.	"	0.016
24" o.c.	"	0.013
25 ga.		
12" o.c.	SF	0.020
16" o.c.	"	0.016
24" o.c.	"	0.013
4", 20 ga.		
12" o.c.	SF	0.020
16" o.c.	"	0.016
24" o.c.	"	0.013
25 ga.		
12" o.c.	SF	0.020
16" o.c.	"	0.016
24" o.c.	"	0.013
6", 20 ga.		
12" o.c.	SF	0.025
16" o.c.	"	0.020
24" o.c.	"	0.017
25 ga.		
12" o.c.	SF	0.025
16" o.c.	"	0.020
24" o.c.	"	0.017
Load bearing studs, galvanized		
3-5/8", 16 ga.		
12" o.c.	SF	0.020
16" o.c.	"	0.016
18 ga.		
12" o.c.	SF	0.013
16" o.c.	"	0.016
4", 16 ga.		
12" o.c.	SF	0.020
16" o.c.	"	0.016
6", 16 ga.		
12" o.c.	SF	0.025
16" o.c.	"	0.020
Furring		
On beams and columns		
7/8" channel	LF	0.053
1-1/2" channel	"	0.062
On ceilings		
3/4" furring channels		
12" o.c.	SF	0.033
16" o.c.	"	0.032

Support Systems	UNIT	MAN/ HOURS
09110.10 **Metal Studs** *(Cont.)*		
24" o.c.	SF	0.029
1-1/2" furring channels		
12" o.c.	SF	0.036
16" o.c.	"	0.033
24" o.c.	"	0.031
On walls		
3/4" furring channels		
12" o.c.	SF	0.027
16" o.c.	"	0.025
24" o.c.	"	0.024
1-1/2" furring channels		
12" o.c.	SF	0.029
16" o.c.	"	0.027
24" o.c.	"	0.025

Lath And Plaster	UNIT	MAN/ HOURS
09205.10 **Gypsum Lath**		
Gypsum lath, 1/2" thick		
Clipped	SY	0.044
Nailed	"	0.050
09205.20 **Metal Lath**		
Diamond expanded, galvanized		
2.5 lb., on walls		
Nailed	SY	0.100
Wired	"	0.114
On ceilings		
Nailed	SY	0.114
Wired	"	0.133
3.4 lb., on walls		
Nailed	SY	0.100
Wired	"	0.114
On ceilings		
Nailed	SY	0.114
Wired	"	0.133
Flat rib		
2.75 lb., on walls		
Nailed	SY	0.100
Wired	"	0.114
On ceilings		
Nailed	SY	0.114
Wired	"	0.133
3.4 lb., on walls		
Nailed	SY	0.100
Wired	"	0.114

Lath And Plaster	UNIT	MAN/HOURS
09205.20　Metal Lath　(Cont.)		
On ceilings		
Nailed	SY	0.114
Wired	"	0.133
Stucco lath		
1.8 lb.	SY	0.100
3.6 lb.	"	0.100
Paper backed		
Minimum	SY	0.080
Maximum	"	0.114
09205.60　Plaster Accessories		
Expansion joint, 3/4", 26 ga., galv.	LF	0.020
Plaster corner beads, 3/4", galvanized	"	0.023
Casing bead, expanded flange, galvanized	"	0.020
Expanded wing, 1-1/4" wide, galvanized	"	0.020
Joint clips for lath	EA	0.004
Metal base, galvanized, 2-1/2" high	LF	0.027
Stud clips for gypsum lath	EA	0.004
Sound deadening board, 1/4"	SF	0.013
09210.10　Plaster		
Gypsum plaster, trowel finish, 2 coats		
Ceilings	SY	0.250
Walls	"	0.235
3 coats		
Ceilings	SY	0.348
Walls	"	0.308
Vermiculite plaster		
2 coats		
Ceilings	SY	0.381
Walls	"	0.348
3 coats		
Ceilings	SY	0.471
Walls	"	0.421
Keenes cement plaster		
2 coats		
Ceilings	SY	0.308
Walls	"	0.267
3 coats		
Ceilings	SY	0.348
Walls	"	0.308
On columns, add to installation, 50%	"	
Chases, fascia, and soffits, add to installation, 50%	"	
Beams, add to installation, 50%	"	
Patch holes, average size holes		
1 sf to 5 sf		
Minimum	SF	0.133
Average	"	0.160
Maximum	"	0.200
Over 5 sf		
Minimum	SF	0.080

Lath And Plaster	UNIT	MAN/HOURS
09210.10　Plaster　(Cont.)		
Average	SF	0.114
Maximum	"	0.133
Patch cracks		
Minimum	SF	0.027
Average	"	0.040
Maximum	"	0.080
09220.10　Portland Cement Plaster		
Stucco, portland, gray, 3 coat, 1" thick		
Sand finish	SY	0.348
Trowel finish	"	0.364
White cement		
Sand finish	SY	0.364
Trowel finish	"	0.400
Scratch coat		
For ceramic tile	SY	0.080
For quarry tile	"	0.080
Portland cement plaster		
2 coats, 1/2"	SY	0.160
3 coats, 7/8"	"	0.200
09250.10　Gypsum Board		
Drywall, plasterboard, 3/8" clipped to		
Metal furred ceiling	SF	0.009
Columns and beams	"	0.020
Walls	"	0.008
Nailed or screwed to		
Wood or metal framed ceiling	SF	0.008
Columns and beams	"	0.018
Walls	"	0.007
1/2", clipped to		
Metal furred ceiling	SF	0.009
Columns and beams	"	0.020
Walls	"	0.008
Nailed or screwed to		
Wood or metal framed ceiling	SF	0.008
Columns and beams	"	0.018
Walls	"	0.007
5/8", clipped to		
Metal furred ceiling	SF	0.010
Columns and beams	"	0.022
Walls	"	0.009
Nailed or screwed to		
Wood or metal framed ceiling	SF	0.010
Columns and beams	"	0.022
Walls	"	0.009
Vinyl faced, clipped to metal studs		
1/2"	SF	0.010
5/8"	"	0.010
Taping and finishing joints		
Minimum	SF	0.005
Average	"	0.007
Maximum	"	0.008

Lath And Plaster	UNIT	MAN/ HOURS
09250.10 **Gypsum Board** *(Cont.)*		
Casing bead		
Minimum	LF	0.023
Average	"	0.027
Maximum	"	0.040
Corner bead		
Minimum	LF	0.023
Average	"	0.027
Maximum	"	0.040

Tile	UNIT	MAN/ HOURS
09310.10 **Ceramic Tile**		
Glazed wall tile, 4-1/4" x 4-1/4"		
Minimum	SF	0.057
Average	"	0.067
Maximum	"	0.080
Base, 4-1/4" high		
Minimum	LF	0.100
Average	"	0.100
Maximum	"	0.100
Unglazed floor tile		
Portland cem., cushion edge, face mtd		
1" x 1"	SF	0.073
2" x 2"	"	0.067
4" x 4"	"	0.067
6" x 6"	"	0.057
12" x 12"	"	0.050
16" x 16"	"	0.044
18" x 18"	"	0.040
Adhesive bed, with white grout		
1" x 1"	SF	0.073
2" x 2"	"	0.067
4" x 4"	"	0.067
6" x 6"	"	0.057
12" x 12"	"	0.050
16" x 16"	"	0.044
18" x 18"	"	0.040
Organic adhesive bed, thin set, back mounted		
1" x 1"	SF	0.073
2" x 2"	"	0.067
Porcelain floor tile		
1" x 1"	SF	0.073
2" x 2"	"	0.070
4" x 4"	"	0.067
6" x 6"	"	0.057
12" x 12"	"	0.050

Tile	UNIT	MAN/ HOURS
09310.10 **Ceramic Tile** *(Cont.)*		
16" x 16"	SF	0.044
18" x 18"	"	0.040
Unglazed wall tile		
Organic adhesive, face mounted cushion edge		
1" x 1"		
Minimum	SF	0.067
Average	"	0.073
Maximum	"	0.080
2" x 2"		
Minimum	SF	0.062
Average	"	0.067
Maximum	"	0.073
Back mounted		
1" x 1"		
Minimum	SF	0.067
Average	"	0.073
Maximum	"	0.080
2" x 2"		
Minimum	SF	0.062
Average	"	0.067
Maximum	"	0.073
Ceramic accessories		
Towel bar, 24" long		
Minimum	EA	0.320
Average	"	0.400
Maximum	"	0.533
Soap dish		
Minimum	EA	0.533
Average	"	0.667
Maximum	"	0.800

09330.10 **Quarry Tile**		
Floor		
4 x 4 x 1/2"	SF	0.107
6 x 6 x 1/2"	"	0.100
6 x 6 x 3/4"	"	0.100
12 x 12 x 3/4"	"	0.089
16 x 16 x 3/4"	"	0.080
18 x 18 x 3/4"	"	0.067
Medallion		
36" dia.	EA	2.000
48" dia.	"	2.000
Wall, applied to 3/4" portland cement bed		
4 x 4 x 1/2"	SF	0.160
6 x 6 x 3/4"	"	0.133
Cove base		
5 x 6 x 1/2" straight top	LF	0.133
6 x 6 x 3/4" round top	"	0.133
Moldings		
2 x 12	LF	0.080
4 x 12	"	0.080
Stair treads 6 x 6 x 3/4"	"	0.200
Window sill 6 x 8 x 3/4"	"	0.160

Tile	UNIT	MAN/HOURS
09330.10 **Quarry Tile** *(Cont.)*		
For abrasive surface, add to material, 25%		
09410.10 **Terrazzo**		
Floors on concrete, 1-3/4" thick, 5/8" topping		
Gray cement	SF	0.114
White cement	"	0.114
Sand cushion, 3" thick, 5/8" top, 1/4"		
Gray cement	SF	0.133
White cement	"	0.133
Monolithic terrazzo, 3-1/2" base slab, 5/8" topping	"	0.100
Terrazzo wainscot, cast-in-place, 1/2" thick	"	0.200
Base, cast in place, terrazzo cove type, 6" high	LF	0.114
Curb, cast in place, 6" wide x 6" high, polished top	"	0.400
Stairs, cast-in-place, topping on concrete or metal		
1-1/2" thick treads, 12" wide	LF	0.400
Combined tread and riser	"	1.000
Precast terrazzo, thin set		
Terrazzo tiles, non-slip surface		
9" x 9" x 1" thick	SF	0.114
12" x 12"		
1" thick	SF	0.107
1-1/2" thick	"	0.114
18" x 18" x 1-1/2" thick	"	0.114
24" x 24" x 1-1/2" thick	"	0.094
Terrazzo wainscot		
12" x 12" x 1" thick	SF	0.200
18" x 18" x 1-1/2" thick	"	0.229
Base		
6" high		
Straight	LF	0.062
Coved	"	0.062
8" high		
Straight	LF	0.067
Coved	"	0.067
Terrazzo curbs		
8" wide x 8" high	LF	0.320
6" wide x 6" high	"	0.267
Precast terrazzo stair treads, 12" wide		
1-1/2" thick		
Diamond pattern	LF	0.145
Non-slip surface	"	0.145
2" thick		
Diamond pattern	LF	0.145
Non-slip surface	"	0.160
Stair risers, 1" thick to 6" high		
Straight sections	LF	0.080
Cove sections	"	0.080
Combined tread and riser		
Straight sections		
1-1/2" tread, 3/4" riser	LF	0.229
3" tread, 1" riser	"	0.229
Curved sections		
2" tread, 1" riser	LF	0.267

Tile	UNIT	MAN/HOURS
09410.10 **Terrazzo** *(Cont.)*		
3" tread, 1" riser	LF	0.267
Stair stringers, notched for treads and risers		
1" thick	LF	0.200
2" thick	"	0.267
Landings, structural, nonslip		
1-1/2" thick	SF	0.133
3" thick	"	0.160

Acoustical Treatment	UNIT	MAN/HOURS
09510.10 **Ceilings And Walls**		
Acoustical panels, suspension system not included		
Fiberglass panels		
5/8" thick		
2' x 2'	SF	0.011
2' x 4'	"	0.009
3/4" thick		
2' x 2'	SF	0.011
2' x 4'	"	0.009
Glass cloth faced fiberglass panels		
3/4" thick	SF	0.013
1" thick	"	0.013
Mineral fiber panels		
5/8" thick		
2' x 2'	SF	0.011
2' x 4'	"	0.009
3/4" thick		
2' x 2'	SF	0.011
2' x 4'	"	0.009
Wood fiber panels		
1/2" thick		
2' x 2'	SF	0.011
2' x 4'	"	0.009
5/8" thick		
2' x 2'	SF	0.011
2' x 4'	"	0.009
Acoustical tiles, suspension system not included		
Fiberglass tile, 12" x 12"		
5/8" thick	SF	0.015
3/4" thick	"	0.018
Glass cloth faced fiberglass tile		
3/4" thick	SF	0.018
3" thick	"	0.020
Mineral fiber tile, 12" x 12"		
5/8" thick		
Standard	SF	0.016

Acoustical Treatment	UNIT	MAN/HOURS
09510.10 **Ceilings And Walls** *(Cont.)*		
Vinyl faced	SF	0.016
3/4" thick		
Standard	SF	0.016
Vinyl faced	"	0.016
Ceiling suspension systems		
T bar system		
2' x 4'	SF	0.008
2' x 2'	"	0.009
Concealed Z bar suspension system, 12" module	"	0.013

Flooring	UNIT	MAN/HOURS
09550.10 **Wood Flooring**		
Wood strip flooring, unfinished		
Fir floor		
C and better		
Vertical grain	SF	0.027
Flat grain	"	0.027
Oak floor		
Minimum	SF	0.038
Average	"	0.038
Maximum	"	0.038
Maple floor		
25/32" x 2-1/4"		
Minimum	SF	0.038
Maximum	"	0.038
33/32" x 3-1/4"		
Minimum	SF	0.038
Maximum	"	0.038
Wood block industrial flooring		
Creosoted		
2" thick	SF	0.021
2-1/2" thick	"	0.025
3" thick	"	0.027
Parquet, 5/16", white oak		
Finished	SF	0.040
Unfinished	"	0.040
Gym floor, 2 ply felt, 25/32" maple, finished, in mastic	"	0.044
Over wood sleepers	"	0.050
Finishing, sand, fill, finish, and wax	"	0.020
Refinish sand, seal, and 2 coats of polyurethane	"	0.027
Clean and wax floors	"	0.004

Flooring	UNIT	MAN/HOURS
09630.10 **Unit Masonry Flooring**		
Clay brick		
9 x 4-1/2 x 3" thick		
Glazed	SF	0.067
Unglazed	"	0.067
8 x 4 x 3/4" thick		
Glazed	SF	0.070
Unglazed	"	0.070
09660.10 **Resilient Tile Flooring**		
Solid vinyl tile, 1/8" thick, 12" x 12"		
Marble patterns	SF	0.020
Solid colors	"	0.020
Travertine patterns	"	0.020
Conductive resilient flooring, vinyl tile		
1/8" thick, 12" x 12"	SF	0.023
09665.10 **Resilient Sheet Flooring**		
Vinyl sheet flooring		
Minimum	SF	0.008
Average	"	0.010
Maximum	"	0.013
Cove, to 6"	LF	0.016
Fluid applied resilient flooring		
Polyurethane, poured in place, 3/8" thick	SF	0.067
Vinyl sheet goods, backed		
0.070" thick	SF	0.010
0.093" thick	"	0.010
0.125" thick	"	0.010
0.250" thick	"	0.010
09678.10 **Resilient Base And Accessories**		
Wall base, vinyl		
4" high	LF	0.027
6" high	"	0.027

Carpet	UNIT	MAN/HOURS
09682.10 **Carpet Padding**		
Carpet padding		
Foam rubber, waffle type, 0.3" thick	SY	0.040
Jute padding		
Minimum	SY	0.036
Average	"	0.040
Maximum	"	0.044
Sponge rubber cushion		
Minimum	SY	0.036

Carpet	UNIT	MAN/ HOURS
09682.10 **Carpet Padding** *(Cont.)*		
Average	SY	0.040
Maximum	"	0.044
Urethane cushion, 3/8" thick		
Minimum	SY	0.036
Average	"	0.040
Maximum	"	0.044
09685.10 **Carpet**		
Carpet, acrylic		
24 oz., light traffic	SY	0.089
28 oz., medium traffic	"	0.089
Nylon		
15 oz., light traffic	SY	0.089
28 oz., medium traffic	"	0.089
Nylon		
28 oz., medium traffic	SY	0.089
35 oz., heavy traffic	"	0.089
Wool		
30 oz., medium traffic	SY	0.089
36 oz., medium traffic	"	0.089
42 oz., heavy traffic	"	0.089
Carpet tile		
Foam backed		
Minimum	SF	0.016
Average	"	0.018
Maximum	"	0.020
Tufted loop or shag		
Minimum	SF	0.016
Average	"	0.018
Maximum	"	0.020
Clean and vacuum carpet		
Minimum	SY	0.004
Average	"	0.005
Maximum	"	0.008

Paint	UNIT	MAN/ HOURS
09905.10 **Painting Preparation**		
Dropcloths		
Minimum	SF	0.001
Average	"	0.001
Maximum	"	0.001
Masking		
Paper and tape		
Minimum	LF	0.008
Average	"	0.010

Paint	UNIT	MAN/ HOURS
09905.10 **Painting Preparation** *(Cont.)*		
Maximum	LF	0.013
Doors		
Minimum	EA	0.100
Average	"	0.133
Maximum	"	0.178
Windows		
Minimum	EA	0.100
Average	"	0.133
Maximum	"	0.178
Sanding		
Walls and flat surfaces		
Minimum	SF	0.005
Average	"	0.007
Maximum	"	0.008
Doors and windows		
Minimum	EA	0.133
Average	"	0.200
Maximum	"	0.267
Trim		
Minimum	LF	0.010
Average	"	0.013
Maximum	"	0.018
Puttying		
Minimum	SF	0.012
Average	"	0.016
Maximum	"	0.020
09910.05 **Ext. Painting, Sitework**		
Concrete Block		
Roller		
First Coat		
Minimum	SF	0.004
Average	"	0.005
Maximum	"	0.008
Second Coat		
Minimum	SF	0.003
Average	"	0.004
Maximum	"	0.007
Spray		
First Coat		
Minimum	SF	0.002
Average	"	0.003
Maximum	"	0.003
Second Coat		
Minimum	SF	0.001
Average	"	0.002
Maximum	"	0.003
Fences, Chain Link		
Roller		
First Coat		
Minimum	SF	0.006
Average	"	0.007
Maximum	"	0.008

Paint	UNIT	MAN/HOURS
09910.05 **Ext. Painting, Sitework** *(Cont.)*		
Second Coat		
Minimum	SF	0.003
Average	"	0.004
Maximum	"	0.005
Spray		
First Coat		
Minimum	SF	0.003
Average	"	0.003
Maximum	"	0.003
Second Coat		
Minimum	SF	0.002
Average	"	0.002
Maximum	"	0.003
Fences, Wood or Masonry		
Brush		
First Coat		
Minimum	SF	0.008
Average	"	0.010
Maximum	"	0.013
Second Coat		
Minimum	SF	0.005
Average	"	0.006
Maximum	"	0.008
Roller		
First Coat		
Minimum	SF	0.004
Average	"	0.005
Maximum	"	0.006
Second Coat		
Minimum	SF	0.003
Average	"	0.004
Maximum	"	0.005
Spray		
First Coat		
Minimum	SF	0.003
Average	"	0.004
Maximum	"	0.005
Second Coat		
Minimum	SF	0.002
Average	"	0.003
Maximum	"	0.003
09910.15 **Ext. Painting, Buildings**		
Decks, Wood, Stained		
Brush		
First Coat		
Minimum	SF	0.004
Average	"	0.004
Maximum	"	0.005
Second Coat		
Minimum	SF	0.003
Average	"	0.003
Maximum	"	0.003

Paint	UNIT	MAN/HOURS
09910.15 **Ext. Painting, Buildings** *(Cont.)*		
Roller		
First Coat		
Minimum	SF	0.003
Average	"	0.003
Maximum	"	0.003
Second Coat		
Minimum	SF	0.003
Average	"	0.003
Maximum	"	0.003
Spray		
First Coat		
Minimum	SF	0.003
Average	"	0.003
Maximum	"	0.003
Second Coat		
Minimum	SF	0.002
Average	"	0.002
Maximum	"	0.003
Doors, Wood		
Brush		
First Coat		
Minimum	SF	0.012
Average	"	0.016
Maximum	"	0.020
Second Coat		
Minimum	SF	0.010
Average	"	0.011
Maximum	"	0.013
Roller		
First Coat		
Minimum	SF	0.005
Average	"	0.007
Maximum	"	0.010
Second Coat		
Minimum	SF	0.004
Average	"	0.004
Maximum	"	0.007
Spray		
First Coat		
Minimum	SF	0.003
Average	"	0.003
Maximum	"	0.004
Second Coat		
Minimum	SF	0.002
Average	"	0.002
Maximum	"	0.003
Gutters and Downspouts		
Brush		
First Coat		
Minimum	LF	0.010
Average	"	0.011
Maximum	"	0.013
Second Coat		

Paint		UNIT	MAN/HOURS
09910.15	**Ext. Painting, Buildings** *(Cont.)*		
Minimum		LF	0.007
Average		"	0.008
Maximum		"	0.010
Siding, Wood			
Roller			
First Coat			
Minimum		SF	0.003
Average		"	0.003
Maximum		"	0.004
Second Coat			
Minimum		SF	0.003
Average		"	0.004
Maximum		"	0.004
Spray			
First Coat			
Minimum		SF	0.003
Average		"	0.003
Maximum		"	0.003
Second Coat			
Minimum		SF	0.002
Average		"	0.003
Maximum		"	0.004
Stucco			
Roller			
First Coat			
Minimum		SF	0.004
Average		"	0.004
Maximum		"	0.005
Second Coat			
Minimum		SF	0.003
Average		"	0.003
Maximum		"	0.004
Spray			
First Coat			
Minimum		SF	0.003
Average		"	0.003
Maximum		"	0.003
Second Coat			
Minimum		SF	0.002
Average		"	0.002
Maximum		"	0.003
Trim			
Brush			
First Coat			
Minimum		LF	0.003
Average		"	0.004
Maximum		"	0.005
Second Coat			
Minimum		LF	0.003
Average		"	0.003
Maximum		"	0.005
Walls			
Roller			

Paint		UNIT	MAN/HOURS
09910.15	**Ext. Painting, Buildings** *(Cont.)*		
First Coat			
Minimum		SF	0.003
Average		"	0.003
Maximum		"	0.003
Second Coat			
Minimum		SF	0.003
Average		"	0.003
Maximum		"	0.003
Spray			
First Coat			
Minimum		SF	0.001
Average		"	0.002
Maximum		"	0.002
Second Coat			
Minimum		SF	0.001
Average		"	0.001
Maximum		"	0.002
Windows			
Brush			
First Coat			
Minimum		SF	0.013
Average		"	0.016
Maximum		"	0.020
Second Coat			
Minimum		SF	0.011
Average		"	0.013
Maximum		"	0.016
09910.25	**Ext. Painting, Misc.**		
Shakes			
Spray			
First Coat			
Minimum		SF	0.003
Average		"	0.004
Maximum		"	0.004
Second Coat			
Minimum		SF	0.003
Average		"	0.003
Maximum		"	0.004
Shingles, Wood			
Roller			
First Coat			
Minimum		SF	0.004
Average		"	0.005
Maximum		"	0.006
Second Coat			
Minimum		SF	0.003
Average		"	0.003
Maximum		"	0.004
Spray			
First Coat			
Minimum		LF	0.003
Average		"	0.003

Paint	UNIT	MAN/HOURS
09910.25 **Ext. Painting, Misc.** *(Cont.)*		
Maximum	LF	0.004
Second Coat		
Minimum	LF	0.002
Average	"	0.003
Maximum	"	0.003
Shutters and Louvres		
Brush		
First Coat		
Minimum	EA	0.160
Average	"	0.200
Maximum	"	0.267
Second Coat		
Minimum	EA	0.100
Average	"	0.123
Maximum	"	0.160
Spray		
First Coat		
Minimum	EA	0.053
Average	"	0.064
Maximum	"	0.080
Second Coat		
Minimum	EA	0.040
Average	"	0.053
Maximum	"	0.064
Stairs, metal		
Brush		
First Coat		
Minimum	SF	0.009
Average	"	0.010
Maximum	"	0.011
Second Coat		
Minimum	SF	0.005
Average	"	0.006
Maximum	"	0.007
Spray		
First Coat		
Minimum	SF	0.004
Average	"	0.006
Maximum	"	0.006
Second Coat		
Minimum	SF	0.003
Average	"	0.004
Maximum	"	0.005
09910.35 **Int. Painting, Buildings**		
Acoustical Ceiling		
Roller		
First Coat		
Minimum	SF	0.005
Average	"	0.007
Maximum	"	0.010
Second Coat		
Minimum	SF	0.004

Paint	UNIT	MAN/HOURS
09910.35 **Int. Painting, Buildings** *(Cont.)*		
Average	SF	0.005
Maximum	"	0.007
Spray		
First Coat		
Minimum	SF	0.002
Average	"	0.003
Maximum	"	0.003
Second Coat		
Minimum	SF	0.002
Average	"	0.002
Maximum	"	0.002
Cabinets and Casework		
Brush		
First Coat		
Minimum	SF	0.008
Average	"	0.009
Maximum	"	0.010
Second Coat		
Minimum	SF	0.007
Average	"	0.007
Maximum	"	0.008
Spray		
First Coat		
Minimum	SF	0.004
Average	"	0.005
Maximum	"	0.006
Second Coat		
Minimum	SF	0.003
Average	"	0.003
Maximum	"	0.004
Ceilings		
Roller		
First Coat		
Minimum	SF	0.003
Average	"	0.004
Maximum	"	0.004
Second Coat		
Minimum	SF	0.003
Average	"	0.003
Maximum	"	0.003
Spray		
First Coat		
Minimum	SF	0.002
Average	"	0.002
Maximum	"	0.003
Second Coat		
Minimum	SF	0.002
Average	"	0.002
Maximum	"	0.002
Doors, Wood		
Brush		
First Coat		
Minimum	SF	0.011

Paint	UNIT	MAN/ HOURS
09910.35 **Int. Painting, Buildings** *(Cont.)*		
Average	SF	0.015
Maximum	"	0.018
Second Coat		
Minimum	SF	0.009
Average	"	0.010
Maximum	"	0.011
Spray		
First Coat		
Minimum	SF	0.002
Average	"	0.003
Maximum	"	0.004
Second Coat		
Minimum	SF	0.002
Average	"	0.002
Maximum	"	0.003
Trim		
Brush		
First Coat		
Minimum	LF	0.003
Average	"	0.004
Maximum	"	0.004
Second Coat		
Minimum	LF	0.002
Average	"	0.003
Maximum	"	0.004
Walls		
Roller		
First Coat		
Minimum	SF	0.003
Average	"	0.003
Maximum	"	0.003
Second Coat		
Minimum	SF	0.003
Average	"	0.003
Maximum	"	0.003
Spray		
First Coat		
Minimum	SF	0.001
Average	"	0.002
Maximum	"	0.002
Second Coat		
Minimum	SF	0.001
Average	"	0.001
Maximum	"	0.002
09955.10 **Wall Covering**		
Vinyl wall covering		
Medium duty	SF	0.011
Heavy duty	"	0.013
Over pipes and irregular shapes		
Lightweight, 13 oz.	SF	0.016
Medium weight, 25 oz.	"	0.018
Heavy weight, 34 oz.	"	0.020

Paint	UNIT	MAN/ HOURS
09955.10 **Wall Covering** *(Cont.)*		
Cork wall covering		
1' x 1' squares		
1/4" thick	SF	0.020
1/2" thick	"	0.020
3/4" thick	"	0.020
Wall fabrics		
Natural fabrics, grass cloths		
Minimum	SF	0.012
Average	"	0.013
Maximum	"	0.016
Flexible gypsum coated wall fabric, fire resistant	"	0.008
Vinyl corner guards		
3/4" x 3/4" x 8'	EA	0.100
2-3/4" x 2-3/4" x 4'	"	0.100

10   SPECIALTIES

Specialties	UNIT	MAN/HOURS
10185.10 **Shower Stalls**		
Shower receptors		
Precast, terrazzo		
32" x 32"	EA	0.667
32" x 48"	"	0.800
Concrete		
32" x 32"	EA	0.667
48" x 48"	"	0.889
Shower door, trim and hardware		
Economy, 24" wide, chrome, tempered glass	EA	0.800
Porcelain enameled steel, flush	"	0.800
Baked enameled steel, flush	"	0.800
Aluminum, tempered glass, 48" wide, sliding	"	1.000
Folding	"	1.000
Aluminum and tempered glass, molded plastic		
Complete with receptor and door		
32" x 32"	EA	2.000
36" x 36"	"	2.000
40" x 40"	"	2.286
10210.10 **Vents And Wall Louvers**		
Block vent, 8"x16"x4" alum., w/screen, mill finish	EA	0.267
Standard	"	0.250
Vents w/screen, 4" deep, 8" wide, 5" high		
Modular	EA	0.250
Aluminum gable louvers	SF	0.133
Vent screen aluminum, 4" wide, continuous	LF	0.027
Aluminum louvers		
Residential use, fixed type, with screen		
8" x 8"	EA	0.400
12" x 12"	"	0.400
12" x 18"	"	0.400
14" x 24"	"	0.400
18" x 24"	"	0.400
30" x 24"	"	0.444
10290.10 **Pest Control**		
Termite control		
Under slab spraying		
Minimum	SF	0.002
Average	"	0.004
Maximum	"	0.008
10350.10 **Flagpoles**		
Installed in concrete base		
Fiberglass		
25' high	EA	5.333
50' high	"	13.333
Aluminum		
25' high	EA	5.333
50' high	"	13.333
Bonderized steel		
25' high	EA	6.154
50' high	"	16.000

Specialties	UNIT	MAN/HOURS
10350.10 **Flagpoles** *(Cont.)*		
Freestanding tapered, fiberglass		
30' high	EA	5.714
40' high	"	7.273
10800.10 **Bath Accessories**		
Grab bar, 1-1/2" dia., stainless steel, wall mounted		
24" long	EA	0.400
36" long	"	0.421
1" dia., stainless steel		
12" long	EA	0.348
24" long	"	0.400
36" long	"	0.444
Medicine cabinet, 16 x 22, baked enamel, lighted	"	0.320
With mirror, lighted	"	0.533
Mirror, 1/4" plate glass, up to 10 sf	SF	0.080
Mirror, stainless steel frame		
18"x24"	EA	0.267
18"x32"	"	0.320
24"x30"	"	0.400
24"x60"	"	0.800
Soap dish, stainless steel, wall mounted	"	0.533
Toilet tissue dispenser, stainless, wall mounted		
Single roll	EA	0.200
Towel bar, stainless steel		
18" long	EA	0.320
24" long	"	0.364
30" long	"	0.400
36" long	"	0.444
Toothbrush and tumbler holder	"	0.267

Architectural Equipment	UNIT	MAN/ HOURS
11010.10 **Maintenance Equipment**		
Vacuum cleaning system		
3 valves		
1.5 hp	EA	8.889
2.5 hp	"	11.429
5 valves	"	16.000
7 valves	"	20.000
11450.10 **Residential Equipment**		
Compactor, 4 to 1 compaction	EA	2.000
Dishwasher, built-in		
2 cycles	EA	4.000
4 or more cycles	"	4.000
Disposal		
Garbage disposer	EA	2.667
Heaters, electric, built-in		
Ceiling type	EA	2.667
Wall type		
Minimum	EA	2.000
Maximum	"	2.667
Hood for range, 2-speed, vented		
30" wide	EA	2.667
42" wide	"	2.667
Ice maker, automatic		
30 lb per day	EA	1.143
50 lb per day	"	4.000
Folding access stairs, disappearing metal stair		
8' long	EA	1.143
11' long	"	1.143
12' long	"	1.143
Wood frame, wood stair		
22" x 54" x 8'9" long	EA	0.800
25" x 54" x 10' long	"	0.800
Ranges electric		
Built-in, 30", 1 oven	EA	2.667
2 oven	"	2.667
Counter top, 4 burner, standard	"	2.000
With grill	"	2.000
Free standing, 21", 1 oven	"	2.667
30", 1 oven	"	1.600
2 oven	"	1.600
Water softener		
30 grains per gallon	EA	2.667
70 grains per gallon	"	4.000

12 FURNISHINGS

Casework	UNIT	MAN/HOURS
12302.10 — **Wood Casework**		
Kitchen base cabinet, standard, 24" deep, 35" high		
12"wide	EA	0.800
18" wide	"	0.800
24" wide	"	0.889
27" wide	"	0.889
36" wide	"	1.000
48" wide	"	1.000
Drawer base, 24" deep, 35" high		
15"wide	EA	0.800
18" wide	"	0.800
24" wide	"	0.889
27" wide	"	0.889
30" wide	"	0.889
Sink-ready, base cabinet		
30" wide	EA	0.889
36" wide	"	0.889
42" wide	"	0.889
60" wide	"	1.000
Corner cabinet, 36" wide	"	1.000
Wall cabinet, 12" deep, 12" high		
30" wide	EA	0.800
36" wide	"	0.800
15" high		
30" wide	EA	0.889
36" wide	"	0.889
24" high		
30" wide	EA	0.889
36" wide	"	0.889
30" high		
12" wide	EA	1.000
18" wide	"	1.000
24" wide	"	1.000
27" wide	"	1.000
30" wide	"	1.143
36" wide	"	1.143
Corner cabinet, 30" high		
24" wide	EA	1.333
30" wide	"	1.333
36" wide	"	1.333
Wardrobe	"	2.000
Vanity with top, laminated plastic		
24" wide	EA	2.000
30" wide	"	2.000
36" wide	"	2.667
48" wide	"	3.200
12390.10 — **Countertops**		
Stainless steel, counter top, with backsplash	SF	0.200
Acid-proof, kemrock surface	"	0.133

Casework	UNIT	MAN/HOURS
12500.10 — **Window Treatment**		
Drapery tracks, wall or ceiling mounted		
Basic traverse rod		
50 to 90"	EA	0.400
84 to 156"	"	0.444
136 to 250"	"	0.444
165 to 312"	"	0.500
Traverse rod with stationary curtain rod		
30 to 50"	EA	0.400
50 to 90"	"	0.400
84 to 156"	"	0.444
136 to 250"	"	0.500
Double traverse rod		
30 to 50"	EA	0.400
50 to 84"	"	0.400
84 to 156"	"	0.444
136 to 250"	"	0.500
12510.10 — **Blinds**		
Venetian blinds		
2" slats	SF	0.020
1" slats	"	0.020

13 SPECIAL CONSTRUCTION

Construction		UNIT	MAN/ HOURS
13056.10	**Vaults**		
Floor safes			
1.0 cf		EA	0.667
1.3 cf		"	1.000
13121.10	**Pre-engineered Buildings**		
Pre-engineered metal building, 40'x100'			
14' eave height		SF	0.032
16' eave height		"	0.037
13200.10	**Storage Tanks**		
Oil storage tank, underground, single wall, no excv.			
Steel			
500 gals		EA	3.000
1,000 gals		"	4.000
Fiberglass, double wall			
550 gals		EA	4.000
1,000 gals		"	4.000
Above ground			
Steel, single wall			
275 gals		EA	2.400
500 gals		"	4.000
1,000 gals		"	4.800
Fill cap		"	0.800
Vent cap		"	0.800
Level indicator		"	0.800

Lifts	UNIT	MAN/ HOURS
14410.10 **Personnel Lifts**		
Residential stair climber, per story	EA	6.667
14410.20 **Wheelchair Lifts**		
600 lb, Residential	EA	8.000

Basic Materials	UNIT	MAN/ HOURS
15100.10 **Specialties**		
Wall penetration		
Concrete wall, 6" thick		
2" dia.	EA	0.267
4" dia.	"	0.400
12" thick		
2" dia.	EA	0.364
4" dia.	"	0.571
15120.10 **Backflow Preventers**		
Backflow preventer, flanged, cast iron, with valves		
3" pipe	EA	4.000
4" pipe	"	4.444
Threaded		
3/4" pipe	EA	0.500
2" pipe	"	0.800
15140.11 **Pipe Hangers, Light**		
A band, black iron		
1/2"	EA	0.057
1"	"	0.059
1-1/4"	"	0.062
1-1/2"	"	0.067
2"	"	0.073
2-1/2"	"	0.080
3"	"	0.089
4"	"	0.100
Copper		
1/2"	EA	0.057
3/4"	"	0.059
1"	"	0.059
1-1/4"	"	0.062
1-1/2"	"	0.067
2"	"	0.073
2-1/2"	"	0.080
3"	"	0.089
4"	"	0.100
2 hole clips, galvanized		
3/4"	EA	0.053
1"	"	0.055
1-1/4"	"	0.057
1-1/2"	"	0.059
2"	"	0.062
2-1/2"	"	0.064
3"	"	0.067
4"	"	0.073
Perforated strap		
3/4"		
Galvanized, 20 ga.	LF	0.040
Copper, 22 ga.	"	0.040
J-Hooks		
1/2"	EA	0.036
3/4"	"	0.036
1"	"	0.038

Basic Materials	UNIT	MAN/ HOURS
15140.11 **Pipe Hangers, Light** *(Cont.)*		
1-1/4"	EA	0.039
1-1/2"	"	0.040
2"	"	0.040
3"	"	0.042
4"	"	0.042
PVC coated hangers, galvanized, 28 ga.		
1-1/2" x 12"	EA	0.053
2" x 12"	"	0.057
3" x 12"	"	0.062
4" x 12"	"	0.067
Copper, 30 ga.		
1-1/2" x 12"	EA	0.053
2" x 12"	"	0.057
3" x 12"	"	0.062
4" x 12"	"	0.067
Wire hook hangers		
Black wire, 1/2" x		
4"	EA	0.040
6"	"	0.042
Copper wire hooks		
1/2" x		
4"	EA	0.040
6"	"	0.042
15240.10 **Vibration Control**		
Vibration isolator, in-line, stainless connector		
1/2"	EA	0.444
3/4"	"	0.471
1"	"	0.500
1-1/4"	"	0.533
1-1/2"	"	0.571
2"	"	0.615
2-1/2"	"	0.667
3"	"	0.727
4"	"	0.800

Insulation	UNIT	MAN/ HOURS
15290.10 **Ductwork Insulation**		
Fiberglass duct insulation, plain blanket		
1-1/2" thick	SF	0.010
2" thick	"	0.013
With vapor barrier		
1-1/2" thick	SF	0.010
2" thick	"	0.013
Rigid with vapor barrier		

Insulation	UNIT	MAN/HOURS
15290.10 **Ductwork Insulation** *(Cont.)*		
2" thick	SF	0.027

Facility Water Distribution	UNIT	MAN/HOURS
15410.05 **C.I. Pipe, Above Ground**		
No hub pipe		
1-1/2" pipe	LF	0.057
2" pipe	"	0.067
3" pipe	"	0.080
4" pipe	"	0.133
No hub fittings, 1-1/2" pipe		
1/4 bend	EA	0.267
1/8 bend	"	0.267
Sanitary tee	"	0.400
Sanitary cross	"	0.400
Wye	"	0.400
Tapped tee	"	0.267
P-trap	"	0.267
Tapped cross	"	0.267
2" pipe		
1/4 bend	EA	0.320
1/8 bend	"	0.320
Sanitary tee	"	0.533
Sanitary cross	"	0.533
Wye	"	0.667
Double wye	"	0.667
2x1-1/2" wye & 1/8 bend	"	0.500
Double wye & 1/8 bend	"	0.667
Test tee less 2" plug	"	0.320
Tapped tee		
2"x2"	EA	0.320
2"x1-1/2"	"	0.320
P-trap		
2"x2"	EA	0.320
Tapped cross		
2"x1-1/2"	EA	0.320
3" pipe		
1/4 bend	EA	0.400
1/8 bend	"	0.400
Sanitary tee	"	0.500
3"x2" sanitary tee	"	0.500
3"x1-1/2" sanitary tee	"	0.500
Sanitary cross	"	0.667
3x2" sanitary cross	"	0.667
Wye	"	0.667
3x2" wye	"	0.667

Facility Water Distribution	UNIT	MAN/HOURS
15410.05 **C.I. Pipe, Above Ground** *(Cont.)*		
Double wye	EA	0.667
3x2" double wye	"	0.667
3x2" wye & 1/8 bend	"	0.571
3x1-1/2" wye & 1/8 bend	"	0.571
Double wye & 1/8 bend	"	0.667
3x2" double wye & 1/8 bend	"	0.667
3x2" reducer	"	0.364
Test tee, less 3" plug	"	0.400
3x3" tapped tee	"	0.400
3x2" tapped tee	"	0.400
3x1-1/2" tapped tee	"	0.400
P-trap	"	0.400
3x2" tapped cross	"	0.400
3x1-1/2" tapped cross	"	0.400
Closet flange, 3-1/2" deep	"	0.200
4" pipe		
1/4 bend	EA	0.400
1/8 bend	"	0.400
Sanitary tee	"	0.667
4x3" sanitary tee	"	0.667
4x2" sanitary tee	"	0.667
Sanitary cross	"	0.800
4x3" sanitary cross	"	0.800
4x2" sanitary cross	"	0.800
Wye	"	0.667
4x3" wye	"	0.667
4x2" wye	"	0.667
Double wye	"	0.800
4x3" double wye	"	0.800
4x2" double wye	"	0.800
Wye & 1/8 bend	"	0.667
4x3" wye & 1/8 bend	"	0.667
4x2" wye & 1/8 bend	"	0.667
Double wye & 1/8 bend	"	0.800
4x3" double wye & 1/8 bend	"	0.800
4x2" double wye & 1/8 bend	"	0.800
4x3" reducer	"	0.400
4x2" reducer	"	0.400
Test tee, less 4" plug	"	0.400
4x2" tapped tee	"	0.400
4x1-1/2" tapped tee	"	0.400
P-trap	"	0.400
4x2" tapped cross	"	0.400
4x1-1/2" tapped cross	"	0.400
Closet flange		
3" deep	EA	0.400
8" deep	"	0.400

15 MECHANICAL

Facility Water Distribution	UNIT	MAN/HOURS
15410.06 **C.I. Pipe, Below Ground**		
No hub pipe		
1-1/2" pipe	LF	0.040
2" pipe	"	0.044
3" pipe	"	0.050
4" pipe	"	0.067
Fittings, 1-1/2"		
1/4 bend	EA	0.229
1/8 bend	"	0.229
Wye	"	0.320
Wye & 1/8 bend	"	0.229
P-trap	"	0.229
2"		
1/4 bend	EA	0.267
1/8 bend	"	0.267
Double wye	"	0.500
Wye & 1/8 bend	"	0.400
Double wye & 1/8 bend	"	0.500
P-trap	"	0.267
3"		
1/4 bend	EA	0.320
1/8 bend	"	0.320
Wye	"	0.500
3x2" wye	"	0.500
Wye & 1/8 bend	"	0.500
Double wye & 1/8 bend	"	0.500
3x2" double wye & 1/8 bend	"	0.500
3x2" reducer	"	0.320
P-trap	"	0.320
4"		
1/4 bend	EA	0.320
1/8 bend	"	0.320
Wye	"	0.500
4x3" wye	"	0.500
4x2" wye	"	0.500
Double wye	"	0.667
4x3" double wye	"	0.667
4x2" double wye	"	0.667
Wye & 1/8 bend	"	0.500
4x3" wye & 1/8 bend	"	0.500
4x2" wye & 1/8 bend	"	0.500
Double wye & 1/8 bend	"	0.667
4x3" double wye & 1/8 bend	"	0.667
4x2" double wye & 1/8 bend	"	0.667
4x3" reducer	"	0.320
4x2" reducer	"	0.320
15410.10 **Copper Pipe**		
Type "K" copper		
1/2"	LF	0.025
3/4"	"	0.027
1"	"	0.029
DWV, copper		
1-1/4"	LF	0.033

Facility Water Distribution	UNIT	MAN/HOURS
15410.10 **Copper Pipe** *(Cont.)*		
1-1/2"	LF	0.036
2"	"	0.040
3"	"	0.044
4"	"	0.050
6"	"	0.057
Refrigeration tubing, copper, sealed		
1/8"	LF	0.032
3/16"	"	0.033
1/4"	"	0.035
Type "L" copper		
1/4"	LF	0.024
3/8"	"	0.024
1/2"	"	0.025
3/4"	"	0.027
1"	"	0.029
Type "M" copper		
1/2"	LF	0.025
3/4"	"	0.027
1"	"	0.029
15410.11 **Copper Fittings**		
Coupling, with stop		
1/4"	EA	0.267
3/8"	"	0.320
1/2"	"	0.348
5/8"	"	0.400
3/4"	"	0.444
1"	"	0.471
Reducing coupling		
1/4" x 1/8"	EA	0.320
3/8" x 1/4"	"	0.348
1/2" x		
3/8"	EA	0.400
1/4"	"	0.400
1/8"	"	0.400
3/4" x		
3/8"	EA	0.444
1/2"	"	0.444
1" x		
3/8"	EA	0.500
1" x 1/2"	"	0.500
1" x 3/4"	"	0.500
Slip coupling		
1/4"	EA	0.267
1/2"	"	0.320
3/4"	"	0.400
1"	"	0.444
Coupling with drain		
1/2"	EA	0.400
3/4"	"	0.444
1"	"	0.500
Reducer		
3/8" x 1/4"	EA	0.320

Facility Water Distribution	UNIT	MAN/HOURS	Facility Water Distribution	UNIT	MAN/HOURS
15410.11 Copper Fittings *(Cont.)*			**15410.11** Copper Fittings *(Cont.)*		
1/2" x 3/8"	EA	0.320	1-1/4"	EA	0.421
3/4" x			1" x 1-1/4"	"	0.444
1/4"	EA	0.364	Reducing male adapters		
3/8"	"	0.364	1/2" x		
1/2"	"	0.364	1/4"	EA	0.400
1" x			3/8"	"	0.400
1/2"	EA	0.400	3/4" x 1/2"	"	0.421
3/4"	"	0.400	1" x		
Female adapters			1/2"	EA	0.444
1/4"	EA	0.320	3/4"	"	0.444
3/8"	"	0.364	Fitting x male adapters		
1/2"	"	0.400	1/2"	EA	0.400
3/4"	"	0.444	3/4"	"	0.421
1"	"	0.444	1"	"	0.444
Increasing female adapters			90 ells		
1/8" x			1/8"	EA	0.320
3/8"	EA	0.320	1/4"	"	0.320
1/2"	"	0.320	3/8"	"	0.364
1/4" x 1/2"	"	0.348	1/2"	"	0.400
3/8" x 1/2"	"	0.364	3/4"	"	0.421
1/2" X			1"	"	0.444
3/4"	EA	0.400	Reducing 90 ell		
1"	"	0.400	3/8" x 1/4"	EA	0.364
3/4" X			1/2" x		
1"	EA	0.444	1/4"	EA	0.400
1-1/4"	"	0.444	3/8"	"	0.400
1" x			3/4" x 1/2"	"	0.421
1-1/4"	EA	0.444	1" x		
1-1/2"	"	0.444	1/2"	EA	0.444
Reducing female adapters			3/4"	"	0.444
3/8" x 1/4"	EA	0.364	Street ells, copper		
1/2" x			1/4"	EA	0.320
1/4"	EA	0.400	3/8"	"	0.364
3/8"	"	0.400	1/2"	"	0.400
3/4" x 1/2"	"	0.444	3/4"	"	0.421
1" x			1"	"	0.444
1/2"	EA	0.444	Female, 90 ell		
3/4"	"	0.444	1/2"	EA	0.400
Female fitting adapters			3/4"	"	0.421
1/2"	EA	0.400	1"	"	0.444
3/4"	"	0.400	Female increasing, 90 ell		
3/4" x 1/2"	"	0.421	3/8" x 1/2"	EA	0.364
1"	"	0.444	1/2" x		
Male adapters			3/4"	EA	0.400
1/4"	EA	0.364	1"	"	0.400
3/8"	"	0.364	3/4" x 1"	"	0.421
Increasing male adapters			1" x 1-1/4"	"	0.444
3/8" x 1/2"	EA	0.364	Female reducing, 90 ell		
1/2" x			1/2" x 3/8"	EA	0.400
3/4"	EA	0.400	3/4" x 1/2"	"	0.421
1"	"	0.400	1" x		
3/4" x			1/2"	EA	0.444
1"	EA	0.421	3/4"	"	0.444

Facility Water Distribution	UNIT	MAN/ HOURS	Facility Water Distribution	UNIT	MAN/ HOURS
15410.11 Copper Fittings *(Cont.)*			**15410.11** Copper Fittings *(Cont.)*		
Male, 90 ell			1/2"	EA	0.421
1/4"	EA	0.320	1" x		
3/8"	"	0.364	1/2"	EA	0.444
1/2"	"	0.400	3/4"	"	0.444
3/4"	"	0.421	Female flush bushing		
1"	"	0.444	1/2" x		
Male, increasing 90 ell			1/2" x 1/8"	EA	0.400
1/2" x			1/4"	"	0.400
3/4"	EA	0.400	Union		
1"	"	0.400	1/4"	EA	0.320
3/4" x 1"	"	0.421	3/8"	"	0.364
1" x 1-1/4"	"	0.444	Female		
Male, reducing 90 ell			1/2"	EA	0.400
1/2" x 3/8"	EA	0.400	3/4"	"	0.421
3/4" x 1/2"	"	0.421	Male		
1" x			1/2"	EA	0.400
1/2"	EA	0.444	3/4"	"	0.421
3/4"	"	0.444	1"	"	0.444
Drop ear ells			45 degree wye		
1/2"	EA	0.400	1/2"	EA	0.400
Female drop ear ells			3/4"	"	0.421
1/2"	FA	0.400	1"	"	0.444
1/2" x 3/8"	"	0.400	1" x 3/4" x 3/4"	"	0.444
3/4"	"	0.421	Twin ells		
Female flanged sink ell			1" x 3/4" x 3/4"	EA	0.444
1/2"	EA	0.400	1" x 1" x 1"	"	0.444
45 ells			90 union ells, male		
1/4"	EA	0.320	1/2"	EA	0.400
3/8"	"	0.364	3/4"	"	0.421
45 street ell			1"	"	0.444
1/4"	EA	0.320	DWV fittings, coupling with stop		
3/8"	"	0.364	1-1/4"	EA	0.471
1/2"	"	0.400	1-1/2"	"	0.500
3/4"	"	0.421	1-1/2" x 1-1/4"	"	0.500
1"	"	0.444	2"	"	0.533
Tee			2" x 1-1/4"	"	0.533
1/8"	EA	0.320	2" x 1-1/2"	"	0.533
1/4"	"	0.320	3"	"	0.667
3/8"	"	0.364	3" x 1-1/2"	"	0.667
Caps			3" x 2"	"	0.667
1/4"	EA	0.320	4"	"	0.800
3/8"	"	0.364	Slip coupling		
Test caps			1-1/2"	EA	0.500
1/2"	EA	0.400	2"	"	0.533
3/4"	"	0.421	3"	"	0.667
1"	"	0.444	90 ells		
Flush bushing			1-1/2"	EA	0.500
1/4" x 1/8"	EA	0.320	1-1/2" x 1-1/4"	"	0.500
1/2" x			2"	"	0.533
1/4"	EA	0.400	2" x 1-1/2"	"	0.533
3/8"	"	0.400	3"	"	0.667
3/4" x			4"	"	0.800
3/8"	EA	0.421	Street, 90 elbows		

Facility Water Distribution	UNIT	MAN/ HOURS	Facility Water Distribution	UNIT	MAN/ HOURS
15410.11 Copper Fittings *(Cont.)*			**15410.11** Copper Fittings *(Cont.)*		
1-1/2"	EA	0.500	2" x 1-1/2" x 1-1/2"	EA	0.533
2"	"	0.533	2" x 1-1/2" x 2"	"	0.533
3"	"	0.667	2" x 2" x 1-1/2"	"	0.533
4"	"	0.800	3"	"	0.667
Female, 90 elbows			3" x 3" x 1-1/2"	"	0.667
1-1/2"	EA	0.500	3" x 3" x 2"	"	0.667
2"	"	0.533	4"	"	0.800
Male, 90 elbows			4" x 4" x 3"	"	0.800
1-1/2"	EA	0.500	Female sanitary tee		
2"	"	0.533	1-1/2"	EA	0.500
90 with side inlet			Long turn tee		
3" x 3" x 1"	EA	0.667	1-1/2"	EA	0.500
3" x 3" x 1-1/2"	"	0.667	2"	"	0.533
3" x 3" x 2"	"	0.667	3" x 1-1/2"	"	0.667
45 ells			Double wye		
1-1/4"	EA	0.471	1-1/2"	EA	0.500
1-1/2"	"	0.500	2"	"	0.533
2"	"	0.533	2" x 2" x 1-1/2" x 1-1/2"	"	0.533
3"	"	0.667	3"	"	0.667
4"	"	0.800	3" x 3" x 1-1/2" x 1-1/2"	"	0.667
Street, 45 ell			3" x 3" x 2" x 2"	"	0.667
1-1/2"	EA	0.500	4" x 4" x 1-1/2" x 1-1/2"	"	0.800
2"	"	0.533	Double sanitary tee		
3"	"	0.667	1-1/2"	EA	0.500
60 ell			2"	"	0.533
1-1/2"	EA	0.500	2" x 2" x 1-1/2"	"	0.533
2"	"	0.533	3"	"	0.667
3"	"	0.667	3" x 3" x 1-1/2" x 1-1/2"	"	0.667
22-1/2 ell			3" x 3" x 2" x 2"	"	0.667
1-1/2"	EA	0.500	4" x 4" x 1-1/2" x 1-1/2"	"	0.800
2"	"	0.533	Long		
3"	"	0.667	2" x 1-1/2"	EA	0.533
11-1/4 ell			Twin elbow		
1-1/2"	EA	0.500	1-1/2"	EA	0.500
2"	"	0.533	2"	"	0.533
3"	"	0.667	2" x 1-1/2" x 1-1/2"	"	0.533
Wye			Spigot adapter, manoff		
1-1/4"	EA	0.471	1-1/2" x 2"	EA	0.500
1-1/2"	"	0.500	1-1/2" x 3"	"	0.500
2"	"	0.533	2"	"	0.533
2" x 1-1/2" x 1-1/2"	"	0.533	2" x 3"	"	0.533
2" x 1-1/2" x 2"	"	0.533	2" x 4"	"	0.533
2" x 1-1/2" x 2-1/2"	"	0.533	3"	"	0.667
3"	"	0.667	3" x 4"	"	0.667
3" x 3" x 1-1/2"	"	0.667	4"	"	0.800
3" x 3" x 2"	"	0.667	No-hub adapters		
4"	"	0.800	1-1/2" x 2"	EA	0.500
4" x 4" x 2"	"	0.800	2"	"	0.533
4" x 4" x 3"	"	0.800	2" x 3"	"	0.533
Sanitary tee			3"	"	0.667
1-1/4"	EA	0.471	3" x 4"	"	0.667
1-1/2"	"	0.500	4"	"	0.800
2"	"	0.533	Fitting reducers		

Facility Water Distribution	UNIT	MAN/HOURS	Facility Water Distribution	UNIT	MAN/HOURS
15410.11 Copper Fittings *(Cont.)*			**15410.11** Copper Fittings *(Cont.)*		
1-1/2" x 1-1/4"	EA	0.500	1-1/2"	EA	0.500
2" x 1-1/2"	"	0.533	2"	"	0.533
3" x 1-1/2"	"	0.667	Closet flanges		
3" x 2"	"	0.667	3"	EA	0.667
Slip joint (Desanco)			4"	"	0.800
1-1/4"	EA	0.471	Drum traps, with cleanout		
1-1/2"	"	0.500	1-1/2" x 3" x 6"	EA	0.500
1-1/2" x 1-1/4"	"	0.500	P-trap, swivel, with cleanout		
Street x slip joint (Desanco)			1-1/2"	EA	0.500
1-1/2"	EA	0.500	P-trap, solder union		
1-1/2" x 1-1/4"	"	0.500	1-1/2"	EA	0.500
Flush bushing			2"	"	0.533
1-1/2" x 1-1/4"	EA	0.500	With cleanout		
2" x 1-1/2"	"	0.533	1-1/2"	EA	0.500
3" x 1-1/2"	"	0.667	2"	"	0.533
3" x 2"	"	0.667	2" x 1-1/2"	"	0.533
Male hex trap bushing			Swivel joint, with cleanout		
1-1/4" x 1-1/2"	EA	0.471	1-1/2" x 1-1/4"	EA	0.500
1-1/2"	"	0.500	1-1/2"	"	0.500
1-1/2" x 2"	"	0.500	2" x 1-1/2"	"	0.533
2"	"	0.533	Estabrook TY, with inlets		
Round trap bushing			3", with 1-1/2" inlet	EA	0.667
1-1/2"	EA	0.500	Fine thread adapters		
2"	"	0.533	1/2"	EA	0.400
Female adapter			1/2" x 1/2" IPS	"	0.400
1-1/4"	EA	0.471	1/2" x 3/4" IPS	"	0.400
1-1/2"	"	0.500	1/2" x male	"	0.400
1-1/2" x 2"	"	0.500	1/2" x female	"	0.400
2"	"	0.533	Copper pipe fittings		
2" x 1-1/2"	"	0.533	1/2"		
3"	"	0.667	90 deg ell	EA	0.178
Fitting x female adapter			45 deg ell	"	0.178
1-1/2"	EA	0.500	Tee	"	0.229
2"	"	0.533	Cap	"	0.089
Male adapters			Coupling	"	0.178
1-1/4"	EA	0.471	Union	"	0.200
1-1/4" x 1-1/2"	"	0.471	3/4"		
1-1/2"	"	0.500	90 deg ell	EA	0.200
1-1/2" x 2"	"	0.500	45 deg ell	"	0.200
2"	"	0.533	Tee	"	0.267
2" x 1-1/2"	"	0.533	Cap	"	0.094
3"	"	0.667	Coupling	"	0.200
Male x slip joint adapters			Union	"	0.229
1-1/2" x 1-1/4"	EA	0.500	1"		
Dandy cleanout			90 deg ell	EA	0.267
1-1/2"	EA	0.500	45 deg ell	"	0.267
2"	"	0.533	Tee	"	0.320
3"	"	0.667	Cap	"	0.133
End cleanout, flush pattern			Coupling	"	0.267
1-1/2" x 1"	EA	0.500	Union	"	0.267
2" x 1-1/2"	"	0.533			
3" x 2-1/2"	"	0.667			
Copper caps					

Facility Water Distribution		UNIT	MAN/HOURS
15410.14	**Brass I.P.S. Fittings**		
Fittings, iron pipe size, 45 deg ell			
1/8"		EA	0.320
1/4"		"	0.320
3/8"		"	0.364
1/2"		"	0.400
3/4"		"	0.421
1"		"	0.444
90 deg ell			
1/8"		EA	0.320
1/4"		"	0.320
3/8"		"	0.364
1/2"		"	0.400
3/4"		"	0.421
1"		"	0.444
90 deg ell, reducing			
1/4" x 1/8"		EA	0.320
3/8" x 1/8"		"	0.364
3/8" x 1/4"		"	0.364
1/2" x 1/4"		"	0.400
1/2" x 3/8"		"	0.400
3/4" x 1/2"		"	0.421
1" x 3/8"		"	0.444
1" x 1/2"		"	0.444
1" x 3/4"		"	0.444
Street ell, 45 deg			
1/2"		EA	0.400
3/4"		"	0.421
90 deg			
1/8"		EA	0.320
1/4"		"	0.320
3/8"		"	0.364
1/2"		"	0.400
3/4"		"	0.421
1"		"	0.444
Tee, 1/8"		"	0.320
1/4"		"	0.320
3/8"		"	0.364
1/2"		"	0.400
3/4"		"	0.421
1"		"	0.444
Tee, reducing, 3/8" x			
1/4"		EA	0.364
1/2"		"	0.364
1/2" x			
1/4"		EA	0.400
3/8"		"	0.400
3/4"		"	0.400
3/4" x			
1/4"		EA	0.421
1/2"		"	0.421
1"		"	0.421
1" x			
1/2"		EA	0.444

Facility Water Distribution		UNIT	MAN/HOURS
15410.14	**Brass I.P.S. Fittings** *(Cont.)*		
3/4"		EA	0.444
Tee, reducing			
1/2" x 3/8" x 1/2"		EA	0.400
3/4" x 1/2" x 1/2"		"	0.421
3/4" x 1/2" x 3/4"		"	0.421
1" x 1/2" x 1/2"		"	0.444
1" x 1/2" x 3/4"		"	0.444
1" x 3/4" x 1/2"		"	0.444
1" x 3/4" x 3/4"		"	0.444
Union			
1/8"		EA	0.320
1/4"		"	0.320
3/8"		"	0.364
1/2"		"	0.400
3/4"		"	0.421
1"		"	0.444
Brass face bushing			
3/8" x 1/4"		EA	0.364
1/2" x 3/8"		"	0.400
3/4" x 1/2"		"	0.421
1" x 3/4"		"	0.444
Hex bushing, 1/4" x 1/8"		"	0.320
1/2" x			
1/4"		EA	0.400
3/8"		"	0.400
5/8" x			
1/8"		EA	0.400
1/4"		"	0.400
3/4" x			
1/8"		EA	0.421
1/4"		"	0.421
3/8"		"	0.421
1/2"		"	0.421
1" x			
1/4"		EA	0.444
3/8"		"	0.444
1/2"		"	0.444
3/4"		"	0.444
Caps			
1/8"		EA	0.320
1/4"		"	0.320
3/8"		"	0.364
1/2"		"	0.400
3/4"		"	0.421
1"		"	0.444
Couplings			
1/8"		EA	0.320
1/4"		"	0.320
3/8"		"	0.364
1/2"		"	0.400
3/4"		"	0.421
1"		"	0.444
Couplings, reducing, 1/4" x 1/8"		"	0.320

Facility Water Distribution	UNIT	MAN/HOURS
15410.14 **Brass I.P.S. Fittings** *(Cont.)*		
3/8" x		
1/8"	EA	0.364
1/4"	"	0.364
1/2" x		
1/8"	EA	0.400
1/4"	"	0.400
3/8"	"	0.400
3/4" x		
1/4"	EA	0.421
3/8"	"	0.421
1/2"	"	0.421
1" x		
1/2"	EA	0.421
3/4"	"	0.421
Square head plug, solid		
1/8"	EA	0.320
1/4"	"	0.320
3/8"	"	0.364
1/2"	"	0.400
3/4"	"	0.421
Cored		
1/2"	FA	0.400
3/4"	"	0.421
1"	"	0.444
Countersunk		
1/2"	EA	0.400
3/4"	"	0.421
Locknut		
3/4"	EA	0.421
1"	"	0.444
Close standard red nipple, 1/8"	"	0.320
1/8" x		
1-1/2"	EA	0.320
2"	"	0.320
2-1/2"	"	0.320
3"	"	0.320
3-1/2"	"	0.320
4"	"	0.320
4-1/2"	"	0.320
5"	"	0.320
5-1/2"	"	0.320
6"	"	0.320
1/4" x close	"	0.320
1/4" x		
1-1/2"	EA	0.320
2"	"	0.320
2-1/2"	"	0.320
3"	"	0.320
3-1/2"	"	0.320
4"	"	0.320
4-1/2"	"	0.320
5"	"	0.320
5-1/2"	"	0.320

Facility Water Distribution	UNIT	MAN/HOURS
15410.14 **Brass I.P.S. Fittings** *(Cont.)*		
6"	EA	0.320
3/8" x close	"	0.364
3/8" x		
1-1/2"	EA	0.364
2"	"	0.364
2-1/2"	"	0.364
3"	"	0.364
3-1/2"	"	0.364
4"	"	0.364
4-1/2"	"	0.364
5"	"	0.364
5-1/2"	"	0.364
6"	"	0.364
1/2" x close	"	0.400
1/2" x		
1-1/2"	EA	0.400
2"	"	0.400
2-1/2"	"	0.400
3"	"	0.400
3-1/2"	"	0.400
4"	"	0.400
4-1/2"	"	0.400
5"	"	0.400
5-1/2"	"	0.400
6"	"	0.400
7-1/2"	"	0.400
8"	"	0.400
3/4" x close	"	0.421
3/4" x		
1-1/2"	EA	0.421
2"	"	0.421
2-1/2"	"	0.421
3"	"	0.421
3-1/2"	"	0.421
4"	"	0.421
4-1/2"	"	0.421
5"	"	0.421
5-1/2"	"	0.421
6"	"	0.421
1" x close	"	0.444
1" x		
2"	EA	0.444
2-1/2"	"	0.444
3"	"	0.444
3-1/2"	"	0.444
4"	"	0.444
4-1/2"	"	0.444
5"	"	0.444
5-1/2"	"	0.444
6"	"	0.444

Facility Water Distribution	UNIT	MAN/ HOURS
15410.15 **Brass Fittings**		
Compression fittings, union		
3/8"	EA	0.133
1/2"	"	0.133
5/8"	"	0.133
Union elbow		
3/8"	EA	0.133
1/2"	"	0.133
5/8"	"	0.133
Union tee		
3/8"	EA	0.133
1/2"	"	0.133
5/8"	"	0.133
Male connector		
3/8"	EA	0.133
1/2"	"	0.133
5/8"	"	0.133
Female connector		
3/8"	EA	0.133
1/2"	"	0.133
5/8"	"	0.133
Brass flare fittings, union		
3/8"	EA	0.129
1/2"	"	0.129
5/8"	"	0.129
90 deg elbow union		
3/8"	EA	0.129
1/2"	"	0.129
5/8"	"	0.129
Three way tee		
3/8"	EA	0.216
1/2"	"	0.216
5/8"	"	0.216
Cross		
3/8"	EA	0.286
1/2"	"	0.286
5/8"	"	0.286
Male connector, half union		
3/8"	EA	0.129
1/2"	"	0.129
5/8"	"	0.129
Female connector, half union		
3/8"	EA	0.129
1/2"	"	0.129
5/8"	"	0.129
Long forged nut		
3/8"	EA	0.129
1/2"	"	0.129
5/8"	"	0.129
Short forged nut		
3/8"	EA	0.129
1/2"	"	0.129
5/8"	"	0.129
Sleeve		

Facility Water Distribution	UNIT	MAN/ HOURS
15410.15 **Brass Fittings** *(Cont.)*		
1/8"	EA	0.160
1/4"	"	0.160
5/16"	"	0.160
3/8"	"	0.160
1/2"	"	0.160
5/8"	"	0.160
Tee		
1/4"	EA	0.229
5/16"	"	0.229
Male tee		
5/16" x 1/8"	EA	0.229
Female union		
1/8" x 1/8"	EA	0.200
1/4" x 3/8"	"	0.200
3/8" x 1/4"	"	0.200
3/8" x 1/2"	"	0.200
5/8" x 1/2"	"	0.229
Male union, 1/4"		
1/4" x 1/4"	EA	0.200
3/8"	"	0.200
1/2"	"	0.200
5/16" x		
1/8"	EA	0.200
1/4"	"	0.200
3/8"	"	0.200
3/8" x		
1/8"	EA	0.200
1/4"	"	0.200
1/2"	"	0.200
5/8" x		
3/8"	EA	0.229
1/2"	"	0.229
Female elbow, 1/4" x 1/4"	"	0.229
5/16" x		
1/8"	EA	0.229
1/4"	"	0.229
3/8" x		
3/8"	EA	0.229
1/2"	"	0.229
Male elbow, 1/8" x 1/8"	"	0.229
3/16" x 1/4"	"	0.229
1/4" x		
1/8"	EA	0.229
1/4"	"	0.229
3/8"	"	0.229
5/16" x		
1/8"	EA	0.229
1/4"	"	0.229
3/8"	"	0.229
3/8" x		
1/8"	EA	0.229
1/4"	"	0.229
3/8"	"	0.229

Facility Water Distribution	UNIT	MAN/HOURS
15410.15 **Brass Fittings** *(Cont.)*		
1/2"	EA	0.229
1/2" x		
1/4"	EA	0.267
3/8"	"	0.267
1/2"	"	0.267
5/8" x		
3/8"	EA	0.267
1/2"	"	0.267
3/4"	"	0.267
Union		
1/8"	EA	0.229
3/16"	"	0.229
1/4"	"	0.229
5/16"	"	0.229
3/8"	"	0.229
Reducing union		
3/8" x 1/4"	EA	0.267
5/8" x		
3/8"	EA	0.267
1/2"	"	0.267
15410.17 **Chrome Plated Fittings**		
Fittings		
90 ell		
3/8"	EA	0.200
1/2"	"	0.200
45 ell		
3/8"	EA	0.200
1/2"	"	0.200
Tee		
3/8"	EA	0.267
1/2"	"	0.267
Coupling		
3/8"	EA	0.200
1/2"	"	0.200
Union		
3/8"	EA	0.200
1/2"	"	0.200
Tee		
1/2" x 3/8" x 3/8"	EA	0.267
1/2" x 3/8" x 1/2"	"	0.267
15410.30 **PVC/CPVC Pipe**		
PVC schedule 40		
1/2" pipe	LF	0.033
3/4" pipe	"	0.036
1" pipe	"	0.040
1-1/4" pipe	"	0.044
1-1/2" pipe	"	0.050
2" pipe	"	0.057
2-1/2" pipe	"	0.067
3" pipe	"	0.080
4" pipe	"	0.100

Facility Water Distribution	UNIT	MAN/HOURS
15410.30 **PVC/CPVC Pipe** *(Cont.)*		
Fittings, 1/2"		
90 deg ell	EA	0.100
45 deg ell	"	0.100
Tee	"	0.114
Reducing insert	"	0.133
Threaded	"	0.100
Male adapter	"	0.133
Female adapter	"	0.100
Union	"	0.160
Cap	"	0.133
Flange	"	0.160
3/4"		
90 deg elbow	EA	0.133
45 deg elbow	"	0.133
Tee	"	0.160
Reducing insert	"	0.114
Threaded	"	0.133
1"		
90 deg elbow	EA	0.160
45 deg elbow	"	0.160
Tee	"	0.178
Reducing insert	"	0.160
Threaded	"	0.178
Male adapter	"	0.200
Female adapter	"	0.200
Union	"	0.267
Cap	"	0.160
Flange	"	0.267
1-1/4"		
90 deg elbow	EA	0.229
45 deg elbow	"	0.229
Tee	"	0.267
Reducing insert	"	0.267
Threaded	"	0.267
Female adapter	"	0.267
Union	"	0.320
Cap	"	0.267
Flange	"	0.320
1-1/2"		
90 deg elbow	EA	0.229
45 deg elbow	"	0.229
Tee	"	0.267
Reducing insert	"	0.267
Threaded	"	0.267
Male adapter	"	0.267
Female adapter	"	0.267
Union	"	0.400
Cap	"	0.267
Flange	"	0.400
2"		
90 deg elbow	EA	0.267
45 deg elbow	"	0.267
Tee	"	0.320

Facility Water Distribution	UNIT	MAN/ HOURS	Facility Water Distribution	UNIT	MAN/ HOURS
15410.30 PVC/CPVC Pipe *(Cont.)*			**15410.30** PVC/CPVC Pipe *(Cont.)*		
Reducing insert	EA	0.320	Union	EA	0.400
Threaded	"	0.320	Cap	"	0.267
Male adapter	"	0.320	Flange	"	0.400
Female adapter	"	0.320	2"		
Union	"	0.500	90 deg elbow	EA	0.320
Cap	"	0.320	45 deg elbow	"	0.320
Flange	"	0.500	Tee	"	0.500
2-1/2"			Reducing insert	"	0.320
90 deg elbow	EA	0.500	Threaded	"	0.320
45 deg elbow	"	0.500	Male adapter	"	0.320
Tee	"	0.533	Female adapter	"	0.320
Reducing insert	"	0.533	2-1/2"		
Threaded	"	0.533	90 deg elbow	EA	0.500
Male adapter	"	0.533	45 deg elbow	"	0.500
Female adapter	"	0.533	Tee	"	0.667
Union	"	0.667	Reducing insert	"	0.500
Cap	"	0.500	Threaded	"	0.500
Flange	"	0.667	Male adapter	"	0.500
3"			Female adapter	"	0.500
90 deg elbow	EA	0.667	Union	"	0.667
45 deg elbow	"	0.667	Cap	"	0.500
Tee	"	0.727	Flange	"	0.667
Reducing insert	"	0.667	3"		
Threaded	"	0.667	90 deg elbow	EA	0.667
Male adapter	"	0.667	45 deg elbow	"	0.667
Female adapter	"	0.667	Tee	"	0.800
Union	"	0.800	Reducing insert	"	0.667
Cap	"	0.667	Threaded	"	0.667
Flange	"	0.800	Male adapter	"	0.667
4"			Female adapter	"	0.667
90 deg elbow	EA	0.800	Union	"	0.800
45 deg elbow	"	0.800	Cap	"	0.667
Tee	"	0.889	Flange	"	0.800
Reducing insert	"	0.800	4"		
Threaded	"	0.800	90 deg elbow	EA	0.800
Male adapter	"	0.800	45 deg elbow	"	0.800
Female adapter	"	0.800	Tee	"	1.000
Union	"	1.000	Reducing insert	"	0.800
Cap	"	0.800	Threaded	"	0.800
Flange	"	1.000	Male adapter	"	0.800
PVC schedule 80 pipe			Union	"	1.000
1-1/2" pipe	LF	0.050	Cap	"	0.800
2" pipe	"	0.057	Flange	"	1.000
3" pipe	"	0.080	CPVC schedule 40		
4" pipe	"	0.100	1/2" pipe	LF	0.033
Fittings, 1-1/2"			3/4" pipe	"	0.036
90 deg elbow	EA	0.267	1" pipe	"	0.040
45 deg elbow	"	0.267	1-1/4" pipe	"	0.044
Tee	"	0.400	1-1/2" pipe	"	0.050
Reducing insert	"	0.267	2" pipe	"	0.057
Threaded	"	0.267	Fittings, CPVC, schedule 80		
Male adapter	"	0.267	1/2", 90 deg ell	EA	0.080
Female adapter	"	0.267	Tee	"	0.133

15410.30 PVC/CPVC Pipe (Cont.)

Facility Water Distribution	UNIT	MAN/ HOURS
3/4", 90 deg ell	EA	0.080
Tee	"	0.133
1", 90 deg ell	"	0.089
Tee	"	0.145
1-1/4", 90 deg ell	"	0.089
Tee	"	0.145
1-1/2", 90 deg ell	"	0.160
Tee	"	0.200
2", 90 deg ell	"	0.160
Tee	"	0.200

15410.33 ABS DWV Pipe

Facility Water Distribution	UNIT	MAN/ HOURS
Schedule 40 ABS		
1-1/2" pipe	LF	0.040
2" pipe	"	0.044
3" pipe	"	0.057
4" pipe	"	0.080
Fittings		
1/8 bend		
1-1/2"	EA	0.160
2"	"	0.200
3"	"	0.267
4"	"	0.320
Tee, sanitary		
1-1/2"	EA	0.267
2"	"	0.320
3"	"	0.400
4"	"	0.500
Tee, sanitary reducing		
2 x 1-1/2 x 1-1/2	EA	0.320
2 x 1-1/2 x 2	"	0.333
2 x 2 x 1-1/2	"	0.364
3 x 3 x 1-1/2	"	0.400
3 x 3 x 2	"	0.444
4 x 4 x 1-1/2	"	0.500
4 x 4 x 2	"	0.571
4 x 4 x 3	"	0.615
Wye		
1-1/2"	EA	0.229
2"	"	0.320
3"	"	0.400
4"	"	0.500
Reducer		
2 x 1-1/2	EA	0.200
3 x 1-1/2	"	0.267
3 x 2	"	0.267
4 x 2	"	0.320
4 x 3	"	0.320
P-trap		
1-1/2"	EA	0.267
2"	"	0.296
3"	"	0.348
4"	"	0.400

15410.33 ABS DWV Pipe (Cont.)

Facility Water Distribution	UNIT	MAN/ HOURS
Double sanitary, tee		
1-1/2"	EA	0.320
2"	"	0.400
3"	"	0.500
4"	"	0.667
Long sweep, 1/4 bend		
1-1/2"	EA	0.160
2"	"	0.200
3"	"	0.267
4"	"	0.400
Wye, standard		
1-1/2"	EA	0.267
2"	"	0.320
3"	"	0.400
4"	"	0.500
Wye, reducing		
2 x 1-1/2 x 1-1/2	EA	0.267
2 x 2 x 1-1/2	"	0.320
4 x 4 x 2	"	0.500
4 x 4 x 3	"	0.533
Double wye		
1-1/2"	EA	0.320
2"	"	0.400
3"	"	0.500
4"	"	0.667
2 x 2 x 1-1/2 x 1-1/2	"	0.400
3 x 3 x 2 x 2	"	0.500
4 x 4 x 3 x 3	"	0.667
Combination wye and 1/8 bend		
1-1/2"	EA	0.267
2"	"	0.320
3"	"	0.400
4"	"	0.500
2 x 2 x 1-1/2	"	0.320
3 x 3 x 1-1/2	"	0.400
3 x 3 x 2	"	0.400
4 x 4 x 2	"	0.500
4 x 4 x 3	"	0.500

15410.80 Steel Pipe

Facility Water Distribution	UNIT	MAN/ HOURS
Black steel, extra heavy pipe, threaded		
1/2" pipe	LF	0.032
3/4" pipe	"	0.032
Fittings, malleable iron, threaded, 1/2" pipe		
90 deg ell	EA	0.267
45 deg ell	"	0.267
Tee	"	0.400
Reducing tee	"	0.400
Cap	"	0.160
Coupling	"	0.320
Union	"	0.267
Nipple, 4" long	"	0.267
3/4" pipe		

Facility Water Distribution	UNIT	MAN/ HOURS
15410.80 **Steel Pipe** *(Cont.)*		
90 deg ell	EA	0.267
45 deg ell	"	0.400
Tee	"	0.400
Reducing tee	"	0.267
Cap	"	0.160
Coupling	"	0.267
Union	"	0.267
Nipple, 4" long	"	0.267
Cast iron fittings		
1/2" pipe		
90 deg. ell	EA	0.267
45 deg. ell	"	0.267
Tee	"	0.400
Reducing tee	"	0.400
3/4" pipe		
90 deg. ell	EA	0.267
45 deg. ell	"	0.267
Tee	"	0.400
Reducing tee	"	0.400
15410.82 **Galvanized Steel Pipe**		
Galvanized pipe		
1/2" pipe	LF	0.080
3/4" pipe	"	0.100
90 degree ell, 150 lb malleable iron, galvanized		
1/2"	EA	0.160
3/4"	"	0.200
45 degree ell, 150 lb m.i., galv.		
1/2"	EA	0.160
3/4"	"	0.200
Tees, straight, 150 lb m.i., galv.		
1/2"	EA	0.200
3/4"	"	0.229
Tees, reducing, out, 150 lb m.i., galv.		
1/2"	EA	0.200
3/4"	"	0.229
Couplings, straight, 150 lb m.i., galv.		
1/2"	EA	0.160
3/4"	"	0.178
Couplings, reducing, 150 lb m.i., galv		
1/2"	EA	0.160
3/4"	"	0.178
Caps, 150 lb m.i., galv.		
1/2"	EA	0.080
3/4"	"	0.084
Unions, 150 lb m.i., galv.		
1/2"	EA	0.200
3/4"	"	0.229
Nipples, galvanized steel, 4" long		
1/2"	EA	0.100
3/4"	"	0.107

Facility Water Distribution	UNIT	MAN/ HOURS
15410.82 **Galvanized Steel Pipe** *(Cont.)*		
90 degree reducing ell, 150 lb m.i., galv.		
3/4" x 1/2"	EA	0.160
1" x 3/4"	"	0.178
Square head plug (C.I.)		
1/2"	EA	0.089
3/4"	"	0.100
15430.23 **Cleanouts**		
Cleanout, wall		
2"	EA	0.533
3"	"	0.533
4"	"	0.667
Floor		
2"	EA	0.667
3"	"	0.667
4"	"	0.800
15430.25 **Hose Bibbs**		
Hose bibb		
1/2"	EA	0.267
3/4"	"	0.267
15430.60 **Valves**		
Gate valve, 125 lb, bronze, soldered		
1/2"	EA	0.200
3/4"	"	0.200
Threaded		
1/4", 125 lb	EA	0.320
1/2"		
125 lb	EA	0.320
150 lb	"	0.320
300 lb	"	0.320
3/4"		
125 lb	EA	0.320
150 lb	"	0.320
300 lb	"	0.320
Check valve, bronze, soldered, 125 lb		
1/2"	EA	0.200
3/4"	"	0.200
Threaded		
1/2"		
125 lb	EA	0.267
150 lb	"	0.267
200 lb	"	0.267
3/4"		
125 lb	EA	0.320
150 lb	"	0.320
200 lb	"	0.320
Vertical check valve, bronze, 125 lb, threaded		
1/2"	EA	0.320
3/4"	"	0.364
Globe valve, bronze, soldered, 125 lb		
1/2"	EA	0.229

Facility Water Distribution	UNIT	MAN/HOURS
15430.60 **Valves** *(Cont.)*		
3/4"	EA	0.250
Threaded		
1/2"		
125 lb	EA	0.267
150 lb	"	0.267
300 lb	"	0.267
3/4"		
125 lb	EA	0.320
150 lb	"	0.320
300 lb	"	0.320
Ball valve, bronze, 250 lb, threaded		
1/2"	EA	0.320
3/4"	"	0.320
Angle valve, bronze, 150 lb, threaded		
1/2"	EA	0.286
3/4"	"	0.320
Balancing valve, meter connections, circuit setter		
1/2"	EA	0.320
3/4"	"	0.364
Balancing valve, straight type		
1/2"	EA	0.320
3/4"	"	0.320
Angle type		
1/2"	EA	0.320
3/4"	"	0.320
Square head cock, 125 lb, bronze body		
1/2"	EA	0.267
3/4"	"	0.320
Radiator temp control valve, with control and sensor		
1/2" valve	EA	0.500
Pressure relief valve, 1/2", bronze		
Low pressure	EA	0.320
High pressure	"	0.320
Pressure and temperature relief valve		
Bronze, 3/4"	EA	0.320
Cast iron, 3/4"		
High pressure	EA	0.320
Temperature relief	"	0.320
Pressure & temp relief valve	"	0.320
Pressure reducing valve, bronze, threaded, 250 lb		
1/2"	EA	0.500
3/4"	"	0.500
Solar water temperature regulating valve		
3/4"	EA	0.667
Tempering valve, threaded		
3/4"	EA	0.267
Thermostatic mixing valve, threaded		
1/2"	EA	0.286
3/4"	"	0.320
Sweat connection		
1/2"	EA	0.286
3/4"	"	0.320
Mixing valve, sweat connection		

Facility Water Distribution	UNIT	MAN/HOURS
15430.60 **Valves** *(Cont.)*		
1/2"	EA	0.286
3/4"	"	0.320
Liquid level gauge, aluminum body		
3/4"	EA	0.320
125 psi, pvc body		
3/4"	EA	0.320
150 psi, crs body		
3/4"	EA	0.320
175 psi, bronze body, 1/2"	"	0.286
15430.65 **Vacuum Breakers**		
Vacuum breaker, atmospheric, threaded connection		
3/4"	EA	0.320
Anti-siphon, brass		
3/4"	EA	0.320
15430.68 **Strainers**		
Strainer, Y pattern, 125 psi, cast iron body, threaded		
3/4"	EA	0.286
250 psi, brass body, threaded		
3/4"	EA	0.320
Cast iron body, threaded		
3/4"	EA	0.320
15430.70 **Drains, Roof & Floor**		
Floor drain, cast iron, with cast iron top		
2"	EA	0.667
3"	"	0.667
4"	"	0.667
Roof drain, cast iron		
2"	EA	0.667
3"	"	0.667
4"	"	0.667
15430.80 **Traps**		
Bucket trap, threaded		
3/4"	EA	0.500
Inverted bucket steam trap, threaded		
3/4"	EA	0.500
With stainless interior		
1/2"	EA	0.500
3/4"	"	0.500
Brass interior		
3/4"	EA	0.500
Cast steel body, threaded, high temperature		
3/4"	EA	0.500
Float trap, 15 psi		
3/4"	EA	0.500
Float and thermostatic trap, 15 psi		
3/4"	EA	0.500
Steam trap, cast iron body, threaded, 125 psi		

15 MECHANICAL

Facility Water Distribution	UNIT	MAN/HOURS
15430.80 **Traps** *(Cont.)*		
3/4"	EA	0.500
Thermostatic trap, low pressure, angle type, 25 psi		
1/2"	EA	0.500
3/4"	"	0.500
Cast iron body, threaded, 125 psi		
3/4"	EA	0.500

Plumbing Fixtures	UNIT	MAN/HOURS
15440.10 **Baths**		
Bath tub, 5' long		
Minimum	EA	2.667
Average	"	4.000
Maximum	"	8.000
6' long		
Minimum	EA	2.667
Average	"	4.000
Maximum	"	8.000
Square tub, whirlpool, 4'x4'		
Minimum	EA	4.000
Average	"	8.000
Maximum	"	10.000
5'x5'		
Minimum	EA	4.000
Average	"	8.000
Maximum	"	10.000
6'x6'		
Minimum	EA	4.000
Average	"	8.000
Maximum	"	10.000
For trim and rough-in		
Minimum	EA	2.667
Average	"	4.000
Maximum	"	8.000
15440.12 **Disposals & Accessories**		
Disposal, continuous feed		
Minimum	EA	1.600
Average	"	2.000
Maximum	"	2.667
Batch feed, 1/2 hp		
Minimum	EA	1.600
Average	"	2.000
Maximum	"	2.667
Hot water dispenser		
Minimum	EA	1.600

Plumbing Fixtures	UNIT	MAN/HOURS
15440.12 **Disposals & Accessories** *(Cont.)*		
Average	EA	2.000
Maximum	"	2.667
Epoxy finish faucet	"	1.600
Lock stop assembly	"	1.000
Mounting gasket	"	0.667
Tailpipe gasket	"	0.667
Stopper assembly	"	0.800
Switch assembly, on/off	"	1.333
Tailpipe gasket washer	"	0.400
Stop gasket	"	0.444
Tailpipe flange	"	0.400
Tailpipe	"	0.500
15440.15 **Faucets**		
Kitchen		
Minimum	EA	1.333
Average	"	1.600
Maximum	"	2.000
Bath		
Minimum	EA	1.333
Average	"	1.600
Maximum	"	2.000
Lavatory, domestic		
Minimum	EA	1.333
Average	"	1.600
Maximum	"	2.000
Washroom		
Minimum	EA	1.333
Average	"	1.600
Maximum	"	2.000
Handicapped		
Minimum	EA	1.600
Average	"	2.000
Maximum	"	2.667
Shower		
Minimum	EA	1.333
Average	"	1.600
Maximum	"	2.000
For trim and rough-in		
Minimum	EA	1.600
Average	"	2.000
Maximum	"	4.000
15440.18 **Hydrants**		
Wall hydrant		
8" thick	EA	1.333
12" thick	"	1.600

Plumbing Fixtures	UNIT	MAN/ HOURS
15440.20 **Lavatories**		
Lavatory, counter top, porcelain enamel on cast iron		
Minimum	EA	1.600
Average	"	2.000
Maximum	"	2.667
Wall hung, china		
Minimum	EA	1.600
Average	"	2.000
Maximum	"	2.667
Handicapped		
Minimum	EA	2.000
Average	"	2.667
Maximum	"	4.000
For trim and rough-in		
Minimum	EA	2.000
Average	"	2.667
Maximum	"	4.000
15440.30 **Showers**		
Shower, fiberglass, 36"x34"x84"		
Minimum	EA	5.714
Average	"	8.000
Maximum	"	8.000
Steel, 1 piece, 36"x36"		
Minimum	EA	5.714
Average	"	8.000
Maximum	"	8.000
Receptor, molded stone, 36"x36"		
Minimum	EA	2.667
Average	"	4.000
Maximum	"	6.667
For trim and rough-in		
Minimum	EA	3.636
Average	"	4.444
Maximum	"	8.000
15440.40 **Sinks**		
Service sink, 24"x29"		
Minimum	EA	2.000
Average	"	2.667
Maximum	"	4.000
Kitchen sink, single, stainless steel, single bowl		
Minimum	EA	1.600
Average	"	2.000
Maximum	"	2.667
Double bowl		
Minimum	EA	2.000
Average	"	2.667
Maximum	"	4.000
Porcelain enamel, cast iron, single bowl		
Minimum	EA	1.600
Average	"	2.000
Maximum	"	2.667
Double bowl		

Plumbing Fixtures	UNIT	MAN/ HOURS
15440.40 **Sinks** *(Cont.)*		
Minimum	EA	2.000
Average	"	2.667
Maximum	"	4.000
Mop sink, 24"x36"x10"		
Minimum	EA	1.600
Average	"	2.000
Maximum	"	2.667
Washing machine box		
Minimum	EA	2.000
Average	"	2.667
Maximum	"	4.000
For trim and rough-in		
Minimum	EA	2.667
Average	"	4.000
Maximum	"	5.333
15440.60 **Water Closets**		
Water closet flush tank, floor mounted		
Minimum	EA	2.000
Average	"	2.667
Maximum	"	4.000
Handicapped		
Minimum	EA	2.667
Average	"	4.000
Maximum	"	8.000
For trim and rough-in		
Minimum	EA	2.000
Average	"	2.667
Maximum	"	4.000
15440.70 **Domestic Water Heaters**		
Water heater, electric		
6 gal	EA	1.333
10 gal	"	1.333
15 gal	"	1.333
20 gal	"	1.600
30 gal	"	1.600
40 gal	"	1.600
52 gal	"	2.000
Oil fired		
20 gal	EA	4.000
50 gal	"	5.714
Tankless water heater, natural gas		
Min.	EA	5.333
Ave.	"	8.000
Max.	"	16.000
Propane		
Min.	EA	5.333
Ave.	"	8.000
Max.	"	16.000

15 MECHANICAL

Plumbing Fixtures	UNIT	MAN/HOURS
15440.75 **Solar Water Heaters**		
Hydronic system, 100-120 Gallons including material		
Min.	EA	
Ave.	"	
Max.	"	
Direct-Solar, 100-120 Gallons		
Min.	EA	
Ave.	"	
Max.	"	
Indirect-Solar tank, 50-80 Gallons		
Min.	EA	
Ave.	"	
Max.	"	
100-120 Gallons		
Min.	EA	
Ave.	"	
Max.	"	
Solar water collector panel, 3 x 8		
Min.	EA	1.000
Ave.	"	1.143
Max.	"	1.333
4 x 7		
Min.	EA	1.000
Ave.	"	1.143
Max.	"	1.333
4 x 8		
Min.	EA	1.000
Ave.	"	1.143
Max.	"	1.333
4 x 10		
Min.	EA	1.143
Ave.	"	1.333
Max.	"	1.600
Passive tube tank system, 12 Tube		
Min.	EA	1.000
Ave.	"	1.143
Max.	"	1.333
24 Tube		
Min.	EA	1.000
Ave.	"	1.143
Max.	"	1.333
27 Tube		
Min.	EA	1.000
Ave.	"	1.143
Max.	"	1.333
15440.76 **Solar Water Heaters, Pools**		
Solar Water Heater, 1000 BTU/SF panel, 4x8	EA	1.000
4X10	"	1.143
4X12	"	1.333
Panel Mounting Kit, 4x8	"	0.400
4X10	"	0.444
4X12	"	0.500

Plumbing Fixtures	UNIT	MAN/HOURS
15450.40 **Storage Tanks**		
Hot water storage tank, cement lined		
10 gallon	EA	2.667
70 gallon	"	4.000

HVAC	UNIT	MAN/HOURS
15555.10 **Boilers**		
Cast iron, gas fired, hot water		
115 mbh	EA	20.000
175 mbh	"	21.818
235 mbh	"	24.000
Steam		
115 mbh	EA	20.000
175 mbh	"	21.818
235 mbh	"	24.000
Electric, hot water		
115 mbh	EA	12.000
175 mbh	"	12.000
235 mbh	"	12.000
Steam		
115 mbh	EA	12.000
175 mbh	"	12.000
235 mbh	"	12.000
Oil fired, hot water		
115 mbh	EA	16.000
175 mbh	"	18.462
235 mbh	"	21.818
Steam		
115 mbh	EA	16.000
175 mbh	"	18.462
235 mbh	"	21.818
15610.10 **Furnaces**		
Electric, hot air		
40 mbh	EA	4.000
60 mbh	"	4.211
80 mbh	"	4.444
100 mbh	"	4.706
125 mbh	"	4.848
Gas fired hot air		
40 mbh	EA	4.000
60 mbh	"	4.211
80 mbh	"	4.444
100 mbh	"	4.706
125 mbh	"	4.848

HVAC	UNIT	MAN/ HOURS
15610.10 Furnaces *(Cont.)*		
Oil fired hot air		
40 mbh	EA	4.000
60 mbh	"	4.211
80 mbh	"	4.444
100 mbh	"	4.706
125 mbh	"	4.848

Central Cooling Equipment	UNIT	MAN/ HOURS
15670.10 Condensing Units		
Air cooled condenser, single circuit		
3 ton	EA	1.333
5 ton	"	1.333
With low ambient dampers		
3 ton	EA	2.000
5 ton	"	2.000
15780.20 Rooftop Units		
Packaged, single zone rooftop unit, with roof curb		
2 ton	EA	8.000
3 ton	"	8.000
4 ton	"	10.000
15830.10 Radiation Units		
Baseboard radiation unit		
1.7 mbh/lf	LF	0.320
2.1 mbh/lf	"	0.400
15830.70 Unit Heaters		
Steam unit heater, horizontal		
12,500 btuh, 200 cfm	EA	1.333
17,000 btuh, 300 cfm	"	1.333

Air Handling	UNIT	MAN/ HOURS
15855.10 Air Handling Units		
Air handling unit, medium pressure, single zone		
1500 cfm	EA	5.000
3000 cfm	"	8.889
Rooftop air handling units		
4950 cfm	EA	8.889
7370 cfm	"	11.429
15870.20 Exhaust Fans		
Belt drive roof exhaust fans		
640 cfm, 2618 fpm	EA	1.000
940 cfm, 2604 fpm	"	1.000

Air Distribution	UNIT	MAN/ HOURS
15890.10 Metal Ductwork		
Rectangular duct		
Galvanized steel		
Minimum	LB	0.073
Average	"	0.089
Maximum	"	0.133
Aluminum		
Minimum	LB	0.160
Average	"	0.200
Maximum	"	0.267
Fittings		
Minimum	EA	0.267
Average	"	0.400
Maximum	"	0.800
15890.30 Flexible Ductwork		
Flexible duct, 1.25" fiberglass		
5" dia.	LF	0.040
6" dia.	"	0.044
7" dia.	"	0.047
8" dia.	"	0.050
10" dia.	"	0.057
12" dia.	"	0.062
Flexible duct connector, 3" wide fabric	"	0.133
15910.10 Dampers		
Horizontal parallel aluminum backdraft damper		
12" x 12"	EA	0.200
16" x 16"	"	0.229

Air Distribution	UNIT	MAN/HOURS
15940.10 **Diffusers**		
Ceiling diffusers, round, baked enamel finish		
6" dia.	EA	0.267
8" dia.	"	0.333
10" dia.	"	0.333
12" dia.	"	0.333
Rectangular		
6x6"	EA	0.267
9x9"	"	0.400
12x12"	"	0.400
15x15"	"	0.400
18x18"	"	0.400
15940.40 **Registers And Grilles**		
Lay in flush mounted, perforated face, return		
6x6/24x24	EA	0.320
8x8/24x24	"	0.320
9x9/24x24	"	0.320
10x10/24x24	"	0.320
12x12/24x24	"	0.320
Rectangular, ceiling return, single deflection		
10x10	EA	0.400
12x12	"	0.400
14x14	"	0.400
16x8	"	0.400
16x16	"	0.400
Wall, return air register		
12x12	EA	0.200
16x16	"	0.200
18x18	"	0.200
Ceiling, return air grille		
6x6	EA	0.267
8x8	"	0.320
10x10	"	0.320
Ceiling, exhaust grille, aluminum egg crate		
6x6	EA	0.267
8x8	"	0.320
10x10	"	0.320
12x12	"	0.400

Basic Materials	UNIT	MAN/HOURS
16050.30 **Bus Duct**		
Bus duct, 100a, plug-in		
10', 600v	EA	2.759
With ground	"	4.211
Circuit breakers, with enclosure		
1 pole		
15a-60a	EA	1.000
70a-100a	"	1.250
2 pole		
15a-60a	EA	1.100
70a-100a	"	1.301
Circuit breaker, adapter cubicle		
225a	EA	1.509
400a	"	1.600
Fusible switches, 240v, 3 phase		
30a	EA	1.000
60a	"	1.250
100a	"	1.509
200a	"	2.105
16110.20 **Conduit Specialties**		
Rod beam clamp, 1/2"	EA	0.050
Hanger rod		
3/8"	LF	0.040
1/2"	"	0.050
All thread rod		
1/4"	LF	0.030
3/8"	"	0.040
1/2"	"	0.050
5/8"	"	0.080
Hanger channel, 1-1/2"		
No holes	EA	0.030
Holes	"	0.030
Channel strap		
1/2"	EA	0.050
3/4"	"	0.050
Conduit penetrations, roof and wall, 8" thick		
1/2"	EA	0.615
3/4"	"	0.615
1"	"	0.800
Threaded rod couplings		
1/4"	EA	0.050
3/8"	"	0.050
1/2"	"	0.050
5/8"	"	0.050
3/4"	"	0.050
Hex nuts, 1/4"	"	0.050
3/8"	"	0.050
1/2"	"	0.050
5/8"	"	0.050
3/4"	"	0.050
Square nuts		
1/4"	EA	0.050
3/8"	"	0.050

Basic Materials	UNIT	MAN/HOURS
16110.20 **Conduit Specialties** *(Cont.)*		
1/2"	EA	0.050
5/8"	"	0.050
3/4"	"	0.050
16110.21 **Aluminum Conduit**		
Aluminum conduit		
1/2"	LF	0.030
3/4"	"	0.040
1"	"	0.050
90 deg. elbow		
1/2"	EA	0.190
3/4"	"	0.250
1"	"	0.308
Coupling		
1/2"	EA	0.050
3/4"	"	0.059
1"	"	0.080
16110.22 **EMT Conduit**		
EMT conduit		
1/2"	LF	0.030
3/4"	"	0.040
1"	"	0.050
90 deg. elbow		
1/2"	EA	0.089
3/4"	"	0.100
1"	"	0.107
Connector, steel compression		
1/2"	EA	0.089
3/4"	"	0.089
1"	"	0.089
Coupling, steel, compression		
1/2"	EA	0.059
3/4"	"	0.059
1"	"	0.059
1 hole strap, steel		
1/2"	EA	0.040
3/4"	"	0.040
1"	"	0.040
Connector, steel set screw		
1/2"	EA	0.070
3/4"	"	0.070
1"	"	0.070
Insulated throat		
1/2"	EA	0.070
3/4"	"	0.070
1"	"	0.070
Connector, die cast set screw		
1/2"	EA	0.059
3/4"	"	0.059
1"	"	0.059
Insulated throat		
1/2"	EA	0.059

Basic Materials	UNIT	MAN/HOURS	Basic Materials	UNIT	MAN/HOURS
16110.22 EMT Conduit *(Cont.)*			**16110.22** EMT Conduit *(Cont.)*		
3/4"	EA	0.059	1"	EA	0.500
1"	"	0.059	Type "T" compression condulets		
Coupling, steel set screw			1/2"	EA	0.400
1/2"	EA	0.040	3/4"	"	0.444
3/4"	"	0.040	1"	"	0.615
1"	"	0.040	Condulet covers		
Diecast set screw			1/2"	EA	0.123
1/2"	EA	0.040	3/4"	"	0.123
3/4"	"	0.040	1"	"	0.123
1"	"	0.040	Clamp type entrance caps		
1 hole malleable straps			1/2"	EA	0.250
1/2"	EA	0.040	3/4"	"	0.296
3/4"	"	0.040	1"	"	0.400
1"	"	0.040	Slip fitter type entrance caps		
EMT to rigid compression coupling			1/2"	EA	0.250
1/2"	EA	0.100	3/4"	"	0.296
3/4"	"	0.100	1"	"	0.400
1"	"	0.150	**16110.23** Flexible Conduit		
Set screw couplings			Flexible conduit, steel		
1/2"	EA	0.100	3/8"	LF	0.030
3/4"	"	0.100	1/2	"	0.030
1"	"	0.145	3/4"	"	0.040
Set screw offset connectors			1"	"	0.040
1/2"	EA	0.100	Flexible conduit, liquid tight		
3/4"	"	0.100	3/8"	LF	0.030
1"	"	0.145	1/2"	"	0.030
Compression offset connectors			3/4"	"	0.040
1/2"	EA	0.100	1"	"	0.040
3/4"	"	0.100	Connector, straight		
1"	"	0.145	3/8"	EA	0.080
Type "LB" set screw condulets			1/2"	"	0.080
1/2"	EA	0.229	3/4"	"	0.089
3/4"	"	0.296	1"	"	0.100
1"	"	0.381	Straight insulated throat connectors		
Type "T" set screw condulets			3/8"	EA	0.123
1/2"	EA	0.296	1/2"	"	0.123
3/4"	"	0.400	3/4"	"	0.145
1"	"	0.444	1"	"	0.145
Type "C" set screw condulets			90 deg connectors		
1/2"	EA	0.250	3/8"	EA	0.148
3/4"	"	0.296	1/2"	"	0.148
1"	"	0.381	3/4"	"	0.170
Type "LL" set screw condulets			1"	"	0.182
1/2"	EA	0.250	90 degree insulated throat connectors		
3/4"	"	0.296	3/8"	EA	0.145
1"	"	0.381	1/2"	"	0.145
Type "LR" set screw condulets			3/4"	"	0.170
1/2"	EA	0.250	1"	"	0.178
3/4"	"	0.296	Flexible aluminum conduit		
1"	"	0.381	3/8"	LF	0.030
Type "LB" compression condulets			1/2"	"	0.030
1/2"	EA	0.296	3/4"	"	0.040
3/4"	"	0.500			

Basic Materials	UNIT	MAN/ HOURS
16110.23 **Flexible Conduit** *(Cont.)*		
1"	LF	0.040
Connector, straight		
3/8"	EA	0.100
1/2"	"	0.100
3/4"	"	0.107
1"	"	0.123
Straight insulated throat connectors		
3/8"	EA	0.089
1/2"	"	0.089
3/4"	"	0.089
1"	"	0.100
90 deg connectors		
3/8"	EA	0.145
1/2"	"	0.145
3/4"	"	0.145
1"	"	0.170
90 deg insulated throat connectors		
3/8"	EA	0.145
1/2"	"	0.145
3/4"	"	0.145
1"	"	0.170
16110.24 **Galvanized Conduit**		
Galvanized rigid steel conduit		
1/2"	LF	0.040
3/4"	"	0.050
1"	"	0.059
1-1/4"	"	0.080
1-1/2"	"	0.089
2"	"	0.100
90 degree ell		
1/2"	EA	0.250
3/4"	"	0.308
1"	"	0.381
1-1/4"	"	0.444
1-1/2"	"	0.500
2"	"	0.533
Couplings, with set screws		
1/2"	EA	0.050
3/4"	"	0.059
1"	"	0.080
1-1/4"	"	0.100
1-1/2"	"	0.123
2"	"	0.145
Split couplings		
1/2"	EA	0.190
3/4"	"	0.250
1"	"	0.276
1-1/4"	"	0.308
1-1/2"	"	0.381
2"	"	0.571
Erickson couplings		
1/2"	EA	0.444

Basic Materials	UNIT	MAN/ HOURS
16110.24 **Galvanized Conduit** *(Cont.)*		
3/4"	EA	0.500
1"	"	0.615
1-1/4"	"	0.889
1-1/2"	"	1.000
2"	"	1.333
Seal fittings		
1/2"	EA	0.667
3/4"	"	0.800
1"	"	1.000
1-1/4"	"	1.143
1-1/2"	"	1.333
2"	"	1.600
Entrance fitting, (weather head), threaded		
1/2"	EA	0.444
3/4"	"	0.500
1"	"	0.571
1-1/4"	"	0.727
1-1/2"	"	0.800
2"	"	0.889
Locknuts		
1/2"	EA	0.050
3/4"	"	0.050
1"	"	0.050
1-1/4"	"	0.050
1-1/2"	"	0.059
2"	"	0.059
Plastic conduit bushings		
1/2"	EA	0.123
3/4"	"	0.145
1"	"	0.190
1-1/4"	"	0.222
1-1/2"	"	0.250
2"	"	0.308
Conduit bushings, steel		
1/2"	EA	0.123
3/4"	"	0.145
1"	"	0.190
1-1/4"	"	0.222
1-1/2"	"	0.250
2"	"	0.308
Pipe cap		
1/2"	EA	0.050
3/4"	"	0.050
1"	"	0.050
1-1/4"	"	0.080
1-1/2"	"	0.080
2"	"	0.080
Threaded couplings		
1/2"	EA	0.050
3/4"	"	0.059
1"	"	0.080
1-1/4"	"	0.089
1-1/2"	"	0.100

Basic Materials	UNIT	MAN/ HOURS
16110.24 **Galvanized Conduit** *(Cont.)*		
2"	EA	0.107
Threadless couplings		
1/2"	EA	0.100
3/4"	"	0.123
1"	"	0.145
1-1/4"	"	0.190
1-1/2"	"	0.250
2"	"	0.308
Threadless connectors		
1/2"	EA	0.100
3/4"	"	0.123
1"	"	0.145
1-1/4"	"	0.190
1-1/2"	"	0.250
2"	"	0.308
Setscrew connectors		
1/2"	EA	0.080
3/4"	"	0.089
1"	"	0.100
1-1/4"	"	0.123
1-1/2"	"	0.145
2"	"	0.190
Clamp type entrance caps		
1/2"	EA	0.308
3/4"	"	0.381
1"	"	0.444
1-1/4"	"	0.500
1-1/2"	"	0.615
2"	"	0.727
"LB" condulets		
1/2"	EA	0.308
3/4"	"	0.381
1"	"	0.444
1-1/4"	"	0.500
1-1/2"	"	0.615
2"	"	0.727
"T" condulets		
1/2"	EA	0.381
3/4"	"	0.444
1"	"	0.500
1-1/4"	"	0.571
1-1/2"	"	0.615
2"	"	0.727
"X" condulets		
1/2"	EA	0.444
3/4"	"	0.500
1"	"	0.571
1-1/4"	"	0.615
1-1/2"	"	0.667
2"	"	0.879
Blank steel condulet covers		
1/2"	EA	0.100
3/4"	"	0.100

Basic Materials	UNIT	MAN/ HOURS
16110.24 **Galvanized Conduit** *(Cont.)*		
1"	EA	0.100
1-1/4"	"	0.123
1-1/2"	"	0.123
2"	"	0.123
Solid condulet gaskets		
1/2"	EA	0.050
3/4"	"	0.050
1"	"	0.050
1-1/4"	"	0.080
1-1/2"	"	0.080
2"	"	0.080
One-hole malleable straps		
1/2"	EA	0.040
3/4"	"	0.040
1"	"	0.040
1-1/4"	"	0.050
1-1/2"	"	0.050
2"	"	0.050
One-hole steel straps		
1/2"	EA	0.040
3/4"	"	0.040
1"	"	0.040
1-1/4"	"	0.050
1-1/2"	"	0.050
2"	"	0.050
Grounding locknuts		
1/2"	EA	0.080
3/4"	"	0.080
1"	"	0.080
1-1/4"	"	0.089
1-1/2"	"	0.089
2"	"	0.089
Insulated grounding metal bushings		
1/2"	EA	0.190
3/4"	"	0.222
1"	"	0.250
1-1/4"	"	0.308
1-1/2"	"	0.381
2"	"	0.444
16110.25 **Plastic Conduit**		
PVC conduit, schedule 40		
1/2"	LF	0.030
3/4"	"	0.030
1"	"	0.040
1-1/4"	"	0.040
1-1/2"	"	0.050
2"	"	0.050
Couplings		
1/2"	EA	0.050
3/4"	"	0.050
1"	"	0.050
1-1/4"	"	0.059

Basic Materials		UNIT	MAN/ HOURS
16110.25	**Plastic Conduit** *(Cont.)*		
1-1/2"		EA	0.059
2"		"	0.059
90 degree elbows			
1/2"		EA	0.100
3/4"		"	0.123
1"		"	0.123
1-1/4"		"	0.145
1-1/2"		"	0.190
2"		"	0.222
Terminal adapters			
1/2"		EA	0.100
3/4"		"	0.100
1"		"	0.100
1-1/4"		"	0.160
1-1/2"		"	0.160
2"		"	0.160
LB conduit body			
1/2"		EA	0.190
3/4"		"	0.190
1		"	0.190
1-1/4"		"	0.308
1-1/2"		"	0.308
2"		"	0.308
16110.27	**Plastic Coated Conduit**		
Rigid steel conduit, plastic coated			
1/2"		LF	0.050
3/4"		"	0.059
1"		"	0.080
1-1/4"		"	0.100
1-1/2"		"	0.123
2"		"	0.145
90 degree elbows			
1/2"		EA	0.308
3/4"		"	0.381
1"		"	0.444
1-1/4"		"	0.500
1-1/2"		"	0.615
2"		"	0.800
Couplings			
1/2"		EA	0.059
3/4"		"	0.080
1"		"	0.089
1-1/4"		"	0.107
1-1/2"		"	0.123
2"		"	0.145
1 hole conduit straps			
3/4"		EA	0.050
1"		"	0.050
1-1/4"		"	0.059
1-1/2"		"	0.059
2"		"	0.059

Basic Materials		UNIT	MAN/ HOURS
16110.28	**Steel Conduit**		
Intermediate metal conduit (IMC)			
1/2"		LF	0.030
3/4"		"	0.040
1"		"	0.050
1-1/4"		"	0.059
1-1/2"		"	0.080
2"		"	0.089
90 degree ell			
1/2"		EA	0.250
3/4"		"	0.308
1"		"	0.381
1-1/4"		"	0.444
1-1/2"		"	0.500
2"		"	0.571
Couplings			
1/2"		EA	0.050
3/4"		"	0.059
1"		"	0.080
1-1/4"		"	0.089
1-1/2"		"	0.100
2"		"	0.107
16110.35	**Surface Mounted Raceway**		
Single Raceway			
3/4" x 17/32" Conduit		LF	0.040
Mounting Strap		EA	0.053
Connector		"	0.053
Elbow			
45 degree		EA	0.050
90 degree		"	0.050
internal		"	0.050
external		"	0.050
Switch		"	0.400
Utility Box		"	0.400
Receptacle		"	0.400
3/4" x 21/32" Conduit		LF	0.040
Mounting Strap		EA	0.053
Connector		"	0.053
Elbow			
45 degree		EA	0.050
90 degree		"	0.050
internal		"	0.050
external		"	0.050
Switch		"	0.400
Utility Box		"	0.400
Receptacle		"	0.400
16120.41	**Aluminum Conductors**		
Type XHHW, stranded aluminum, 600v			
#8		LF	0.005
#6		"	0.006
#4		"	0.008
#2		"	0.009

Basic Materials		UNIT	MAN/HOURS
16120.41	**Aluminum Conductors** *(Cont.)*		
1/0		LF	0.011
2/0		"	0.012
3/0		"	0.014
4/0		"	0.015
Type S.E.U. cable			
#8/3		LF	0.025
#6/3		"	0.028
#4/3		"	0.035
#2/3		"	0.038
#1/3		"	0.040
1/0-3		"	0.042
2/0-3		"	0.044
3/0-3		"	0.052
4/0-3		"	0.057
Type S.E.R. cable with ground			
#8/3		LF	0.028
#6/3		"	0.035
16120.43	**Copper Conductors**		
Copper conductors, type THW, solid			
#14		LF	0.004
#12		"	0.005
#10		"	0.006
THHN-THWN, solid			
#14		LF	0.004
#12		"	0.005
#10		"	0.006
Stranded			
#14		LF	0.004
#12		"	0.005
#10		"	0.006
#8		"	0.008
#6		"	0.009
#4		"	0.010
#2		"	0.012
#1		"	0.014
1/0		"	0.016
2/0		"	0.020
3/0		"	0.025
4/0		"	0.028
Bare stranded wire			
#8		LF	0.008
#6		"	0.010
#4		"	0.010
#2		"	0.011
#1		"	0.014
Type "BX" solid armored cable			
#14/2		LF	0.025
#14/3		"	0.028
#14/4		"	0.031
#12/2		"	0.028
#12/3		"	0.031
#12/4		"	0.035

Basic Materials		UNIT	MAN/HOURS
16120.43	**Copper Conductors** *(Cont.)*		
#10/2		LF	0.031
#10/3		"	0.035
#10/4		"	0.040
Steel type, metal clad cable, solid, with ground			
#14/2		LF	0.018
#14/3		"	0.020
#14/4		"	0.023
#12/2		"	0.020
#12/3		"	0.025
#12/4		"	0.030
#10/2		"	0.023
#10/3		"	0.028
#10/4		"	0.033
16120.47	**Sheathed Cable**		
Non-metallic sheathed cable			
Type NM cable with ground			
#14/2		LF	0.015
#12/2		"	0.016
#10/2		"	0.018
#8/2		"	0.020
#6/2		"	0.025
#14/3		"	0.026
#12/3		"	0.027
#10/3		"	0.027
#8/3		"	0.028
#6/3		"	0.028
#4/3		"	0.032
#2/3		"	0.035
Type U.F. cable with ground			
#14/2		LF	0.016
#12/2		"	0.019
#14/3		"	0.020
#12/3		"	0.022
Type S.F.U. cable, 3 conductor			
#8		LF	0.028
#6		"	0.031
Type SER cable, 4 conductor			
#6		LF	0.036
#4		"	0.039
Flexible cord, type STO cord			
#18/2		LF	0.004
#18/3		"	0.005
#18/4		"	0.006
#16/2		"	0.004
#16/3		"	0.004
#16/4		"	0.005
#14/2		"	0.005
#14/3		"	0.006
#14/4		"	0.007
#12/2		"	0.006
#12/3		"	0.007
#12/4		"	0.008

Basic Materials	UNIT	MAN/ HOURS
16120.47 Sheathed Cable *(Cont.)*		
#10/2	LF	0.007
#10/3	"	0.008
#10/4	"	0.009
16130.40 Boxes		
Round cast box, type SEH		
1/2"	EA	0.348
3/4"	"	0.421
SEHC		
1/2"	EA	0.348
3/4"	"	0.421
SEHL		
1/2"	EA	0.348
3/4"	"	0.444
SEHT		
1/2"	EA	0.421
3/4"	"	0.500
SEHX		
1/2"	EA	0.500
3/4"	"	0.615
Blank cover	"	0.145
1/2", hub cover	"	0.145
Cover with gasket	"	0.178
Rectangle, type FS boxes		
1/2"	EA	0.348
3/4"	"	0.400
1"	"	0.500
FSA		
1/2"	EA	0.348
3/4"	"	0.400
FSC		
1/2"	EA	0.348
3/4"	"	0.421
1"	"	0.500
FSL		
1/2"	EA	0.348
3/4"	"	0.400
FSR		
1/2"	EA	0.348
3/4"	"	0.400
FSS		
1/2"	EA	0.348
3/4"	"	0.400
FSLA		
1/2"	EA	0.348
3/4"	"	0.400
FSCA		
1/2"	EA	0.348
3/4"	"	0.400
FSCC		
1/2"	EA	0.400
3/4"	"	0.500
FSCT		

Basic Materials	UNIT	MAN/ HOURS
16130.40 Boxes *(Cont.)*		
1/2"	EA	0.400
3/4"	"	0.500
1"	"	0.571
FST		
1/2"	EA	0.500
3/4"	"	0.571
FSX		
1/2"	EA	0.615
3/4"	"	0.727
FSCD boxes		
1/2"	EA	0.615
3/4"	"	0.727
Rectangle, type FS, 2 gang boxes		
1/2"	EA	0.348
3/4"	"	0.400
1"	"	0.500
FSC, 2 gang boxes		
1/2"	EA	0.348
3/4"	"	0.400
1"	"	0.500
FSS, 2 gang boxes		
3/4"	EA	0.400
FS, tandem boxes		
1/2"	EA	0.400
3/4"	"	0.444
FSC, tandem boxes		
1/2"	EA	0.400
3/4"	"	0.444
FS, three gang boxes		
3/4"	EA	0.444
1"	"	0.500
FSS, three gang boxes, 3/4"	"	0.500
Weatherproof cast aluminum boxes, 1 gang, 3 outlets		
1/2"	EA	0.400
3/4"	"	0.500
2 gang, 3 outlets		
1/2"	EA	0.500
3/4"	"	0.533
1 gang, 4 outlets		
1/2"	EA	0.615
3/4"	"	0.727
2 gang, 4 outlets		
1/2"	EA	0.615
3/4"	"	0.727
1 gang, 5 outlets		
1/2"	EA	0.727
3/4"	"	0.800
2 gang, 5 outlets		
1/2"	EA	0.727
3/4"	"	0.800
2 gang, 6 outlets		
1/2"	EA	0.851
3/4"	"	0.899

Basic Materials	UNIT	MAN/HOURS
16130.40 **Boxes** *(Cont.)*		
2 gang, 7 outlets		
1/2"	EA	1.000
3/4"	"	1.096
Weatherproof and type FS box covers, blank, 1 gang	"	0.145
Tumbler switch, 1 gang	"	0.145
1 gang, single recept	"	0.145
Duplex recept	"	0.145
Despard	"	0.145
Red pilot light	"	0.145
SW and		
Single recept	EA	0.200
Duplex recept	"	0.200
2 gang		
Blank	EA	0.182
Tumbler switch	"	0.182
Single recept	"	0.182
Duplex recept	"	0.182
3 gang		
Blank	EA	0.200
Tumbler switch	"	0.200
4 gang		
Tumbler switch	EA	0.250
Box covers		
Surface	EA	0.200
Sealing	"	0.200
Dome	"	0.200
1/2" nipple	"	0.200
3/4" nipple	"	0.200
16130.60 **Pull And Junction Boxes**		
4"		
Octagon box	EA	0.114
Box extension	"	0.059
Plaster ring	"	0.059
Cover blank	"	0.059
Square box	"	0.114
Box extension	"	0.059
Plaster ring	"	0.059
Cover blank	"	0.059
Switch and device boxes		
2 gang	EA	0.114
3 gang	"	0.114
4 gang	"	0.160
Device covers		
2 gang	EA	0.059
3 gang	"	0.059
4 gang	"	0.059
Handy box	"	0.114
Extension	"	0.059
Switch cover	"	0.059
Switch box with knockout	"	0.145
Weatherproof cover, spring type	"	0.080
Cover plate, dryer receptacle 1 gang plastic	"	0.100

Basic Materials	UNIT	MAN/HOURS
16130.60 **Pull And Junction Boxes** *(Cont.)*		
For 4" receptacle, 2 gang	EA	0.100
Duplex receptacle cover plate, plastic	"	0.059
4", vertical bracket box, 1-1/2" with		
RMX clamps	EA	0.145
BX clamps	"	0.145
4", octagon device cover		
1 switch	EA	0.059
1 duplex recept	"	0.059
4", octagon swivel hanger box, 1/2" hub	"	0.059
3/4" hub	"	0.059
4" octagon adjustable bar hangers		
18-1/2"	EA	0.050
26-1/2"	"	0.050
With clip		
18-1/2"	EA	0.050
26-1/2"	"	0.050
4", square face bracket boxes, 1-1/2"		
RMX	EA	0.145
BX	"	0.145
4" square to round plaster rings	"	0.059
2 gang device plaster rings	"	0.059
Surface covers		
1 gang switch	EA	0.059
2 gang switch	"	0.059
1 single recept	"	0.059
1 20a twist lock recept	"	0.059
1 30a twist lock recept	"	0.059
1 duplex recept	"	0.059
2 duplex recept	"	0.059
Switch and duplex recept	"	0.059
4" plastic round boxes, ground straps		
Box only	EA	0.145
Box w/clamps	"	0.200
Box w/16" bar	"	0.229
Box w/24" bar	"	0.250
4" plastic round box covers		
Blank cover	EA	0.059
Plaster ring	"	0.059
4" plastic square boxes		
Box only	EA	0.145
Box w/clamps	"	0.200
Box w/hanger	"	0.250
Box w/nails and clamp	"	0.250
4" plastic square box covers		
Blank cover	EA	0.059
1 gang ring	"	0.059
2 gang ring	"	0.059
Round ring	"	0.059

Basic Materials

Basic Materials	UNIT	MAN/HOURS
16130.80 **Receptacles**		
Contractor grade duplex receptacles, 15a 120v		
Duplex	EA	0.200
125 volt, 20a, duplex, standard grade	"	0.200
Ground fault interrupter type	"	0.296
250 volt, 20a, 2 pole, single, ground type	"	0.200
120/208v, 4 pole, single receptacle, twist lock		
20a	EA	0.348
50a	"	0.348
125/250v, 3 pole, flush receptacle		
30a	EA	0.296
50a	"	0.296
60a	"	0.348
Dryer receptacle, 250v, 30a/50a, 3 wire	"	0.296
Clock receptacle, 2 pole, grounding type	"	0.200
125v, 20a single recept. grounding type		
Standard grade	EA	0.200
125/250v, 3 pole, 3 wire surface recepts		
30a	EA	0.296
50a	"	0.296
60a	"	0.348
Cord set, 3 wire, 6' cord		
30a	EA	0.296
50a	"	0.296
125/250v, 3 pole, 3 wire cap		
30a	EA	0.400
50a	"	0.400
60a	"	0.444
16199.10 **Utility Poles & Fittings**		
Wood pole, creosoted		
25'	EA	2.353
30'	"	2.963
Treated, wood preservative, 6"x6"		
8'	EA	0.500
10'	"	0.800
12'	"	0.889
14'	"	1.333
16'	"	1.600
18'	"	2.000
20'	"	2.000
Aluminum, brushed, no base		
8'	EA	2.000
10'	"	2.667
15'	"	2.759
20'	"	3.200
25'	"	3.810
Steel, no base		
10'	EA	2.500
15'	"	2.963
20'	"	3.810
25'	"	4.520
Concrete, no base		
13'	EA	5.517

Basic Materials	UNIT	MAN/HOURS
16199.10 **Utility Poles & Fittings** *(Cont.)*		
16'	EA	7.273
18'	"	8.791
25'	"	10.000
16350.10 **Circuit Breakers**		
Load center circuit breakers, 240v		
1 pole, 10-60a	EA	0.250
2 pole		
10-60a	EA	0.400
70-100a	"	0.667
110-150a	"	0.727
Load center, G.F.I. breakers, 240v		
1 pole, 15-30a	EA	0.296
Tandem breakers, 240v		
1 pole, 15-30a	EA	0.400
2 pole, 15-30a	"	0.533
16365.10 **Fuses**		
Fuse, one-time, 250v		
30a	EA	0.050
60a	"	0.050
100a	"	0.050
16395.10 **Grounding**		
Ground rods, copper clad, 1/2" x		
6'	EA	0.667
8'	"	0.727
5/8" x		
6'	EA	0.727
8'	"	1.000
Ground rod clamp		
5/8"	EA	0.123
Ground rod couplings		
1/2"	EA	0.100
5/8"	"	0.100
Ground rod, driving stud		
1/2"	EA	0.100
5/8"	"	0.100
Ground rod clamps, #8-2 to		
1" pipe	EA	0.200
2" pipe	"	0.250

Service And Distribution	UNIT	MAN/ HOURS
16430.20 **Metering**		
Outdoor wp meter sockets, 1 gang, 240v, 1 phase		
Includes sealing ring, 100a	EA	1.509
150a	"	1.778
200a	"	2.000
Die cast hubs, 1-1/4"	"	0.320
1-1/2"	"	0.320
2"	"	0.320
16470.10 **Panelboards**		
Indoor load center, 1 phase 240v main lug only		
30a - 2 spaces	EA	2.000
100a - 8 spaces	"	2.424
150a - 16 spaces	"	2.963
200a - 24 spaces	"	3.478
200a - 42 spaces	"	4.000
Main circuit breaker		
100a - 8 spaces	EA	2.424
100a - 16 spaces	"	2.759
150a - 16 spaces	"	2.963
150a - 24 spaces	"	3.200
200a - 24 spaces	"	3.478
200a - 42 spaces	"	3.636
120/208v, flush, 3 ph., 4 wire, main only		
100a		
12 circuits	EA	5.096
20 circuits	"	6.299
30 circuits	"	7.018
225a		
30 circuits	EA	7.767
42 circuits	"	9.524
16490.10 **Switches**		
Photo electric switches		
1000 watt		
105-135v	EA	0.727
Dimmer switch and switch plate		
600w	EA	0.308
Time clocks with skip, 40a, 120v		
SPST	EA	0.748
Contractor grade wall switch 15a, 120v		
Single pole	EA	0.160
Three way	"	0.200
Four way	"	0.267
Specification grade toggle switches, 20a, 120-277v		
Single pole	EA	0.200
Double pole	"	0.296
3 way	"	0.250
4 way	"	0.296
Combination switch and pilot light, single pole	"	0.296
3 way	"	0.348
Combination switch and receptacle, single pole	"	0.296
3 way	"	0.296
Switch plates, plastic ivory		

Service And Distribution	UNIT	MAN/ HOURS
16490.10 **Switches** *(Cont.)*		
1 gang	EA	0.080
2 gang	"	0.100
3 gang	"	0.119
4 gang	"	0.145
5 gang	"	0.160
6 gang	"	0.182
Stainless steel		
1 gang	EA	0.080
2 gang	"	0.100
3 gang	"	0.123
4 gang	"	0.145
5 gang	"	0.160
6 gang	"	0.182
Brass		
1 gang	EA	0.080
2 gang	"	0.100
3 gang	"	0.123
4 gang	"	0.145
5 gang	"	0.160
6 gang	"	0.182

Lighting	UNIT	MAN/ HOURS
16510.05 **Interior Lighting**		
Recessed fluorescent fixtures, 2'x2'		
2 lamp	EA	0.727
4 lamp	"	0.727
Surface mounted incandescent fixtures		
40w	EA	0.667
75w	"	0.667
100w	"	0.667
150w	"	0.667
Pendant		
40w	EA	0.800
75w	"	0.800
100w	"	0.800
150w	"	0.800
Contractor grade recessed down lights		
100 watt housing only	EA	1.000
150 watt housing only	"	1.000
100 watt trim	"	0.500
150 watt trim	"	0.500
Recessed incandescent fixtures		
40w	EA	1.509
75w	"	1.509
100w	"	1.509

Lighting	UNIT	MAN/HOURS
16510.05 Interior Lighting *(Cont.)*		
150w	EA	1.509
Light track single circuit		
2'	EA	0.500
4'	"	0.500
8'	"	1.000
12'	"	1.509
Fittings and accessories		
Dead end	EA	0.145
Starter kit	"	0.250
Conduit feed	"	0.145
Straight connector	"	0.145
Center feed	"	0.145
L-connector	"	0.145
T-connector	"	0.145
X-connector	"	0.200
Cord and plug	"	0.100
Rigid corner	"	0.145
Flex connector	"	0.145
2 way connector	"	0.200
Spacer clip	"	0.050
Grid box	"	0.145
T-bar clip	"	0.050
Utility hook	"	0.145
Fixtures, square		
R-20	EA	0.145
R-30	"	0.145
40w flood	"	0.145
40w spot	"	0.145
100w flood	"	0.145
100w spot	"	0.145
Mini spot	"	0.145
Mini flood	"	0.145
Quartz, 500w	"	0.145
R-20 sphere	"	0.145
R-30 sphere	"	0.145
R-20 cylinder	"	0.145
R-30 cylinder	"	0.145
R-40 cylinder	"	0.145
R-30 wall wash	"	0.145
R-40 wall wash	"	0.145
16510.10 Industrial Lighting		
Surface mounted fluorescent, wrap around lens		
1 lamp	EA	0.800
2 lamps	"	0.889
Wall mounted fluorescent		
2-20w lamps	EA	0.500
2-30w lamps	"	0.500
2-40w lamps	"	0.667
Strip fluorescent		
4'		
1 lamp	EA	0.667
2 lamps	"	0.667

Lighting	UNIT	MAN/HOURS
16510.10 Industrial Lighting *(Cont.)*		
8'		
1 lamp	EA	0.727
2 lamps	"	0.889
Compact fluorescent		
2-7w	EA	1.000
2-13w	"	1.333
16670.10 Lightning Protection		
Lightning protection		
Copper point, nickel plated, 12'		
1/2" dia.	EA	1.000
5/8" dia.	"	1.000
16750.20 Signaling Systems		
Contractor grade doorbell chime kit		
Chime	EA	1.000
Doorbutton	"	0.320
Transformer	"	0.500

Resistance Heating	UNIT	MAN/HOURS
16850.10 Electric Heating		
Baseboard heater		
2', 375w	EA	1.000
3', 500w	"	1.000
4', 750w	"	1.143
5', 935w	"	1.333
6', 1125w	"	1.600
7', 1310w	"	1.818
8', 1500w	"	2.000
9', 1680w	"	2.222
10', 1875w	"	2.286
Unit heater, wall mounted		
750w	EA	1.600
1500w	"	1.667
Thermostat		
Integral	EA	0.500
Line voltage	"	0.500
Electric heater connection	"	0.250
Fittings		
Inside corner	EA	0.400
Outside corner	"	0.400
Receptacle section	"	0.400
Blank section	"	0.400

Resistance Heating	UNIT	MAN/ HOURS
16850.10 **Electric Heating** *(Cont.)*		
Radiant ceiling heater panels		
500w	EA	1.000
750w	"	1.000
Unit heater thermostat	"	0.533
Mounting bracket	"	0.727
Relay	"	0.615
16910.40 **Control Cable**		
Control cable, 600v, #14 THWN, PVC jacket		
2 wire	LF	0.008
4 wire	"	0.010

Solar	UNIT	MAN/ HOURS
16990.10 **Solar Electrical Systems**		
Photovoltaic, Full Grid-Tie System		
Panel array, inverter, mounts, racks, conduit, etc., 3,000 Watt		
Min.	EA	
Ave.	"	
Max.	"	
4,000 Watt		
Min.	EA	
Ave.	"	
Max.	"	
5,000 Watt		
Min.	EA	
Ave.	"	
Max.	"	
Photovoltaic Components		
Polycristalline Rigid Panel, 200 watt		
Min.	EA	0.286
Ave.	"	0.286
Max.	"	0.286
215 Watt		
Min.	EA	0.286
Ave.	"	0.286
Max.	"	0.286
230 Watt		
Min.	EA	0.286
Ave.	"	0.286
Max.	"	0.286
245 Watt		
Min.	EA	0.286
Ave.	"	0.286
Max.	"	0.286
Anodized aluminum rail, 8'		

Solar	UNIT	MAN/ HOURS
16990.20 **Solar Electrical Systems** *(Cont.)*		
Min.	EA	0.178
Ave.	"	0.178
Max.	"	0.178
10'		
Min.	EA	0.200
Ave.	"	0.200
Max.	"	0.200
12'		
Min.	EA	0.229
Ave.	"	0.229
Max.	"	0.229
Panel mounts, aluminum, mount tile, flush		
Min.	EA	0.533
Ave.	"	0.533
Max.	"	0.533
Standard		
Min.	EA	0.533
Ave.	"	0.533
Max.	"	0.533
Rail clamp, mid-clamp		
Min.	EA	
Ave.	"	
Max.	"	
End-clamp		
Min.	EA	
Ave.	"	
Max.	"	
Anodized aluminum, rail splice kit		
Min.	EA	
Ave.	"	
Max.	"	
Power Distribution Panel		
Min.	EA	2.000
Ave.	"	2.000
Max.	"	2.000
Inverters		
Light capacity, micro inverter, 190 Watt		
Min.	EA	0.267
Ave.	"	0.267
Max.	"	0.267
Medium capacity, inverter, 1000 Watt		
Min.	EA	4.000
Ave.	"	4.000
Max.	"	4.000
2500 Watt		
Min.	EA	4.000
Ave.	"	4.000
Max.	"	4.000
5000 Watt		
Min.	EA	4.000
Ave.	"	4.000
Max.	"	4.000
Circuits		

Solar	UNIT	MAN/ HOURS
16990.40 **Solar Electrical Systems** *(Cont.)*		
Combiner Box, 12 circuit		
Min.	EA	2.667
Ave.	"	2.667
Max.	"	2.667
28 circuit		
Min.	EA	2.667
Ave.	"	2.667
Max.	"	2.667
Solar Circuit Breaker, 15A, 150 VDC		
Min.	EA	
Ave.	"	
Max.	"	
Wires and Conductors		
Cable, bare copper, 10 AWG, 500' coils		
Min.	EA	
Ave.	"	
Max.	"	
Multi-Contact branch connector, MC4		
Min.	EA	0.267
Ave.	"	0.267
Max.	"	0.267

Supporting Construction Reference Data

This section contains information, text, charts and tables on various aspects of construction. The intent is to provide the user with a better understanding of unfamiliar areas in order to be able to estimate better. This information includes actual takeoff data for some areas and also selected explanations of common construction materials, methods and common practices.

CONVERSION CALCULATIONS

Commercial Measure

16 grams .... = 1 ounce
16 ounces .... = 1 pound
2,000 pounds .... = 1 ton

Long Measure

12 inches .... = 1 foot
3 feet .... = 1 yard
16 ½ feet .... = 1 rod
40 rods .... = 1 furlong
8 furlongs (5,280 ft.) .... = 1 mile
3 miles .... = 1 league

Square Measure

144 square inches .... = 1 square foot
9 square feet .... = 1 square yard
30 ¼ square yards .... = 1 square rod
160 square rods .... = 1 acre
4,840 square yards .... = 1 acre
640 acres .... = 1 square mile
36 square miles .... = 1 township

Surveyor's Measure

7.92 inches .... = 1 link
25 links .... = 1 rod
4 rods (66 ft.) .... = 1 chain
10 chains .... = 1 furlong
8 furlongs .... = 1 mile
1 square mile .... = 1 section

Cubic Measure

1728 cubic inches .... = 1 cubic foot
27 cubic feet .... = 1 cubic yard
128 cubic feet .... = 1 cord (wood/stone)
231 cubic inches .... = 1 U.S. gallon
7.48 U.S. Gallons .... = 1 cubic foot
2150.4 cubic inches .... = 1 U.S. bushel

Liquid Measure

4 fluid ounces .... = 1 gill
4 gills .... = 1 pint
2 pints .... = 1 quart
4 quarts .... = 1 gallon
9 gallons .... = 1 firkin
31 ½ gallons .... = 1 barrel
2 barrels .... = 1 hogshead

Dry Measure

2 pints .... = 1 quart
8 quarts .... = 1 peck
4 pecks .... = 1 bushel
2150.42 cubic inches .... = 1 bushel

SQUARE

$$A = a^2$$

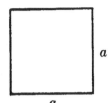

RECTANGLE

$$A = bh$$

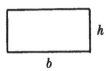

TRIANGLE

$$A = \frac{1}{2}bh$$

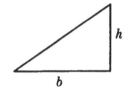
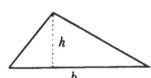

PARALLELOGRAM

$$A = bh = ab \, Sin\phi$$

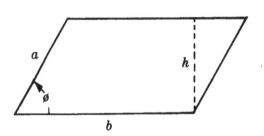

TRAPEZOID

$$A = \left(\frac{a+b}{2}\right)h$$

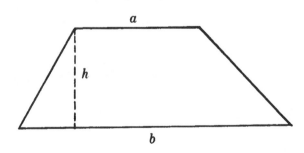

CIRCLE

$$A = \pi r^2 = \frac{\pi d^2}{4}$$

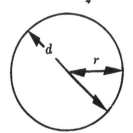

Circumference $= C = 2\pi r = \pi d$

ELLIPSE
$A = \pi ab$

PARABOLA
$A = \frac{2}{3}bh$

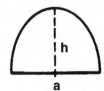

VOLUMES

CUBE
$V = a^3$

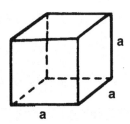

CYLINDER
$V = \pi r^2 h$

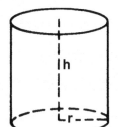

PYRAMID
$V = \frac{1}{3}(Base)\,h$

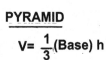
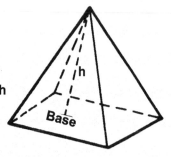

CONE
$V = \frac{1}{3}\pi r^2 h$

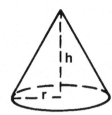

SPHERE
$V = \frac{4}{3}\pi r^3 = \frac{1}{6}\pi d^3$

WEDGE
$V = \frac{1}{2}abc$

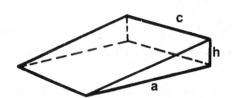

CONVERSION FACTORS
ENGLISH TO SI (SYSTEM INTERNATIONAL)

To Convert from	To	Multiply by
LENGTH		
Inches	Millimetres	25.4[a]
Feet	Metres	0.3048[a]
Yards	Metres	0.9144[a]
Miles (statute)	Kilometres	1.609
AREA		
Square inches	Square millimetres	645.2
Square feet	Square metres	0.0929
Square yards	Square metres	0.8361
VOLUME		
Cubic inches	Cubic millimetres	16.387
Cubic feet	Cubic metres	0.02832
Cubic yards	Cubic metres	0.7646
Gallons (U.S. liquid) [b]	Cubic metres[c]	0.003785
Gallons (Canadian liquid) [b]	Cubic metres[c]	0.004546
Ounces (U.S. liquid) [b]	Millilitres [c,d]	29.57
Quarts (U.S. liquid) [b]	Litres [c,d]	0.9464
Gallons (U.S. liquid) [b]	Litres [c]	3.785
FORCE		
Kilograms force	Newtons	9.807
Pounds force	Newtons	4.448
Pounds force	Kilograms force [d]	0.4536
Kips	Newtons	4448
Kips	Kilograms force [d]	453.6
PRESSURE, STRESS, STRENGTH (FORCE PER UNIT AREA)		
Kilograms force per sq. centimetre	Megapascals	0.09807
Pounds force per square inch (psi)	Megapascals	6895
Kips per square inch	Megapascals	6.895
Pounds force per square inch (psi)	Kilograms force per square centimetre [d]	0.07031
Pounds force per square foot	Pascals	47.88
Pounds force per square foot	Kilograms force per square metre [d]	4.882
SENDING MOMENT OR TORQUE		
Inch-pounds force	Metre-kilog. force [d]	0.01152
Inch-pounds force	Newton-metres	0.1130
Foot-pounds force	Metre-kilog. force [d]	0.1383
Foot-pounds force	Newton-metres	1.356
Metre-kilograms force	Newton-metres	9.807
MASS		
Ounce (avoirdupois)	Grams	28.35
Pounds (avoirdupois)	Kilograms	0.4536
Tons (metric)	Kilograms	1000[a]
Tons, short (2000 pounds)	Kilograms	907.2
Tons, short (2000 pounds)	Megagrams [e]	0.9072
MASS PER UNIT VOLUME		
Pounds mass per cubic foot	Kilog. per cubic metre	16.02
Pounds mass per cubic yard	Kilog. per cubic metre	0.5933
Pds. mass per gallon (U.S. liquid) [b]	Kilog. per cubic metre	119.8
Pds. mass p/gal. (Canadian liquid) [b]	Kilog. per cubic metre	99.78
TEMPERATURE		
Degrees Fahrenheit	Degrees Celsius	$tK = (1F - 32)/1.8$
Degrees Fahrenheit	Degrees Kelvin	$tK = (1F + 459.67)/1.8$
Degree Celsius	Degree Kelvin	$tK = 1C + 273.15$

[a] The factor given is exact.
[b] One U.S. gallon equals 0.8327 Canadian gallon.
[c] 1 litre = 1000 millilitres = 10,000 cubic centimetres = 1 cubic decimetre = 0.001 cubic metre.
[d] Metric but not SI unit.
[e] Called "tonne" in England. Called "metric ton" in other metric systems.

PILES AND PILE DRIVING

General. A pile is a column driven or jetted into the ground which derives its supporting capabilities from end-bearing on the underlying strata, skin friction between the pile surface and the soil, or from a combination of end-bearing and skin friction.

Piles can be divided into two major classes: **Sheet piles** and **load-bearing piles**. Sheet piling is used primarily to restrain lateral forces as in trench sheeting and bulkheads, or to resist the flow of water as in cofferdams. It is prefabricated and is available in steel, wood or concrete. Load-bearing piles are used primarily to transmit loads through soil formations of low bearing values to formations that are capable of supporting the designed loads. If the load is supported predominantly by the action of soil friction on the surface of the pile, it is called a **friction pile**. If the load is transmitted to the soil primarily through the lower tip, it is called an **end-bearing pile**.

There are several load-bearing pile types, which can be classified according to the material from which they are fabricated:
- Timber (Treated and untreated)
- Concrete (Precast and cast in place)
- Steel (H-Section and steel pipe)
- Composite (A combination of two or more materials)

Some of the additional uses of piling are to: eliminate or control settlement of structures, support bridge piers and abutments and protect them from scour, anchor structures against uplift or overturning, and for numerous marine structures such as docks, wharves, fenders, anchorages, piers, trestles and jetties.

Timber Piles. Timber piles, treated or untreated, are the piles most commonly used throughout the world, primarily because they are readily available, economical, easily handled, can be easily cut off to any desired length after driving and can be easily removed if necessary. On the other hand, they have some serious disadvantages which include: difficulty in securing straight piles of long length, problems in driving them into hard formations and difficulty in splicing to increase their length. They are generally not suitable for use as end-bearing piles under heavy load and they are subject to decay and insect attack. Timber piles are resilient and particularly adaptable for use in waterfront structures such as wharves, docks and piers for anchorages since they will bend or give under load or impact where other materials may break. The ease with which they can be worked and their economy makes them popular for trestle construction and for temporary structures such as falsework or centering. Where timber piles can be driven and cut off below the permanent ground-water level, they will last indefinitely; but above this level in the soil, a timber pile will rot or will be attacked by insects and eventually destroyed. In sea water, marine borers and fungus will act to deteriorate timber piles. Treatment of timber piles increases their life but does not protect them indefinitely.

Concrete Piles. Concrete piles are of two general types, precast and cast-in-place. The advantages in the use of concrete piles are that they can be fabricated to meet the most exacting conditions of design, can be cast in any desired shape or length, possess high strength and have excellent resistance to chemical and biological attack. Certain disadvantages are encountered in the use of precast piles, such as:

(a) Their heavy weight and bulk (which introduces problems in handling and driving).
(b) Problems with hair cracks which often develop in the concrete as a result of shrinkage after curing (which may expose the steel reinforcement to deterioration).
(c) Difficulty encountered in cut-off or splicing.
(d) Susceptibility to damage or breakage in handling and driving.
(e) They are more expensive to fabricate, transport and drive.

Precast piles are fabricated in casting yards. Centrifugally spun piles (or piles with square or octagonal cross-sections) are cast in horizontal forms, while round piles are usually cast in vertical forms. With the exception of relatively short lengths, precast piles must be reinforced to provide the designed column strengths and to resist damage or breakage while being transported or driven.

Precast piles can be tapered or have parallel sides. The reinforcement can be of deformed bars or be prestressed or poststressed with high strength steel tendons. Prestressing or prestressing eliminates the problem of open shrinkage cracks in the concrete. Otherwise, the pile must be protected by coating it with a bituminous or plastic material to prevent ultimate deterioration of the reinforcement. Proper curing of the precast concrete in piles is essential.

Cast-in-place pile types are numerous and vary according to the manufacturer of the shell or inventor of the method. In general, they can be classified into two groups: shell-less types and the shell types. The shell-less type is constructed by driving a steel shell into the ground and filling it with concrete as the shell is pulled from the ground. The shell type is constructed by driving a steel shell into the ground and filling it in place with concrete. Some of the advantages of cast-in-place concrete piles are: lightweight shells are handled and driven easily, lengths of the shell may be increased or decreased easily, shells may be transported in short lengths and quickly assembled, the problem of breakage is eliminated and a driven shell may be inspected for shell damage or an uncased hole for "pinching off." Among the disadvantages are problems encountered in the proper centering of the reinforcement cages, in placing and consolidating the concrete without displacement of the reinforcement steel or segregation of the concrete, and shell damage or "pinching-off" of uncased holes.

Shell type piles are fabricated of heavy gage metal or are fluted, corrugated or spirally reinforced with heavy wire to make them strong enough to be driven without a mandrel.

Other thin-shell types are driven with a collapsible steel mandrel or core inside the casing. In addition to making the driving of a long thin shell possible, the mandrel prevents or minimizes damage to the shell from tearing, buckling, collapsing or from hard objects encountered in driving.

Some shell type piles are fabricated of heavy gauge metal with enlargement at the lower end to increase the end bearing.

These enlargements are formed by withdrawing the casing two to three feet after placing concrete in the lower end of the shell. This wet concrete is then struck by a blow of the pile hammer on a core in the casing and the enlargement is formed. As the shell is withdrawn, the core is used to consolidate the concrete after each batch is placed in the shell. The procedure results in completely filling the hole left by the withdrawal of the shell.

Steel Piles. A steel pile is any pile fabricated entirely of steel. They are usually formed of rolled steel H sections, but heavy steel pipe or box piles (fabricated from sections of steel sheet piles welded together) are also used. The advantages of steel piles are that they are readily available, have a thin uniform section and high strength, will take hard driving, will develop high load-bearing values, are easily cut off or extended, are easily adapted to the structure they are to support, and breakage is eliminated. Some disadvantages are: they will rust and deteriorate unless protected from the elements; acid, soils or water will result in corrosion of the pile; and greater lengths may be required than for other types of piles to achieve the same bearing value unless bearing on rock strata. Pipe pile can either be driven open-end or closed-end and can be unfilled, sand filled or concrete filled. After open-end pipe piles are driven, the material from inside can be removed by an earth auger, air or water jets, or other means, inspected, and then filled with concrete. Concrete filled pipe piles are subject to corrosion on the outside surface only.

Composite Piles. Any pile that is fabricated of two or more materials is called a composite pile. There are three general classes of composite piles: wood with concrete, steel with concrete, and wood with steel. Composite piles are usually used for a special purpose or for reasons of economy.

Where a permanent ground-water table exists and a composite pile is to be used, it will generally be of concrete and wood. The wood portion is driven to below the water table level and the concrete upper portion eliminates problems of decay and insect infestation above the water table. Composite piles of steel and concrete are used where high bearing loads are desired or where driving in hard or rocky soils is expected. Composite wood and steel piles are relatively uncommon.

It is important that the pile design provides for a permanent joint between the two materials used, so constructed that the parts do not separate or shift out of axial alignment during driving operations.

Sheet Piles. Sheet piles are made from the same basic materials as other piling: wood, steel and concrete. They are ordinarily designed so as to interlock along the edges of adjacent piles.

Sheet piles are used where support of a vertical wall of earth is required, such as trench walls, bulkheads, waterfront structures or cofferdams. Wood sheet piling is generally used in temporary installations, but is seldom used where water-tightness is required or hard driving expected. Concrete sheet piling has the capability of resisting much larger lateral loads than wood sheet piling, but considerable difficulty is experienced in securing water-tight joints. The type referred to as "fishmouth" type is designed to permit jetting out the joint and filling with grout, but a seal is not always effected unless the adjacent piles are wedged tightly together. Concrete sheet piling has the advantage that it is the most permanent of all types of sheet piling.

Steel sheet piling is manufactured with a tension-type interlock along its edges. Several different shapes are available to permit versatility in its use. It has the advantages that it can take hard driving, has reasonably water-tight joints and can be easily cut, patched, lengthened or reinforced. It can also be easily extracted and reused. Its principal disadvantage is its vulnerability to corrosion.

Types of Pile Driving Hammers. A pile-driving hammer is used to drive load-bearing or sheet piles. The commonly used types are: drop, single-acting, double-acting, differential acting and diesel hammers. The most recent development is a type of hammer that utilizes high-frequency sound and a dead load as the principal sources of driving energy.

Drop Hammers. These hammers employ the principle of lifting a heavy weight by a cable and releasing it to fall on top of the pile. This type of hammer is rapidly disappearing from use, primarily because other types of pile driving hammers are more efficient. Its disadvantages are that it has a slow rate of driving (four to eight blows per minute), that there is some risk of damaging the pile from excessive impact, that damage may occur in adjacent structures from heavy vibration and that it cannot be used directly for driving piles under water. Drop hammers have the advantages of simplicity of operation, ability to vary the energy by changing the height of fall and they represent a small investment in equipment.

Single-Acting Hammers. These hammers can be operated either on steam or compressed air. The driving energy is provided by a free-falling weight (called a ram) which is raised after each stroke by the action of steam or air on a piston. They are manufactured as either open or closed types. Single-acting hammers are best suited for jobs where dense or elastic soil materials must be penetrated or where long heavy timber or precast concrete piles must be driven. The closed type can be used for underwater pile driving. Its advantages include: faster driving (50 blows or more per minute), reduction in skin friction as a result of more frequent blows, lower velocity of the ram which transmits a greater proportion of its energy to the pile and minimizes piles damage during driving, and it has underwater driving capability. Some of its disadvantages are: requires higher investment in equipment (i.e. steam boiler, air compressor, etc.), higher maintenance costs, greater set-up and moving time required, and a larger operating crew.

Double-Acting Hammers. These hammers are similar to the single-acting hammers except that steam or compressed air is used both to lift the ram and to impart energy to the falling ram. While the action is approximately twice as fast as the single-acting hammer (100 blows per minute or more), the ram is much lighter and operates at a greater velocity, thereby making it particularly useful in high production driving of light or medium-weight piles of moderate lengths in granular soils. The hammer is nearly always fully encased by a steel housing which also permits direct driving of piles under water.

Some of its advantages are: faster driving rate, less static skin friction develops between blows, has underwater driving capability and piles can be driven more easily without leads.

Among its disadvantages are: it is less suitable for driving heavy piles in high-friction soils and the more complicated mechanism results in higher maintenance costs.

Differential-Acting Hammers. This type of hammer is, in effect, a modified double-acting hammer with the actuating mechanism having two different diameters. A large-diameter piston operates in an upper cylinder to accelerate the ram on the downstroke and a small-diameter piston operates in a lower cylinder to raise the ram. The additional energy added to the falling ram is the difference in areas of the two pistons multiplied by the unit pressure of the steam or air used. This hammer is a short-stroke, fast-acting hammer with a cycle rate approximately that of the double-acting hammer.

Its advantages are that it has the speed and characteristics of the double-acting hammer with a ram weight comparable to the single-acting type, and it uses 25 to 35 percent less steam or air. It is also more suitable for driving heavy piles under more difficult driving conditions than the double-acting hammer. It is available in the open or closed-type cases, the latter permitting direct underwater pile driving. Its principal disadvantage is higher maintenance costs.

Diesel Hammers. This hammer is a self-contained driving unit which does not require an auxiliary steam boiler or air compressor. It consists essentially of a ram operating as a piston in a cylinder. When the ram is lifted and allowed to fall in the cylinder, diesel fuel is injected in the compression space between the ram and an anvil placed on top of the pile. The continued downstroke of the ram compresses the air and fuel to ignition heat and the resultant explosion drives the pile downward and the ram upward to start another cycle. This hammer is capable of driving at a rate of from 80 to 100 blows per minute. Its advantages are that it has a low equipment investment cost, is easily moved, requires a small crew, has a high driving rate, does not require a steam boiler or air compressor and can be used with or without leads for most work. Its disadvantages are that it is not self-starting (the ram must be mechanically lifted to start the action) and it does not deliver a uniform blow. The latter disadvantage arises from the fact that as the reaction of the pile to driving increases, the reaction to the ram increases correspondingly. That is, when the pile encounters considerable resistance, the rebound of the ram is higher and the energy is increased automatically. The operator is required to observe the driving operations closely to identify changing driving conditions and compensate for such changes with his controls to avoid damaging the pile.

Diesel hammers can be used on all types of piles and they are best suited to jobs where mobility or frequent relocation of the pile driving equipment is necessary.

BEARING PILES

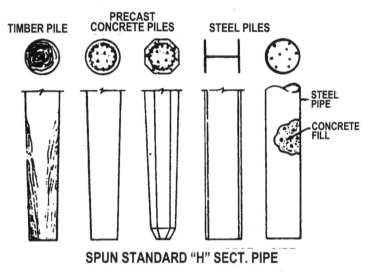

TIMBER PILE

PRECAST CONCRETE PILES

STEEL PILES

SPUN STANDARD "H" SECT. PIPE

STEEL PIPE

CONCRETE FILL

DRIVEN PILES

COMPOSITE WOOD PILE

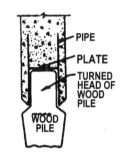

PIPE

PLATE

TURNED HEAD OF WOOD PILE

WOOD PILE

COMPOSITE STEEL PILE

REINF STEEL

WELD LONG. BARS TO PIPE OR H-PIPE

SHELL-TYPE PILES

UNCASED PILES

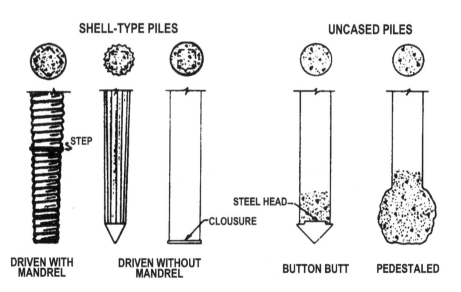

STEP

STEEL HEAD

CLOUSURE

DRIVEN WITH MANDREL

DRIVEN WITHOUT MANDREL

BUTTON BUTT

PEDESTALED

CAST-IN-PLACE PILES

STANDARD NOMENCLATURE
FOR STREET CONSTRUCTION

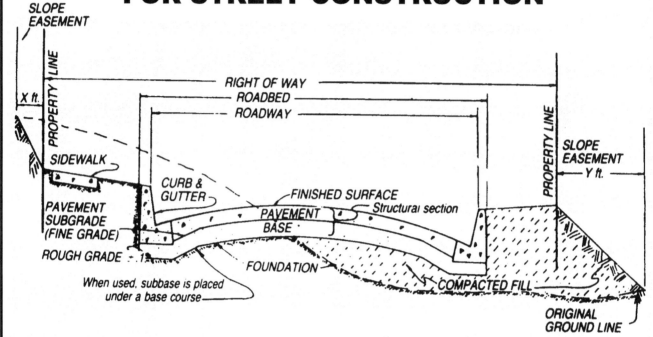

SLOPE EASEMENT

PROPERTY LINE

X ft.

RIGHT OF WAY

ROADBED

ROADWAY

SIDEWALK

CURB & GUTTER

FINISHED SURFACE

Structural section

PAVEMENT SUBGRADE (FINE GRADE)

PAVEMENT BASE

ROUGH GRADE

FOUNDATION

When used, subbase is placed under a base course

COMPACTED FILL

PROPERTY LINE

SLOPE EASEMENT

Y ft.

ORIGINAL GROUND LINE

CONCRETE MASONRY PAVING UNITS

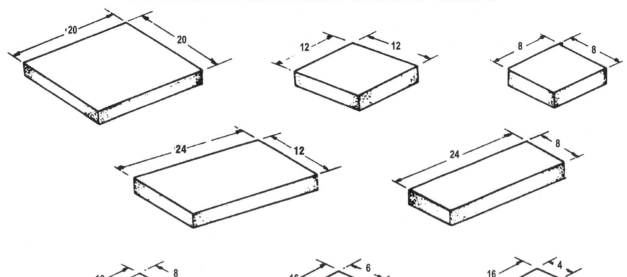

NOTE: Sizes are nominal and will vary by manufacture.

HEXAGON PAVER UNITS
Various Sizes Available

ROUND PAVING UNITS
Various Sizes Available

VEHICULAR PAVING UNITS

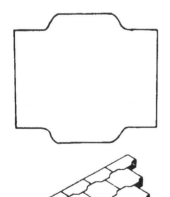

INTERLOCKING PAVER
7¼" x 3" x 8½"

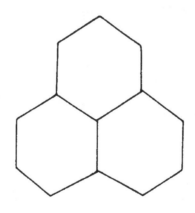

INTERLOCKING PAVER
12" x 3⅝" x 12"

TURF PAVER
24" x 3⅝" x 24"

CAPACITIES FOR SEPTIC TANKS SERVING
AN INDIVIDUAL DWELLING

No. of bedrooms	Capacity of tank (gals.)
2 or less	750
3	900
4	1,000

COMMON TYPES OF STEEL
REINFORCEMENT BARS

ASTM specifications for billet steel reinforcing bars (A 615) require identification marks to be rolled into the surface of one side of the bar to denote the producer's mill designation, bar size and type of steel. For Grade 60 and Grade 75 bars, grade marks indicating yield strength must be show. Grade 40 bars show only three marks (no grade mark) in the following order:

1st — Producing Mill (usually an initial)
2nd — Bar Size Number (#3 through # 18)
3rd — Type (N for New Billet)

NUMBER SYSTEM — GRADE MARKS

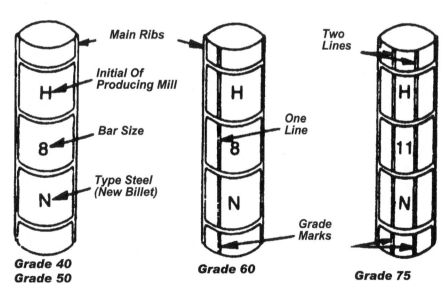

LINE SYSTEM — GRADE MARKS

STANDARD SIZES OF STEEL REINFORCEMENT BARS

STANDARD REINFORCEMENT BARS

Bar Designation Number*	Nominal Weight lb. per ft.	Nominal Dimensions		
		Diameter, in.	Cross Sectional Area, sq. in.	Perimeter, in.
3	0.376	0.375	0.11	1.178
4	0.668	0.500	0.20	1.571
5	1.043	0.625	0.31	1.963
6	1.502	0.750	0.44	2.356
7	2.044	0.875	0.60	2.749
8	2.670	1.000	0.79	3.142
9	3.400	1.128	1.00	3.544
10	4.303	1.270	1.27	3.990
11	5.313	1.410	1.56	4.430
14	7.65	1.693	2.25	5.32
18	13.60	2.257	4.00	7.09

*The bar numbers are based on the number of 1/8 inches included in the nominal diameter of the bar.

Type of Steel & ASTM Specification No.	Size Nos. Inclusive	Grade	Tensile Strength Min., psi.	Yield (a) Min., psi
Billet Steel A 615	3-11	40	70,000	40,000
	3-11 14, 18	60	90,000	60,000
	11, 14, 18	75	100,000	75,000

Style Designation	Steel Area sq. in per ft.		Weight Approx. lbs. per 100 sq. ft.
	Longit.	Transv.	
Rolls			
6×6 — W1.4×W1.4	.03	.03	21
6×8 — W2×W2	.04	.04	29
6×6 — W2.9×W2.9	.08	.06	42
6×6 — W4×W4	.08	.08	58
4×4 — W1.4×W1.4	.04	.04	31
4×4 — W2×W2	.06	.06	43
4×4 — W2.9×W2.9	.09	.09	62
4×4 — W4×W4	.12	.12	
Sheets			
6×6 — W2.9×W2.9	.06	.06	42
6×6 — W4×W4	.08	.08	58
6×6 — W5.5×W5.5	.11	.11	80
4×4 — W4×W4	.12	.12	86

Insofar as is possible, the moisture content should be kept uniform to avoid problems in determining the proper amount of water to be added for mixing. Mixing water must be reduced to compensate for moisture in the aggregate in order to control the slump of the concrete and avoid exceeding the specified water-cement ratio.

Handling Concrete by Pumping Methods. Transportation and placement of concrete by pumping is another method gaining increased popularity. Pumps have several advantages, the primary one being that a pump will high-lift concrete without the need for an expensive crane and bucket. Since the concrete is delivered through pipe and hoses, concrete can be conveyed to remote locations in buildings, in tunnels, to locations otherwise inaccessible on steep hillside slopes for anchor walls, pipe bedding or encasement, or for placing concrete for chain link fence post bases. Concrete pumps have been found to be economical and expedient in the placement of concrete, and this has promoted the use and acceptance of this development. The essence of proper concrete pumping is the placement of the concrete in its final location without segregation.

Modern concrete pumps, depending on the mix design and size of line, can pump to a height of 200 feet or a horizontal distance of 1,000 feet. They can handle, economically, structural mixes, standard mixes, low slump mixes, mixes with two-inch maximum size aggregate and light weight concrete. When a special pump mix is required for structural concrete in a major structure, the mix design must be approved by the Engineer and checked and confirmed by the Supervisor of the Materials Control Group. The Inspector should obtain the pump manufacturer's printed information and evaluate its characteristics and ability to handle the concrete mixture specified for the project.

If concrete is being placed for a major reinforced structure, it is important that the placement continue without interruption. The Inspector should be sure that the contractor has ready access to a back-up pump to be used in the event of a breakdown. In order to further insure the success of the concrete placement by the pumping method, the user should be aware of the following points:

(a) A protective grating over the receiving hopper of the pump is necessary to exclude large pieces of aggregate or foreign material.

(b) The pump and lines require lubrication with a grout of cement and water. All of the excess grout is to be wasted prior to pumping the concrete.

(c) All changes in direction must be made by a large radius bend with a maximum bend of 90 degrees. Wye connections induce segregation and shall not be used.

(d) Pump lines should be made of a material capable of resisting abrasion and with a smooth interior surface having a low coefficient of friction. Steel is commonly used for pump lines, because a chemical reaction occurs between the concrete and the aluminum. Aluminum pipe should not be used for pumping concrete and some of the new plastic or rubber tubing is gaining acceptance. Hydrogen is generated which results in a swelling of the concrete, causing a significant reduction in compressive strength. This reaction is aggravated by any of the following: abrasive coarse aggregate, non-uniformly graded sand, low-slump concrete, low sand-aggregate ratio, high-alkali cement or when no air-entraining agent is used.

(e) During temporary interruptions in pumping, the hopper must remain nearly full, with an occasional turning and pumping to avoid developing a hard slug of concrete in the lines.

(f) Excessive line pressures must be avoided. When this occurs, check these points as the probable cause: segregation caused by too low a slump or too high a slump; large particle contamination caused by large pieces of aggregate or frozen lumps not eliminated by the grating; poor gradation of aggregates or particle shape; rich or lean spots caused by improper mixing.

(g) Corrections must be made to correct excessive slump loss as measured at the transit-mixed concrete truck and as measured at the hose outlet. This may be attributable to porous aggregate, high temperature or rapid setting mixes.

(h) Two transit-mix concrete trucks must be used simultaneously to deliver concrete into the pump hopper. These trucks must be discharged alternately to assure a continuous flow of concrete as trucks are replaced.

(i) Samples of concrete for test specimens prepared to determine the acceptance of the concrete quality are to be taken as required for conventional concrete.

Sampling is done before the concrete is deposited in the pump hopper. However, it is suggested that, where possible, the effect of pumping on the compressive strength be checked by taking companion samples, so identified, from the end of the pump line at the same time. The Record of Test must be properly noted as being a special mix used for pumping purposes. This will enable the Materials Control Group to compile a complete history of mix designs and their respective compressive strengths.

The prudent use of pumped concrete can result in economy and improved quality. However, only the control exercised by the operator will assure continued high standards of quality concrete.

Pump lines must be properly fastened to supports to eliminate excessive vibration. Couplings must be easily and securely fastened in a manner that will prevent mortar leakage. It is preferable to use the flexible hose only at the discharge point. This hose must be moved in such a manner as to avoid kinks or sharp bends. The pump line should be protected from excessive heat during hot weather by water sprinkling or shade.

COMPRESSIVE STRENGTH FOR VARIOUS WATER-CEMENT RATIOS
(The strengths listed are based on the use of normal portland cement)

WATER/CEMENT RATIO		PROBABLE 28-DAY STRENGTH	
WEIGHT	**GALS./100#**	**PSI**	**MEGAPASCALS***
.40	4.8	5000	34
.45	5.4	4500	31
.50	6.0	4000	28
.55	6.6	3500	24
.60	7.2	3000	21
.65	7.8	2500	17
.70	8.4	2000	14

* International system equivalent.

APPROXIMATE CONTENT OF SAND, CEMENT AND WATER PER CUBIC YARD OF CONCRETE

Based on aggregates of average grading and physical characteristics in concrete mixes having a water-cement ratio (W/C) of about .65 by weight (or 7.8 gallons) per sack of cement; 3-in, slump; and a medium natural sand having a fineness modulus of about 2.75.

COURSE AGGREGATE MAX. SIZE	WATER		CEMENT	% SAND
	POUNDS	**GALLONS**		
3/8	385	46	590	57
1/2	365	44	560	50
3/4	340	41	525	43
1	325	39	500	39
1 1/2	300	36	460	37

It can be noted from the above chart that, for a given slump, the amount of mixing water increases as the size of the course aggregate decreases. The size of the course aggregate controls the sand content in the same way; that is, the amount of sand required in the mix increases as the size of the course aggregate decreases.

Other typical examples are contained in the pamphlet published by the Portland Cement Association entitled "Design and Control of Concrete Mixtures."

Effects of Temperature on Concrete. Concrete mixtures gain strength rapidly in the first few days after placement. While the rate of gain in strength diminishes, concrete continues to become stronger with time over a period of many years, so long as drying of the concrete is prevented. Its strength at 28 days is considered to be the compressive strength upon which the Engineer bases his calculations. The temperature of the atmosphere has a significant effect upon the development of strength in concrete. Lower temperatures retard and higher temperatures accelerate the gain in strength.

Most destructive of the natural forces is freezing and thawing action. While the concrete is still wet or moist, expansion of the water as it is converted into ice results in severe damage to the fresh concrete. In situations where freezing may be encountered, high early strength cement may be used. Also, the mixing water or the aggregate (or both) may be preheated before mixing. Covering the concrete, and using steam or salamanders to heat the concrete under the covering, will help prevent freezing. Air-entraining agents help to diminish the effects of freezing of fresh concrete as well as in subsequent freezing and thawing cycles throughout the life of the concrete.

Hot weather will present problems of a different nature in placing concrete. Concrete will set up faster and tend to shrink and crack at the surface. To minimize this problem, the concrete should be placed without delay after mixing. Avoid the use of accelerators (perhaps even use a retarding agent), dampen all subgrade and forms, protect the freshly placed concrete from hot dry winds, and provide for adequate curing. Crushed ice or chilled water can be used as part of the mixing water to reduce the temperature of the mix in extremely hot areas.

Admixture	Purpose	Effects on Concrete	Advantages	Disadvantages
Accelerator	Hasten setting.	Improves cement dispersion and increases early strength.	Permits earlier finishing, form removal, and use of the structure.	Increases shrinkage, decreases sulfate resistance, tends to clog mixing and handling equipment.
Air-Entraining Agent	Increase workability and reduce mixing water.	Reduces segregation, bleeding and increases freeze-thaw resistance. Increases strength	Increases workability and reduces finishing time.	Excess will reduce strength and increase slump. Bulks concrete volume.
Bonding Agent	Increase bond to old concrete.	Produces a non-dusting, slip resistant finish,	Permits a thin topping without roughening old concrete, self-curing, ready in one day.	Quick setting and susceptible to damage from fats, oils and solvents.
Densifier	To obtain dense concrete.	Increased workability and strength.	Increases workability and increases water-proofing characteristics, more impermeable.	Care must be used to reduce mixing water in proportion to amount used.
Foaming Agent	Reduce weight.	Increases insulating properties.	Produces a more plastic mix, reduces dead weight loads.	Its use must be very carefully regulated — following instructions explicitly.
Retarder	Retard setting.	Increases control of setting.	Provides more time to work and finish concrete.	Performance varies with cement used — adds to slump. Requires stronger forms.
Water Reducer and Retarder	Increase compressive and flexural strength.	Reduces segregation, bleeding, absorption, shrinkage, and increases cement dispersion.	Easier to place work, provides better control.	Performance varies with cement. Of no use in cold weather.
Water Reducer, Retarder and Air-Entraining Agent	Increases workability.	Improves cohesiveness. Reduces bleeding and segregation.	Easier to place and work.	Care must be taken to avoid excessive air entrainment.

ARCHITECTURAL WALL PATTERNS (BONDS)

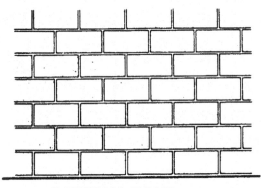

COMMON BOND

8" x 16" UNITS

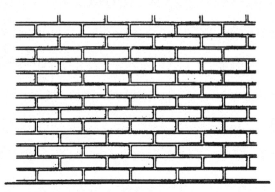

COMMON BOND

4" x 16" UNITS

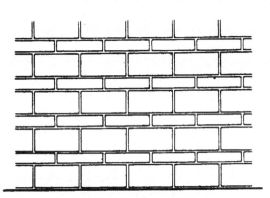

COURSED ASHLER

8" x 16" & 4" x 16" UNITS

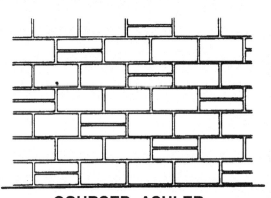

COURSED ASHLER

8" x 16" & 4" x 16" UNITS

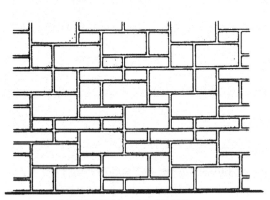

RANDOM ASHLER

8" x 16" & 4" x 16" UNITS
AND 4" x 8" UNITS

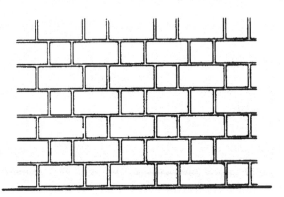

COURSED ASHLER

8" x 16" & 8" x 8" UNITS

ARCHITECTURAL WALL PATTERNS (BONDS) — (Continued)

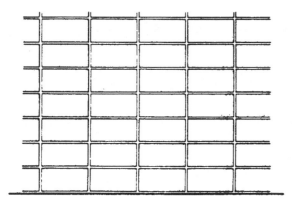

STACKED BOND
8" x 16" UNITS

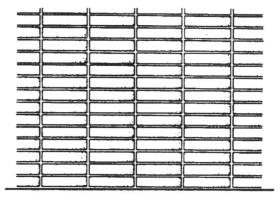

STACKED BOND
8" x 16" & 4" x 16" UNITS

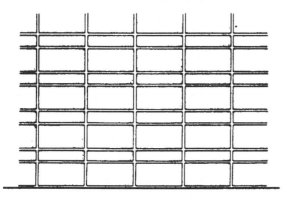

STACKED BOND
8" x 16" & 4" x 16" UNITS

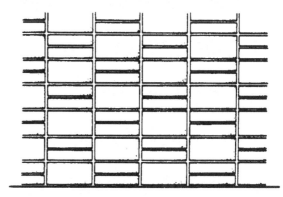

STACKED BOND
8" x 16" & 4" x 16" UNITS

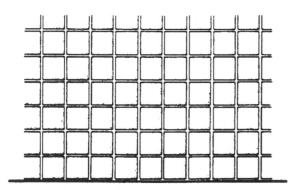

**STACKED BOND VERTICAL
SCORED UNITS**

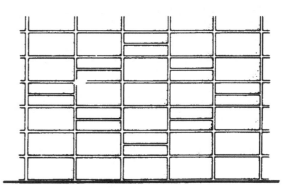

**USE OF BLOCK DESIGN
IN STACKED BOND**

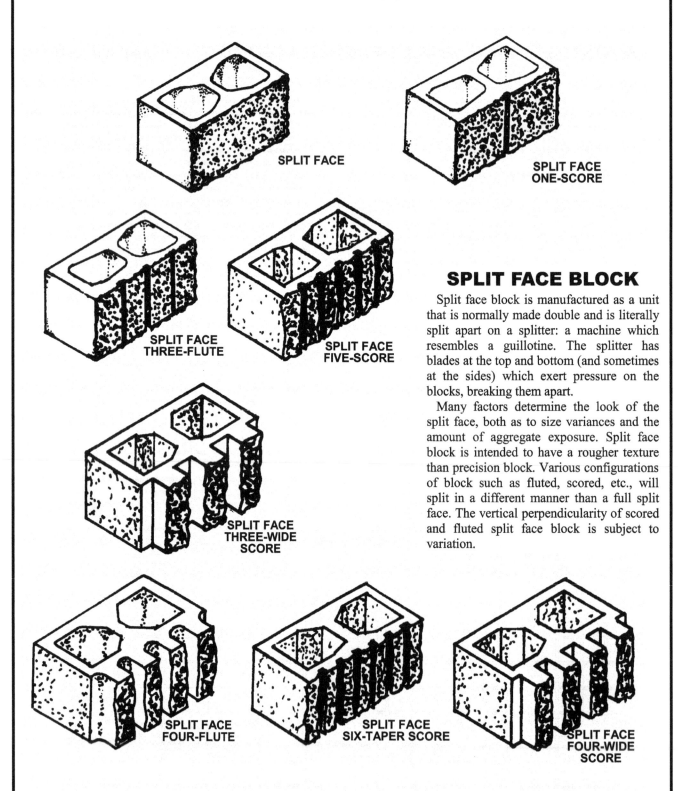

SPLIT FACE

SPLIT FACE
ONE-SCORE

SPLIT FACE
THREE-FLUTE

SPLIT FACE
FIVE-SCORE

SPLIT FACE
THREE-WIDE
SCORE

SPLIT FACE
FOUR-FLUTE

SPLIT FACE
SIX-TAPER SCORE

SPLIT FACE
FOUR-WIDE
SCORE

SPLIT FACE BLOCK

Split face block is manufactured as a unit that is normally made double and is literally split apart on a splitter: a machine which resembles a guillotine. The splitter has blades at the top and bottom (and sometimes at the sides) which exert pressure on the blocks, breaking them apart.

Many factors determine the look of the split face, both as to size variances and the amount of aggregate exposure. Split face block is intended to have a rougher texture than precision block. Various configurations of block such as fluted, scored, etc., will split in a different manner than a full split face. The vertical perpendicularity of scored and fluted split face block is subject to variation.

NOTE: Split face units shown in this manual are a small sampling of the broad range of concrete masonry architectural units available from the industry on special order. Depths and widths of scores vary. Consult a local manufacturer for specific information.

TYPICAL DETAILS — LINTELS AND BOND BEAMS

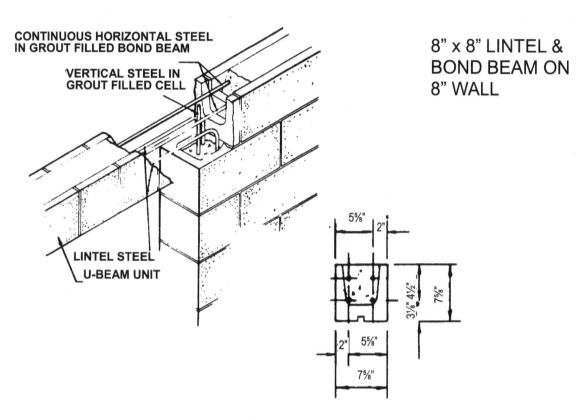

CONTINUOUS HORIZONTAL STEEL
IN GROUT FILLED BOND BEAM

VERTICAL STEEL IN
GROUT FILLED CELL

LINTEL STEEL
U-BEAM UNIT

8" x 8" LINTEL &
BOND BEAM ON
8" WALL

5⅝" 2"

3⅛" 4½" 7⅝"

2" 5⅝"

7⅝"

8" x 16"
BOND BEAM ON
8" WALL

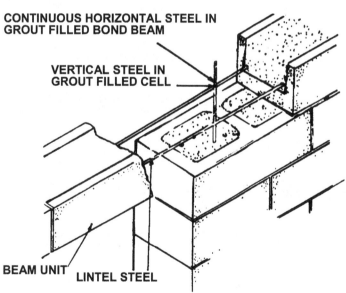

CONTINUOUS HORIZONTAL STEEL IN
GROUT FILLED BOND BEAM

VERTICAL STEEL IN
GROUT FILLED CELL

BEAM UNIT   LINTEL STEEL

BOLTS IN COMMON USAGE

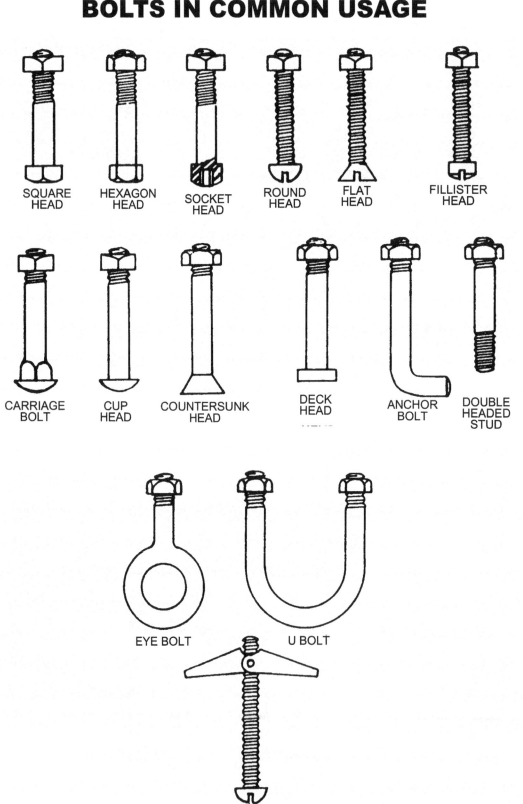

SQUARE HEAD

HEXAGON HEAD

SOCKET HEAD

ROUND HEAD

FLAT HEAD

FILLISTER HEAD

CARRIAGE BOLT

CUP HEAD

COUNTERSUNK HEAD

DECK HEAD

ANCHOR BOLT

DOUBLE HEADED STUD

EYE BOLT

U BOLT

TOGGLE BOLT

COMMON WIRE NAILS (ACTUAL SIZE)

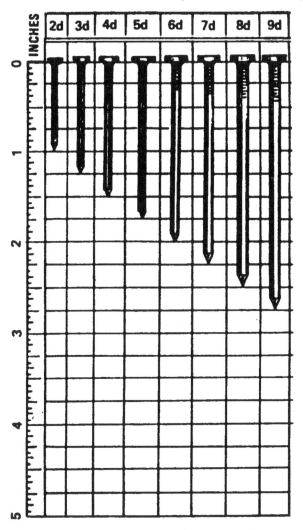

Cut Nails. Cut nails are angular-sided, wedge-shaped with a blunt point.

Wire Nails. Wire nails are round shafted, straight, pointed nails, and are used more generally than cut nails. They are stronger than cut nails and do not buckle as easily when driven into hard wood, but usually split wood more easily than cut nails. Wire nails are available in a variety of sizes varying from two penny to sixty penny.

Nail Finishes. Nails are available with special finishes. Some are galvanized or cadmium plated to resist rust. To increase the resistance to withdrawal, nails are coated with resins or asphalt cement (called cement coated). Nails which are small, sharp-pointed, and often placed in the craftsman's mouth (such as lath or plaster board nails) are generally blued and sterilized.

COMMON WIRE NAILS (ACTUAL SIZE) (Cont.)

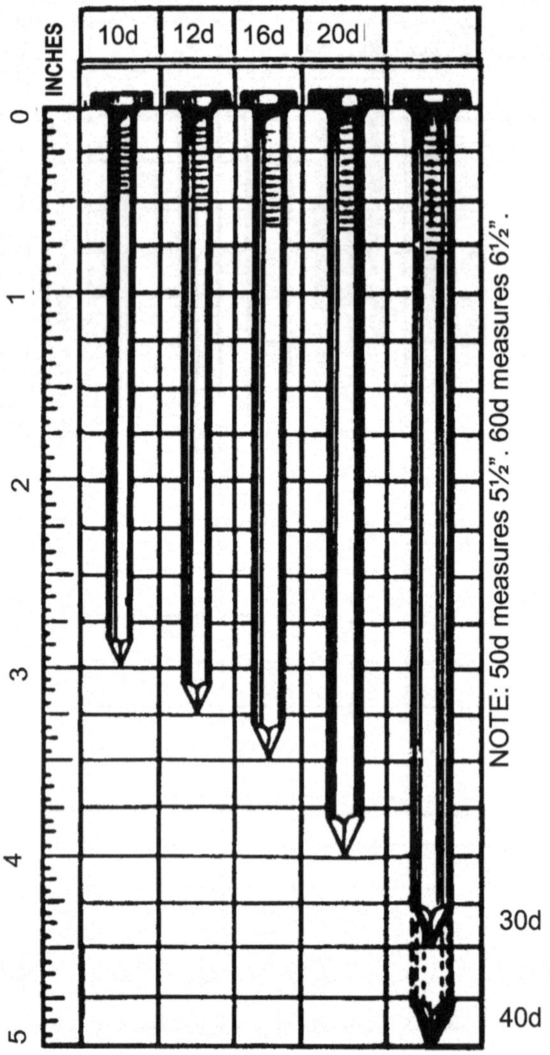

INCHES

10d	12d	16d	20d

NOTE: 50d measures 5½". 60d measures 6½".

30d

40d

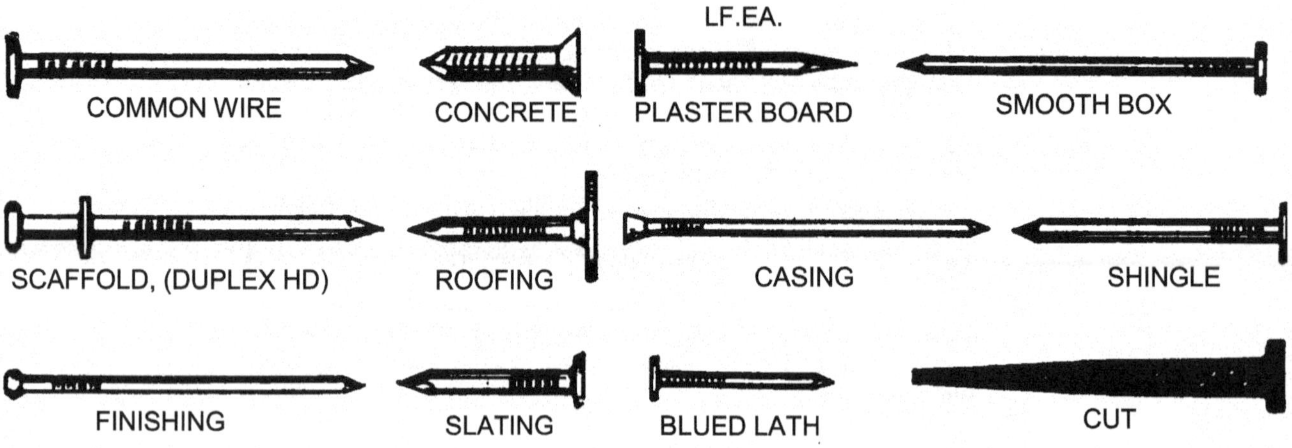

COMMON WIRE

CONCRETE

LF.EA.
PLASTER BOARD

SMOOTH BOX

SCAFFOLD, (DUPLEX HD)

ROOFING

CASING

SHINGLE

FINISHING

SLATING

BLUED LATH

CUT

PLYWOOD — BASIC GRADE MARKS
AMERICAN PLYWOOD ASSOCIATION (APA)

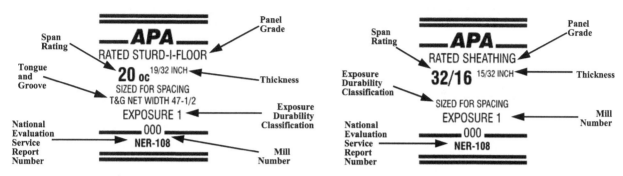

The American Plywood Association's trademarks appear only on products manufactured by APA member mills. The marks signify that the product is manufactured in conformance with APA performance standards and/or U.S. Product Standard PS 1-83 for Construction and Industrial Plywood.

APA A-C
For use where appearance of one side is important in exterior applications such as soffits, fences, structural uses, boxcar and truck linings, farm buildings, tanks, trays, commercial refrigerators, etc. **Exposure Durability Classification: Exterior. Common Thicknesses:** ¼, 11/32, ¾, 15/32, ½, 19/32, 5/8, 23/32, ¾.

APA A-D
For use where appearance of only one side is important in interior applications, such as paneling, built-ins, shelving, partitions, flow racks, etc. **Exposure Durability Classifications: Interior, Exposure 1. Common Thicknesses:** ¼, 11/32, 3/8, 15/32, ½, 19/32, 5/8, 23/32, ¾.

APA B-C
Utility panel for farm service and work buildings, boxcar and truck linings, containers, tanks, agricultural equipment, as a base for exterior coatings and other exterior uses. **Exposure Durability Classification: Exterior. Common Thicknesses:** ¼, 11/32, ¾, 15/32, ½, 19/32, 5/8, 23/32, ¾.

APA B-D
Utility panel for backing, sides or built-ins, industry shelving, slip sheets, separator boards, bins and other interior or protected applications. **Exposure Durability Classifications: Interior, Exposure 1. Common Thicknesses:** ¼, 11/32, 3/8, 15/32, ½, 19/32, 5/8, 23/32, ¾.

APA proprietary concrete form panels designed for high reuse. Sanded both sides and mill-oiled unless otherwise specified. Class I, the strongest, stiffest and more commonly available, is limited to Group 1 faces, Group 1 or 2 crossbands, and Group 1, 2, 3 or 4 inner plies. Class II is limited to Group 1 or 2 faces (Group 3 under certain conditions) and Group 1, 2, 3 or 4 inner plies. Also available in HDO for very smooth concrete finish, in Structural I, and with special overlays. **Exposure Durability Classification: Exterior. Common Thicknesses:** 19/32, ¾, 23/32, ¼.

Plywood panel manufactured with smooth, opaque, resin-treated fiber overlay providing ideal base for paint on one or both sides. Excellent material choice for shelving, factory work surfaces, paneling, built-ins, signs and numerous other construction and industrial applications. Also available as a 303 Siding with texture-embossed or smooth surface on one side only and Structural I. **Exposure Durability Classification: Exterior. Common Thicknesses:** 11/32, 3/8, 19/32, ½, 19/32, 5/8, 23/32, ¾.

SPECIALTY PANELS

HDO · A · A · G-1 · EXT – APA · 000 · PS1 – 83

Plywood panel manufactured with a hard, semi-opaque resin-fiber overlay on both sides. Extremely abrasion resistant and ideally suited to scores of punishing construction and industrial applications, such as concrete forms, industrial tanks, work surfaces, signs, agricultural bins, exhaust ducts, etc. Also available with skid-resistant screen-grid surface and in Structural I. *Exposure Durability Classification:* **Exterior.** *Common Thicknesses:* **3/8, ½, 5/8, 3/4**

MARINE· A · A · EXT – APA · 000 · PS1 – 83

Specialty designed plywood panel made only with Douglas fir or western larch, solid jointed cores, and highly restrictive limitations on core gaps and faces repairs. Ideal for both hulls and other marine applications. Also available with HDO or MDO faces. *Exposure Durability Classification:* **Exterior.** *Common Thicknesses:* **1/4, 3/8, ½, 5/8, 3/4.**

Unsanded and touch-sanded panels, and panels with "B" or better veneer on one side only, usually carry the APA trademark on the panel back. Panels with both sides of "B" or better veneer, or with special overlaid surfaces (such as Medium Density Overlay), carry the APA trademark on the panel edge, like this:

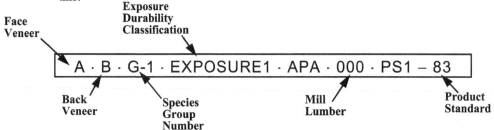

GLOSSARY OF TERMS
Some of the words and terms used in the grading of lumber follow:

Bow. A deviation flatwise from a straight line drawn from end to end of the piece. It is measured at the point of greatest distance from the straight line.

Checks. A separation of the wood which normally occurs across the annual rings and usually as a result of seasoning.

Crook. A deviation edgewise from a straight line drawn from end to end of the piece. It is measured at the point of greatest distance from the straight line.

Cup. A deviation from a straight line drawn across the piece from edge to edge. It is measured at the point of greatest distance from the straight line.

Flat Grain. The annual growth rings pass through the piece at an angle of less than 45 degrees with the flat surface of the piece.

Warp. Any deviation from a true or plane surface, including crook, cup, bow or any combination thereof.

Mixed Grain. The piece may have vertical grain, flat grain or a combination of both vertical and flat grain.

Pitch. An accumulation of resin which occurs in separations in the wood or in the wood cells themselves.

Shake. A separation of the wood which usually occurs between the rings of annual growth.

Splits. A separation of the wood due to tearing apart of the wood cells.

Vertical Grain. The annual growth rings pass through the piece at an angle of 45 degrees or more with the flat surface of the piece.

Wane. Bark or lack of wood from any cause, except eased edges (rounded) on the edge or corner of a piece of lumber.

LUMBER GRADING
GRADING-MARK ABBREVIATIONS

GRADES
(Listed alphabetically — not by quality)

COM	Common
CONST	Construction
ECON	Economy
No. 1	Number One
SEL-MER	Select Merchantable
SEL-STR	Select Structural
STAN	Standard
UTIL	Utility

ALSC TRADEMARKS

CLIS	California Lumber Inspection Service
NELMA	Northeastern Lumber Mfrs. Assoc., Inc.
NH&PMA	Northern Hardwood & Pine Mfrs. Assoc., Inc.
PLIB	Pacific Lumber Inspection Bureau
RIS	Redwood Inspection Service
SPIB	Southern Pine Inspection Bureau
TP	Timber Products Inspection
WCLB	West Coast Lumber Inspection Bureau
WWP	Western Wood Products Association

SPECIES GROUPINGS

AF	Alpine Fir
DF	Douglas Fir
HF	Hem Fir
SP	Sugar Pine
PP	Ponderosa Pipe
LP	Lodgepole Pine
IWP	Idaho White Pine
ES	Engelmann Spruce
WRC	Western Red Cedar
INC CDR	Incense Cedar
L	Larch
LP	Lodgepole Pine
MH	Mountain Hemlock
WW	White Wood

MOISTURE CONTENT

S-GRN	Surfaced at a moisture content of more than 19%.
S-DRY	Surfaced at a moisture content of 19% or less.
MC-15	Surfaced at a moisture content of 15% or less.

FRAMING ESTIMATING RULES OF THUMB

For 16" O.C. stud partitions figure 1 stud for every L.F. of wall; add for top and bottom plates.

For any type of framing, the quantity of basic framing members (in L.F.) can be determined based on spacing and surface area (S.F.):

12" O.C.	1.2 L.F./S.F.
16" O.C.	1.0 L.F./S.F.
24" O.C.	0.8 L.F./S.F.

(Doubled-up members, bands, plates, framed openings, etc., must be added.)

Framing accessories, nails, joist hangers, connectors, etc., should be estimated as separate material costs. Installation should be included with framing. Rule of thumb allowance is 0.5 to 1.5% of lumber cost for rough hardware. Another is 30 to 40 pounds of nails per M.B.F.

BOARD FEET/LINEAR FEET FOR LUMBER

Nominal Size	Actual Size	Board Feet Per Linear Foot	Linear Feet Per 1000 Board Feet
1 x 2	¾ x 1 ½	.167	6000
1 x 3	¾ x 2 ½	.250	4000
1 x 4	¾ x 3 ½	.333	3000
1 x 6	¾ x 5 ½	.500	2000
1 x 8	¾ x 7 ¼	.666	1500
1 x 10	¾ x 9 ¼	.833	1200
1 x 12	¾ x 11 ¼	1.0	1000
2 x 2	1 ½ x 1 ½	.333	3000
2 x 3	1 ½ x 2 ½	.500	2000
2 x 4	1 ½ x 3 ½	.666	1500
2 x 6	1 ½ x 5 ½	1.0	1000
2 x 8	1 ½ x 7 ¼	1.333	750
2 x 10	1 ½ x 9 ¼	1.666	600
2 x 12	1 ½ x 11 ¼	2.0	500

Redwood. Redwood is a fairly strong and moderately lightweight material. The heartwood is red but the sapwood is white. One of the principal advantages of redwood is that the heartwood is highly resistant (but not entirely immune) to decay, fungus and insects. Standard Specifications require that all redwood used in permanent installations shall be "select heart." Grade marking shall be in accordance with the standards established in the California Redwood Association. Grade marking shall be done by, or under the supervision of the Redwood Inspection Service.

Redwood is graded for specific uses as indicated in the following table:

REDWOOD GRADING

Type of Lumber	Grade	Typical Use
Grades for Dimension Only Listed Here	Clear All Heart	Exceptionally fine, knot free, straight-grained timbers. This grade is used primarily for stain finish work of high quality.
	Clear	Same as Clear All Heart except that this grade may contain sound sapwood and medium stain.
	Select Heart	**This grade only is to be used in Agency work, unless otherwise specified in the plans or specifications.** It is sound, live heartwood free from splits or streaks with sound knots. It is generally used where the timber is in contact with the ground, as in posts, mudsills, etc.
	Select Construction Heart	Slightly less quality than Select Heart. It may have some sapwood in the piece. Used for general construction purposes when redwood is needed.
	Construction Common	Same requirement as Construction Heart except that it will contain sapwood and medium stain. Its resistance to decay and insect attack is reduced.
	Merchantable	Used for fence posts, garden stakes, etc.
	Economy	Suitable for crating, bracing and temporary construction.

DOUGLAS FIR GRADING

Type of Lumber	Grade	Typical Use
Select Structural Joists and Planks	Select Structural	Used where strength is the primary consideration, with appearance desirable.
	No. 1	Used where strength is less critical and appearance not a major consideration.
	No. 2	Used for framing elements that will be covered by subsequent construction.
	No. 3	Used for structural framing where strength is required but appearance is not a factor.
Finish Lumber	Superior	For all types of uses as casings, cabinet, exposed members, etc., where a fine appearance is desired.
	Prime	
	E	
Boards (WCLIB)* * Grading is by West Coast Lumber Inspection Bureau rules, but sizes conform to Western Wood Products Assn. rules. These boards are still manufactured by some mills.	Select Merchantable	Intended for use in housing and light construction where a knotty type of lumber with finest appearance is required.
	Construction	Used for sub-flooring, roof and wall sheathing, concrete forms, etc. Has a high degree of serviceability.
	Standard	Used widely for general construction purposes, including subfloors, roof and wall sheathing, concrete forms, etc. Seldom used in exposed construction because appearance.
	Utility	Used in general construction where low cost is a factor and appearance is not important. (Storage shelving, crates, bracing, temporary scaffolding etc.)

BOARD FEET CONVERSION TABLE

Nominal Size (In.)	ACTUAL LENGTH IN FEET								
	8	10	12	14	16	18	20	22	24
1 x 2		1 2/3	2	2 1/3	2 2/3	3	3 ½	3 2/3	4
1 x 3		2 ½	3	3 ½	4	4 ½	5	5 ½	6
1 x 4	2 ¾	3 1/3	4	4 2/3	5 1/3	6	6 2/3	7 1/3	8
1 x 5		4 1/6	5	5 5/6	6 2/3	7 ½	8 1/3	9 1/6	10
1 x 6	4	5	6	7	8	9	10	11	12
1 x 7		5 5/8	7	8 1/6	9 1/3	10 ½	11 2/3	12 5/6	14
1 x 8	5 1/3	6 2/3	8	9 1/3	10 2/3	12	13 1/3	14 2/3	16
1 x 10	6 2/3	8 1/3	10	11 2/3	13 1/3	15	16 2/3	18 1/3	20
1 x 12	8	10	12	14	16	18	20	22	24
1¼ x 4		4 1/6	5	5 5/6	6 2/3	7 ½	8 1/3	9 1/6	10
1¼ x 6		6 ¼	7 ½	8 ¾	10	11 ¼	12 ½	13 ¾	15
1¼ x 8		8 1/3	10	11 2/3	13 1/3	15	16 2/3	18 1/3	20
1¼ x 10		10 5/12	12 ½	14 7/12	16 2/3	18 ¾	20 5/6	22 11/12	25
1¼ x 12		12 ½	15	17 ½	20	22 ½	25	27 ½	30
1½ x 4	4	5	6	7	8	9	10	11	12
1½ x 6	6	7 ½	9	10 ½	12	13 ½	15	16 ½	18
1½ x 8	8	10	12	14	16	18	20	22	24
1½ x 10	10	12 ½	15	17 ½	20	22 ½	25	27 ½	30
1½ x 12	12	15	18	21	24	27	30	33	36
2 x 4	5 1/3	6 2/3	8	9 1/3	10 1/3	12	13 1/3	14 2/3	16
2 x 6	8	10	12	14	16	18	20	22	24
2 x 8	10 2/3	13 1/3	16	18 2/3	21 1/3	24	26 2/3	29 1/3	32
2 x 10	13 1/3	16 2/3	20	23 1/3	26 2/3	30	33 1/3	36 2/3	40
2 x 12	16	20	24	28	32	36	40	44	48
3 x 6	12	15	18	21	24	27	30	33	36
3 x 8	16	20	24	28	32	36	40	44	48
3 x 10	20	25	30	35	40	45	50	55	60
3 x 12	24	30	36	42	48	54	60	66	72
4 x 4	10 2/3	13 1/3	16	18 2/3	21 1/3	24	26 2/3	29 1/3	32
4 x 6	16	20	24	28	32	36	40	44	48
4 x 8	21 1/3	26 2/3	32	37 1/3	42 2/3	48	53 1/3	58 2/3	64
4 x 10	26 2/3	33 1/3	40	46 2/3	53 1/3	60	66 2/3	73 1/3	80
4 x 12	32	40	48	56	64	72	80	88	96

SLOPE AREA CALCULATIONS

Rise and Run	Multiply Flat Area by	LF of Hips or Valleys per LF of Common Run
2 in 12	1.014	1.424
3 in 12	1.031	1.436
4 in 12	1.054	1.453
5 in 12	1.083	1.474
6 in 12	1.118	1.500
7 in 12	1.158	1.530
8 in 12	1.202	1.564
9 in 12	1.250	1.600
10 in 12	1.302	1.641
11 in 12	1.357	1.685
12 in 12	1.413	1.732

DOWNSPOUT/VERTICAL LEADER CALCULATIONS

Roof Type	Slope	S.F. Roof/ Sq. In. Leader
Gravel	Less than ¼" per foot	300
Gravel	Greater than ¼" per foot	250
Metal or Shingle	Any	200

Alternate calculations:

$$\text{Diameter of downspout/leader} = 1.128 \sqrt{\frac{\text{Area of drainage}}{\text{SF Roof/Sq. Inch}}}$$

TYPICAL MINIMUM SIZE OF VERTICAL CONDUCTORS AND LEADERS

Size of leader or conductor (Inches)	Maximum projected roof area (Square feet)
2	544
2 ½	987
3	1,610
4	3,460
5	6,280
6	10,200
8	22,000

TYPICAL MINIMUM SIZE OF ROOF GUTTERS

Diameter gutter (Inches)	MAXIMUM PROJECTED ROOF AREA FOR GUTTERS OF VARIOUS SLOPES			
	1/16 in. Ft. slope (Sq. ft.)	1/8 in. per Ft. slope (Sq. ft.)	¼ in. per Ft. slope (Sq. ft.)	1/2 in. per Ft. /slope (Sq. ft.)
3	170	240	340	480
4	360	510	720	1,020
5	625	880	1,250	1,770
6	960	1,360	1,920	2,770
7	1,380	1,950	2,760	3,900
8	1,990	2,800	3,980	5,600
10	3,600	5,100	7,200	10,000

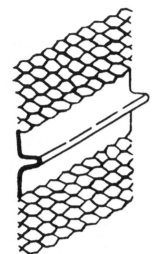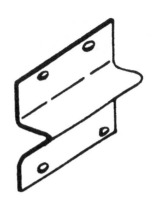

BASE OR PARTING SCREEDS

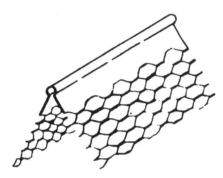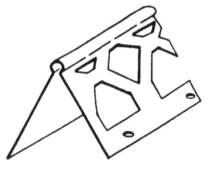

SMALL NOSE CORNER BEADS

WIRE BULL NOSE CORNER BEADS

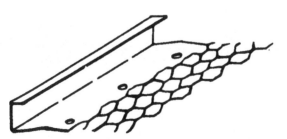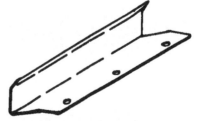

SQUARE CASING BEADS

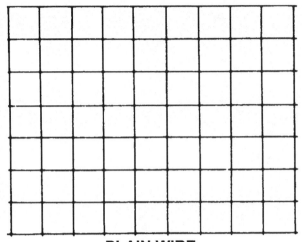

**PLAIN WIRE
FABRIC LATH**

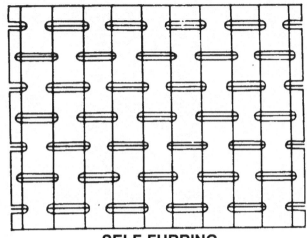

**SELF-FURRING
WIRE FABRIC LATH**

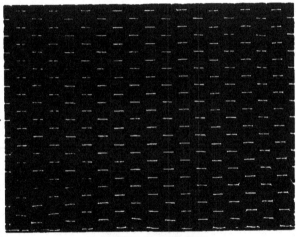

**PAPER BACKED
WOVEN WIRE
FABRIC LATH**

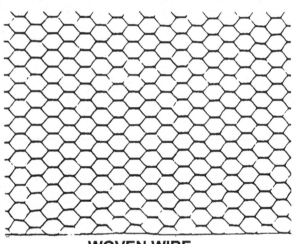

**WOVEN WIRE
FABRIC LATH**
(Also Available Self-Furred)

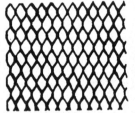

**FLAT
DIAMOND MESH
METAL LATH**

**SELF-
FURRING
METAL LATH**

**FLAT RIB
METAL LATH**

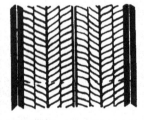

**RIB
METAL LATH**

**RIB
METAL LATH**

FINISHES / LATHING

MAXIMUM SPACING OF SUPPORTS FOR METAL LATH (Inches)

Type Of Lath	Weight of Lath Lb. Per Sq. Yd.	WALLS AND PARTITIONS			CEILINGS	
		Wood Studs	Solid Partitions	Steel Studs Wall Furring,	Wood or Concrete	Metal
Diamond Mesh (flat expanded)	2.5	16	16	13 1/2	12	12
	3.4	16	16	16	16	
Flat Rib	2.75	16	16	16	16	16
	3.4	19	24 (3)	19	19	19
3/8" Rib (1) (2)	3.4	24	(4)	24	24	24
	4.0	24	(4)	24	24	24
3/4" Rib	5.4	-	(4)	24 (5)	36 (6)	36 (6)
Sheet Lath	4.5	24	(4)	24	24	24

NOTE: Weights are exclusive of paper, fiber or other backing.
(1) 3.4 lb. 3/8" Rib Lath is permissible under Concrete Joists at 27" c.c.
(2) These spacings are based on a narrow bearing surface for the lath. When supports with a relatively wide bearing surface are used, these spacings may be increased accordingly, and still assure satisfactory work.
(3) This spacing permissible for Solid Partitions not exceeding 16' in height. For greater heights, permanent horizontal stiffener channels or rods must be provided on channel side of partitions, every 6' vertically, or else spacing shall be reduced 25%.
(4) For studless solid partitions, lath erected vertically.
(5) For interior wall furring or for application over solid surfaces for stucco.
(6) For contact or ceilings only.

TYPES OF LATH-ATTACHMENT TO WOOD AND METAL SUPPORTS

TYPE OF LATH	NAILS — Type & Size	NAILS MAX. SPACING Vertical (In.)	NAILS MAX. SPACING Horizontal (In.)	SCREWS MAX. SPACING Vertical (In.)	SCREWS MAX. SPACING Horizontal (In.)	STAPLES Wire Gauge No.	STAPLES Crown	STAPLES Leg	STAPLES MAX. SPACING Vertical (In.)	STAPLES MAX. SPACING Horizontal (In.)
1. Diamond Mesh Expanded Metal Lath and Flat Rib Metal Lath	4d blued smooth box 1 1/2 No. 14 gauge 7/32" head (clinched); 1" No.11 gauge 7/16" head, barbed; 1 1/2" No.11 gauge 7/16" head, barbed	6 / 6 / 6	– / – / 6	6	6	16	3/4	7/8	6	6
2. 3/8" Rib Metal Lath and Sheet Lath	1 1/2" No. 11 ga. 7/16" head, barbed	6	6	6	6	16	3/4	1 1/2	At Ribs	At Ribs
3. 3/4" Rib Metal Lath	4d common 1 1/2" No.12 1/2 gauge 1/4" head; 2" No.11 gauge 7/16" head, barbed	At Ribs	– / At Ribs	At Ribs	At Ribs	16	3/4	1 5/8	At Ribs	At Ribs
4. Wire Fabric Lath	4d blued smooth box (clinched); 1" No.11 gauge 7/16" head, barbed; 1 1/2" No.11 gauge 7/16" head, barbed; 1 1/4" No. 12 ga. 3/8" head, furring; 1" No.12 gauge 3/8" head	6 / 6 / 6 / 6 / 6	– / – / 6 / 6	6	6	16	34 / 7/16	7/8 / 7/8	6	6
5. 3/8" Gypsum Lath	1 1/8" No.13 gauge 12/61" head, blued	8	8	8	8	16	3/4	7/8	8	8
6. 1/2" Gypsum Lath	1 1/4" No.13 gauge 12/61" head, blued	8	8 / 6	8	8 / 6	16	3/4	1 1/8	8	8 / 6

STRESS RELIEF (CONTROL JOINTS)

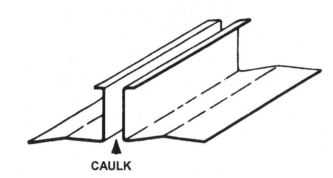

CAULK

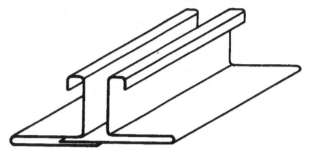

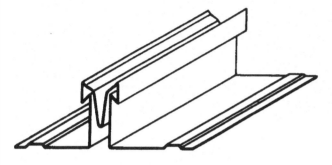

REVEALS

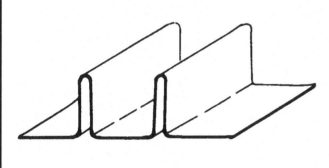

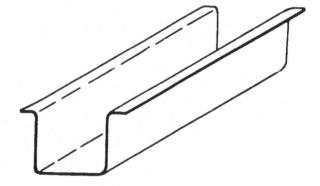

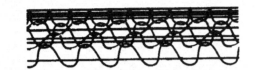

**CORNER REINFORCEMENT
(EXTERIOR) WIRE**

**CORNER REINFORCEMENT
(EXTERIOR) EXPANDED METAL**

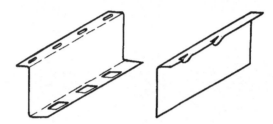

**PARTITION RUNNERS
(Z AND L SHAPE)**

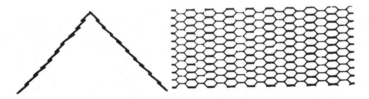

EXPANDED METAL CORNERITE

WIRE CORNERITE

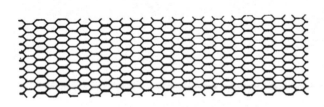

**STRIP REINFORCEMENT
(EXPANDED METAL)**

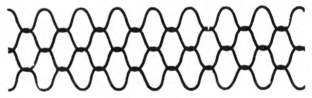

**STRIP REINFORCEMENT
(WIRE)**

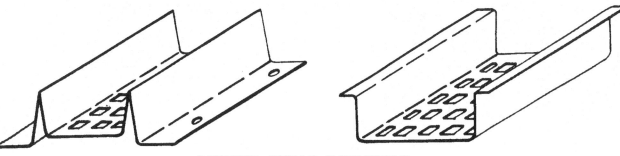

VENTILATING SCREEDS

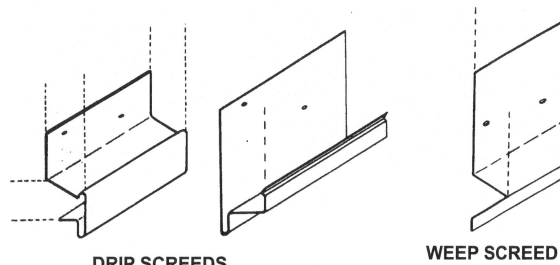

DRIP SCREEDS

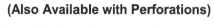

WEEP SCREED
(Also Available with Perforations)

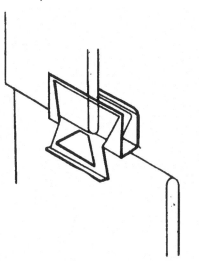

GYPSUM LATH ATTACHMENTS CLIPS

VERTICAL FURRING

VERTICAL FURRING MEMBER	UNBRACED				BRACED			
	STUD SPACING				STUD SPACING			
	24"	19"	16"	12"	24"	19"	16"	12"
	Maximum Furring Heights				Maximum Vertical Distance Between Braces			
3/4" Channel	6'	7'	8'	9'	5'	5'	6'	7'
1 1/2" Channel	8'	9'	10'	12'	6'	7'	8'	9'
2" Channel	9'	10'	11'	13'	7'	8'	9'	10'
2" Prefab. Stud	8'	9'	10'	11'	6'	7'	8'	9'
2 1/2" Prefab. Stud	10'	11'	12'	14'	8'	9'	10'	11'
3 1/4" Prefab.Stud	14'	16'	17'	20'	11'	13'	14'	16'

TYPES OF LATH—MAXIMUM SPACING OF SUPPORTS

TYPE OF LATH	Minimum Weight (psy), Gauge & Mesh Size	VERTICAL			HORIZONTAL	
		WOOD	METAL		Wood or Concrete	Metal
			Solid Plaster Partitions	Other		
Expanded Metal Lath (Diamond Mesh)	2.5 / 3.4	16" / 16"	16" / 16"	12" / 16"	12" / 16"	12" / 16"
Flat rib Expanded Metal Lath	2.75 / 3.4	16" / 19"	16" / 24"	16" / 19"	16" / 19"	16" / 19"
Stucco Mesh Expanded Metal Lath	1.8 and 3.6	16"	-	-	-	-
3/8" Rib Expanded Metal Lath	3.4 / 4.0	24" / 24"	-	24" / 24"	24" / 24"	24" / 24"
Sheet Lath	4.5	24"	-	24"	24"	24"
3/4" Rib Expanded Metal Lath (Not manufactured in West)	5.4	-	-	-	36"	36"
Wire Fabric Lath — Welded	1.95 lbs.,11 ga.,2"x2" / 1.4 lbs.,16 ga.,2"x2" / 1.4 lbs.,18 ga.,1"x1"	24" / 16" / 16"	24" / 16" / -	24" / 16" / -	24" / 16" / -	24" / 16" / -
Wire Fabric Lath — Woven	1.4 lbs.,17 ga.,1 1/2" Hex. / 1.4 lbs.,18 ga.,1" Hex.	24" / 24"	16" / 16"	16" / 16"	24" / 24"	16" / 16"
3/8" Gypsum Lath (plain)	-	16"	-	16"	16"	16"
(Large Size)	-	16"	-	16"	16"	16"
1/2" Gypsum Lath (plain)	-	24"	-	24"	24"	24"
(Large Size)	-	24"	No supports; Erected vertically	24"	24"	16"
5/8" Gypsum Lath (Large Size)	-	24"	No supports; Erected vertically	24"	24"	16"

PLASTERING TABLES

THICKNESS OF PLASTER

PLASTER	FINISHED THICKNESS OF PLASTER FROM FACE OF LATH, MASONRY, CONCRETE	
	Gypsum Plaster	Portland Cement Plaster
Expanded Metal Lath	5/8" minimum	5/8" minimum
Wire Fabric Lath	5/8" minimum	3/4" minimum (interior)
		7/8" minimum (exterior)
Gypsum Lath	1/2" minimum	
Gypsum Veneer Base	1/16" minimum	1/2" minimum
Masonry Walls	1/2" minimum	7/8" maximum
Monolithic Concrete Walls	5/8" maximum	1/2" maximum
Monolithic Concrete Ceilings	3/8" maximum	

GYPSUM PLASTER PROPORTIONS

NUMBER OF COATS	COAT	PLASTER BASE OR LATH	MAXIMUM VOLUME AGGREGATE PER 100# NEAT PLASTER (CUBIC FEET)	
			Damp Loose Sand	Perlite or Vermiculite
Two-Coat Work	Basecoat	Gypsum Lath	2 1/2	2 1/2
	Basecoat	Masonry	3	3
Three-Coat Work	First Coat	Lath	2	2
	Second Coat	Lath	3	3
	First & Second Coat	Masonry	3	3

PORTLAND CEMENT PLASTER

COAT	VOLUME CEMENT	MAXIMUM WEIGHT (OR VOLUME) LIME PER VOLUME CEMENT	MAXIMUM VOLUME SAND PER VOLUME CEMENT	APPROXIMATE MINIMUM THICKNESS	MINIMUM PERIOD MOIST CURING	MINIMUM INTERVAL BETWEEN COATS
First	1	20 lbs.	4	3/8"	48 Hours	48 Hours
Second	1	20 lbs.	5	1st & 2nd Coats total 3/4"	48 Hours	7 Days
Finish	1	1	3	1st, 2nd & Finish Coats total 7/8"	-	

PORTLAND CEMENT - LIME PLASTER

COAT	VOLUME CEMENT	MAXIMUM WEIGHT (OR VOLUME) LIME PER VOLUME CEMENT	MAXIMUM VOLUME SAND PER VOLUME CEMENT	APPROXIMATE MINIMUM THICKNESS	MINIMUM PERIOD MOIST CURING	MINIMUM INTERVAL BETWEEN COATS
First	1	1	4	3/8"	48 Hours	48 Hours
Second	1	1	4 1/2	1st & 2nd Coats total 3/4"	48 Hours	7 Days
Finish	1	1	3	1st, 2nd & Finish Coats total 7/8"	-	

METAL STUD CONSTRUCTION

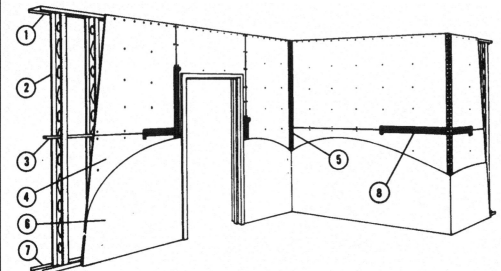

(1) Ceiling Runner Track
(2) Metal Stud (nailable or screw)
(3) Horizontal Stiffener
(4) Large Size Lath
(5) Angle Reinforcement
(6) Veneer Plaster 1/16 to 1/8 inch thick)
(7) Floor Runner Track
(8) Joint Reinforcement

WOOD STUD CONSTRUCTION

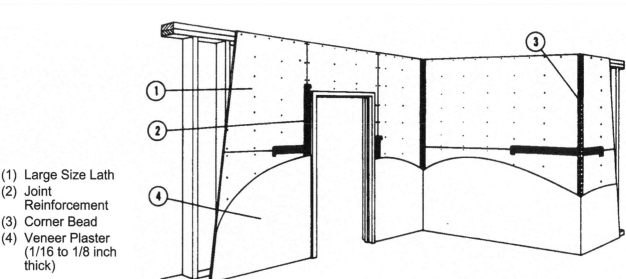

(1) Large Size Lath
(2) Joint Reinforcement
(3) Corner Bead
(4) Veneer Plaster (1/16 to 1/8 inch thick)

EXTERIOR LATH AND PLASTER

OPEN WOOD FRAME CONSTRUCTION

(1) Wire Backing
(2) Building Paper
(3) Wire Fabric Lath
(4) Approved Fasteners
(5) Weep Screed
(6) Three Coats of Plaster (Scratch, Brown, Finish)

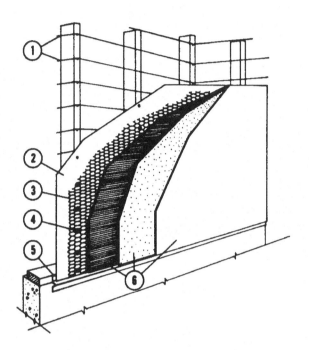

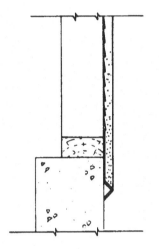

SHEATHED WOOD FRAME CONSTRUCTION

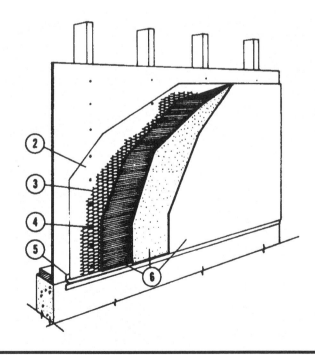

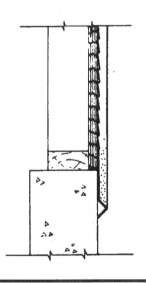

TYPICAL SIZES OF FIXTURE WATER SUPPLY PIPES

Fixture	Nominal pipe size (inches)
Bath tubs	1/2
Combination sink and tray	1/2
Drinking fountain	3/8
Dishwasher (domestic)	1/2
Kitchen sink, residential	1/2
Kitchen sink, commercial	3/4
Lavatory	3/8
Laundry tray, 1, 2 or 3 compartments	1/2
Shower (single head)	1/2
Sinks (service, slop)	1/2
Sinks flushing rim	3/4
Urinal (flash tank)	3/8
Urinal (direct flush valve)	1
Water closet (tank type)	3/8
Water closet (flush valve type)	1
Hose bibs	1/2
Wall hydrant	1/2

AIR CONDITIONING
RECOMMENDED SHEET METAL GAUGES AND CONSTRUCTION FOR RECTANGULAR DUCT

		Steel Metal Gauges		AT JOINTS					Reinforcing Between Joints
PLATE NO.	DIMENSION OF LONGEST SIDE OF DUCT	Steel	Aluminum	Plain "S" Slip (B) / Pocket Lock (K) / Drive Slip (A)	Hemmed "S" Slip (C) / Bar Slip (E) / Seam (1)	Reinforced Bar Slip (G)	Angle Slip (H) / Alternate Bar Slip (F) / Angle RFD Pocket (L)	Companion Angles (M) / Angle Reinforced Standing Seam (J)	
6	Thru 12"	26	24 (.020)	A-B-K	——	——	——	——	
6	13" thru 18"	24	22 (.025)	A-B-K	——	——	——	——	
7 / 7A	19" thru 30"	24	22 (.025)	K @ 5' cc / A	C-E- @ 5' cc / C-E- @ 10' cc	——	——	——	1" x 1" x 1/8" @ 5' cc
8	31" thru 42"	22	20 (.032)	K @ 5' cc	E-G-K @ 5' cc / E-G-K @ 10' cc	——	——	——	1" x 1" x 1/8" @ 5' cc
9	43" thru 54"	22	20 (.032)	K @ 4' cc / K @ 8' cc	E-@ 4' cc / E-@ 8' cc	G- @ 4' cc / G- @ 8' cc	——	——	1½" x 1½" x 1/8" @ 4' cc
9	55" thru 60"	20	18 (.040)	K @ 4' cc / K @ 8' cc	E-@ 4' cc / E-@ 8' cc	G- @ 4' cc / G- @ 8' cc	——	——	1½" x 1½" x 1/8" @ 4' cc
10	61" thru 84"	20	18 (.040)	——	——	G- @ 4' cc / G- @ 5' cc	H- @ 4' cc / F- @ 4' cc / L- @ 4' cc / H- @ 5' cc / F- @ 5' cc / L- @ 5' cc	J- @ 2' cc	1½" x 1½" x 1/8" @ 2' cc / 1½" x 1½" x 1/8" @ 2'- 6" cc
11	85" thru 96"	18	16 (.051)	——	——	——	H- @ 4' cc / L- @ 4' cc / H- @ 5' cc / L- @ 5' cc	M- @ 4' cc / M- @ 5' cc / J- @ 2' cc	1½" x 1½" x 3/16" @ 2' cc / 1½" x 1½" x 3/16" @ 2'- 6" cc / 1½" x 1½" x 3/16" @ 2' cc
12	Over 96"	18	16 (.051)	——	——	——	H- @ 4' cc / L- @ 4' cc / H- @ 5' cc / L- @ 5' cc	M- @ 4' cc / M- @ 5' cc / J- @ 2' cc	2" x 2" x ¼ @ 2' cc / 2" x 2" x ¼ @ 2'- 6" cc / 2" x 2" x ¼ @ 2' cc

LOW PRESSURE — LOW VELOCITY = 2" W.G. MAX

H (height dimension) —— up to 42" = 1"
H (height dimension) —— 43" to 96" = 1½"
H (height dimension) —— over 96" = 2"

305

AIR CONDITIONING (Cont.)
LONGITUDINAL SEAMS
FOR SHEET METAL DUCTWORK

Fig. "N"
PITTSBURGH LOCK

Fig. "Z"
BUTTON PUNCH SNAP LOCK

Fig. "O"
ACME LOCK-GROOVED SEAM

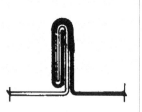

Fig. "T"
DOUBLE SEAM

Approximately 2" Spacing
Between "Buttons"

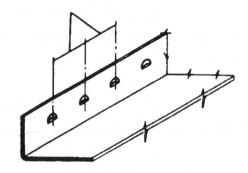

DETAIL NO.1
MALE PIECE-SNAP LOCK

AIR CONDITIONING
TYPICAL DUCT CONNECTIONS
CROSS JOINTS FOR SHEET METAL DUCTWORK
(NOT TO SCALE)

H - HEIGHT REFERRED TO IN DIMESIONS

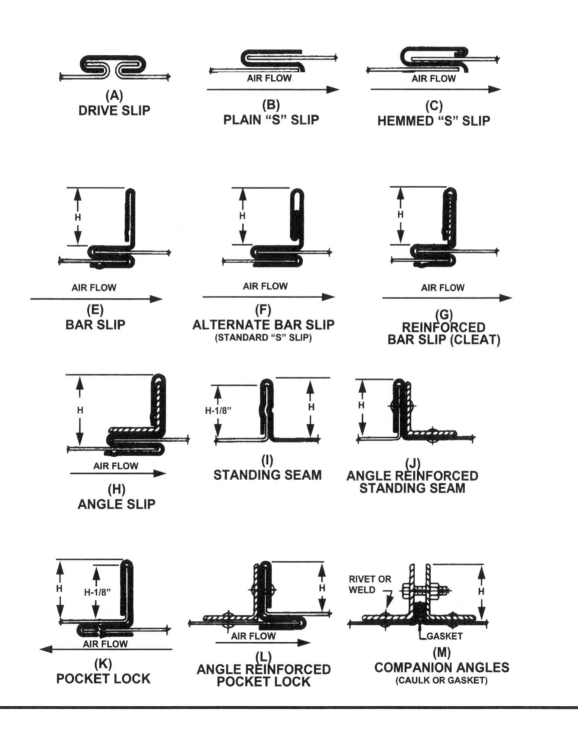

Metro Area Multipliers

The costs presented in this Costbook attempt to represent national averages. Costs, however, vary among regions and states and even between adjacent localities.

In order to more closely approximate the probable costs for specific locations throughout the U.S., this table of Metro Area Multipliers is provided. These adjustment factors can be used to modify costs obtained from this book to help account for regional variations of construction costs and to provide a more accurate estimate for specific areas. The factors are formulated by comparing costs in a specific area to the costs presented in this Costbook. An example of how to use these factors is shown below. Whenever local current costs are known, whether material prices or labor rates, they should be used when more accuracy is required.

| Cost Obtained from Costbook Pages | X | Metro Area Multiplier Divided by 100 | = | **Adjusted Cost** |

For example, a project estimated to cost $1,000,000 using the Costbook can be adjusted to more closely approximate the cost in Los Angeles, where the Multiplier is 133:

$$1{,}000{,}000 \times \frac{133}{100} = \mathbf{1{,}330{,}000}$$

Metro Area Multipliers

State	Metropolitan Area	Multiplier
AK	ANCHORAGE	130
AL	ANNISTON	77
	AUBURN	75
	BIRMINGHAM	78
	DECATUR	75
	DOTHAN	75
	FLORENCE	75
	GADSDEN	77
	HUNTSVILLE	76
	MOBILE	76
	MONTGOMERY	76
	OPELIKA	75
	TUSCALOOSA	79
AR	FAYETTEVILLE	73
	FORT SMITH	73
	JONESBORO	74
	LITTLE ROCK	77
	NORTH LITTLE ROCK	77
	PINE BLUFF	75
	ROGERS	79
	SPRINGDALE	74
	TEXARKANA	75
AZ	FLAGSTAFF	80
	MESA	80
	PHOENIX	82
	TUCSON	76
	YUMA	76
CA	BAKERSFIELD	130
	CHICO	136
	FAIRFIELD	136
	FRESNO	139
	LODI	139
	LONG BEACH	133
	LOS ANGELES	133
	MERCED	139
	MODESTO	139
	NAPA	136
	OAKLAND	139
	ORANGE COUNTY	133
	PARADISE	136
	PORTERVILLE	132
	REDDING	136
	RIVERSIDE	133
	SACRAMENTO	136
	SALINAS	139
	SAN BERNARDINO	133
	SAN DIEGO	130
	SAN FRANCISCO	139

Metro Area Multipliers

State	Metropolitan Area	Multiplier
CA	SAN JOSE	139
	SAN LUIS OBISPO	133
	SANTA BARBARA	133
	SANTA CRUZ	139
	SANTA ROSA	136
	STOCKTON	139
	TULARE	132
	VALLEJO	136
	VENTURA	133
	VISALIA	132
	WATSONVILLE	139
	YOLO	136
	YUBA CITY	136
CO	BOULDER	90
	COLORADO SPRINGS	89
	DENVER	90
	FORT COLLINS	86
	GRAND JUNCTION	87
	GREELEY	88
	LONGMONT	88
	LOVELAND	90
	PUEBLO	87
CT	BRIDGEPORT	127
	DANBURY	127
	HARTFORD	122
	MERIDEN	125
	NEW HAVEN	125
	NEW LONDON	122
	NORWALK	127
	NORWICH	122
	STAMFORD	127
	WATERBURY	125
DC	WASHINGTON	102
DE	DOVER	123
	NEWARK	122
	WILMINGTON	114
FL	BOCA RATON	84
	BRADENTON	82
	CAPE CORAL	81
	CLEARWATER	83
	DAYTONA BEACH	81
	FORT LAUDERDALE	85
	FORT MYERS	79
	FORT PIERCE	81
	FORT WALTON BEACH	79
	GAINESVILLE	79
	JACKSONVILLE	79
	LAKELAND	79

Metro Area Multipliers

State	Metropolitan Area	Multiplier
FL	MELBOURNE	81
	MIAMI	85
	NAPLES	82
	OCALA	80
	ORLANDO	81
	PALM BAY	82
	PANAMA CITY	79
	PENSACOLA	78
	PORT ST. LUCIE	81
	PUNTA GORDA	81
	SARASOTA	81
	ST. PETERSBURG	85
	TALLAHASSEE	80
	TAMPA	84
	TITUSVILLE	94
	WEST PALM BEACH	85
	WINTER HAVEN	85
GA	ALBANY	80
	ATHENS	80
	ATLANTA	77
	AUGUSTA	79
	COLUMBUS	76
	MACON	78
	SAVANNAH	79
HI	HONOLULU	133
IA	CEDAR FALLS	100
	CEDAR RAPIDS	105
	DAVENPORT	111
	DES MOINES	111
	DUBUQUE	104
	IOWA CITY	107
	SIOUX CITY	100
	WATERLOO	100
ID	BOISE CITY	83
	POCATELLO	88
IL	BLOOMINGTON	126
	CHAMPAIGN	121
	CHICAGO	144
	DECATUR	119
	KANKAKEE	126
	NORMAL	126
	PEKIN	123
	PEORIA	123
	ROCKFORD	126
	SPRINGFIELD	119
	URBANA	121
IN	BLOOMINGTON	105
	ELKHART	105

Metro Area Multipliers

State	Metropolitan Area	Multiplier
IN	EVANSVILLE	104
	FORT WAYNE	105
	GARY	119
	GOSHEN	105
	INDIANAPOLIS	105
	KOKOMO	105
	LAFAYETTE	105
	MUNCIE	105
	SOUTH BEND	119
	TERRE HAUTE	104
KS	KANSAS CITY	111
	LAWRENCE	109
	TOPEKA	108
	WICHITA	97
KY	LEXINGTON	100
	LOUISVILLE	100
	OWENSBORO	99
LA	ALEXANDRIA	80
	BATON ROUGE	82
	BOSSIER CITY	82
	HOUMA	81
	LAFAYETTE	81
	LAKE CHARLES	81
	MONROE	80
	NEW ORLEANS	81
	SHREVEPORT	82
MA	BARNSTABLE	139
	BOSTON	139
	BROCKTON	134
	FITCHBURG	133
	LAWRENCE	139
	LEOMINSTER	133
	LOWELL	139
	NEW BEDFORD	139
	PITTSFIELD	133
	SPRINGFIELD	125
	WORCESTER	133
	YARMOUTH	139
MD	BALTIMORE	89
	CUMBERLAND	91
	HAGERSTOWN	91
ME	AUBURN	84
	BANGOR	83
	LEWISTON	84
	PORTLAND	84
MI	ANN ARBOR	118
	BATTLE CREEK	103
	BAY CITY	105

Metro Area Multipliers

State	Metropolitan Area	Multiplier
MI	BENTON HARBOR	118
	DETROIT	118
	EAST LANSING	108
	FLINT	110
	GRAND RAPIDS	87
	HOLLAND	93
	JACKSON	110
	KALAMAZOO	103
	LANSING	108
	MIDLAND	101
	MUSKEGON	93
	SAGINAW	103
MN	DULUTH	122
	MINNEAPOLIS	122
	ROCHESTER	122
	ST. CLOUD	118
	ST. PAUL	123
MO	COLUMBIA	114
	JOPLIN	97
	KANSAS CITY	115
	SPRINGFIELD	98
	ST. JOSEPH	114
	ST. LOUIS	115
MS	BILOXI	69
	GULFPORT	69
	HATTIESBURG	69
	JACKSON	69
	PASCAGOULA	69
MT	BILLINGS	98
	GREAT FALLS	96
	MISSOULA	98
NC	ASHEVILLE	74
	CHAPEL HILL	71
	CHARLOTTE	72
	DURHAM	71
	FAYETTEVILLE	71
	GOLDSBORO	71
	GREENSBORO	71
	GREENVILLE	71
	HICKORY	72
	HIGH POINT	72
	JACKSONVILLE	74
	LENOIR	74
	MORGANTON	72
	RALEIGH	70
	ROCKY MOUNT	72
	WILMINGTON	74
	WINSTON SALEM	71

Metro Area Multipliers

State	Metropolitan Area	Multiplier
ND	BISMARCK	91
	FARGO	94
	GRAND FORKS	92
NE	LINCOLN	92
	OMAHA	92
NH	MANCHESTER	89
	NASHUA	89
	PORTSMOUTH	90
NJ	ATLANTIC CITY	140
	BERGEN	143
	BRIDGETON	138
	CAPE MAY	138
	HUNTERDON	141
	JERSEY CITY	143
	MIDDLESEX	142
	MILLVILLE	138
	MONMOUTH	127
	NEWARK	143
	OCEAN	138
	PASSAIC	144
	SOMERSET	141
	TRENTON	128
	VINELAND	138
NM	ALBUQUERQUE	81
	LAS CRUCES	81
	SANTA FE	80
NV	LAS VEGAS	128
	RENO	125
NY	ALBANY	116
	BINGHAMTON	113
	BUFFALO	111
	DUTCHESS COUNTY	116
	ELMIRA	114
	GLENS FALLS	114
	JAMESTOWN	110
	NASSAU	116
	NEW YORK	164
	NEWBURGH	116
	NIAGARA FALLS	125
	ROCHESTER	114
	ROME	116
	SCHENECTADY	116
	SUFFOLK	154
	SYRACUSE	112
	TROY	116
	UTICA	112

Metro Area Multipliers

State	Metropolitan Area	Multiplier
OH	AKRON	107
	CANTON	102
	CINCINNATI	103
	CLEVELAND	111
	COLUMBUS	104
	DAYTON	102
	ELYRIA	107
	HAMILTON	103
	LIMA	104
	LORAIN	107
	MANSFIELD	104
	MASSILLON	102
	MIDDLETOWN	102
	SPRINGFIELD	103
	STEUBENVILLE	104
	TOLEDO	107
	WARREN	107
	YOUNGSTOWN	107
OK	ENID	80
	LAWTON	83
	OKLAHOMA CITY	80
	TULSA	82
OR	ASHLAND	103
	CORVALLIS	110
	EUGENE	108
	MEDFORD	103
	PORTLAND	113
	SALEM	110
	SPRINGFIELD	108
PA	ALLENTOWN	118
	ALTOONA	112
	BETHLEHEM	120
	CARLISLE	111
	EASTON	120
	ERIE	110
	HARRISBURG	111
	HAZLETON	115
	JOHNSTOWN	111
	LANCASTER	107
	LEBANON	107
	PHILADELPHIA	133
	PITTSBURGH	110
	READING	119
	SCRANTON	114
	SHARON	110
	STATE COLLEGE	112
	WILKES BARRE	115
	WILLIAMSPORT	115

Metro Area Multipliers

State	Metropolitan Area	Multiplier
PA	YORK	109
PR	MAYAGUEZ	72
	PONCE	72
	SAN JUAN	72
RI	PROVIDENCE	125
SC	AIKEN	71
	ANDERSON	69
	CHARLESTON	71
	COLUMBIA	71
	FLORENCE	72
	GREENVILLE	70
	MYRTLE BEACH	69
	NORTH CHARLESTON	71
	SPARTANBURG	71
	SUMTER	71
SD	RAPID CITY	78
	SIOUX FALLS	84
TN	CHATTANOOGA	78
	CLARKSVILLE	77
	JACKSON	76
	JOHNSON CITY	78
	KNOXVILLE	75
	MEMPHIS	78
	NASHVILLE	77
TX	ABILENE	74
	AMARILLO	74
	ARLINGTON	73
	AUSTIN	76
	BEAUMONT	76
	BRAZORIA	76
	BROWNSVILLE	70
	BRYAN	75
	COLLEGE STATION	75
	CORPUS CHRISTI	74
	DALLAS	74
	DENISON	74
	EDINBURG	70
	EL PASO	73
	FORT WORTH	73
	GALVESTON	75
	HARLINGEN	70
	HOUSTON	71
	KILLEEN	73
	LAREDO	72
	LONGVIEW	73
	LUBBOCK	75
	MARSHALL	69
	MCALLEN	70

Metro Area Multipliers

State	Metropolitan Area	Multiplier
TX	MIDLAND	73
	MISSION	70
	ODESSA	73
	PORT ARTHUR	76
	SAN ANGELO	73
	SAN ANTONIO	77
	SAN BENITO	70
	SAN MARCOS	75
	SHERMAN	70
	TEMPLE	73
	TEXARKANA	73
	TEXAS CITY	70
	TYLER	72
	VICTORIA	74
	WACO	73
	WICHITA FALLS	75
UT	OGDEN	77
	OREM	76
	PROVO	76
	SALT LAKE CITY	77
VA	CHARLOTTESVILLE	80
	LYNCHBURG	81
	NEWPORT NEWS	82
	NORFOLK	82
	PETERSBURG	79
	RICHMOND	80
	ROANOKE	83
	VIRGINIA BEACH	82
VT	BURLINGTON	80
WA	BELLEVUE	116
	BELLINGHAM	109
	BREMERTON	111
	EVERETT	115
	KENNEWICK	112
	OLYMPIA	114
	PASCO	112
	RICHLAND	112
	SEATTLE	116
	SPOKANE	93
	TACOMA	116
	YAKIMA	102

Metro Area Multipliers

State	Metropolitan Area	Multiplier
WI	APPLETON	112
	BELOIT	115
	EAU CLAIRE	112
	GREEN BAY	111
	JANESVILLE	115
	KENOSHA	116
	LA CROSSE	112
	MADISON	114
	MILWAUKEE	118
	NEENAH	112
	OSHKOSH	112
	RACINE	117
	SHEBOYGAN	111
	WAUKESHA	118
	WAUSAU	111
WV	CHARLESTON	114
	HUNTINGTON	117
	PARKERSBURG	112
	WHEELING	110
WY	CASPER	85
	CHEYENNE	86

Home Builders
Square Foot Tables

The costs listed in these tables are intended to provide typical, average ranges of costs for the most common types of single family residential construction. Differences in materials and methods of construction will always vary the costs.

Two home types are included: one story and two story. Each is then presented in two quality categories: average, or "contractor grade," and deluxe, or "architectural grade." Low, medium and medium and high ranges of costs are then presented in seven typical square footages.

Each of the examples is further broken down.

The "Frame" includes the foundation, rough framing, siding, roofing, exterior doors and windows. "Interiors" includes all finish carpentry, wall and floor finishes, cabinetry, plumbing, HVAC and electrical. The "Overhead and Profit" includes the markups for insurance, taxes, office overhead and profit for the general contractor.

These costs represent typical averages and do not include land costs, sitework (other than foundation excavation), landscaping or other costs not directly attributable to construction.

ONE STORY AVERAGE

LOW

SQUARE FOOTAGE	1500	1750	2000	2250	2500	2750	3000
COSTS							
FRAME	97.37	95.43	93.48	92.99	90.65	88.34	87.41
INTERIORS	93.14	91.28	89.41	88.95	86.71	84.50	83.61
OVERHEAD & PROFIT	21.17	20.74	20.32	20.22	19.71	19.20	19.00
TOTAL	$211.68	$207.45	$203.21	$202.15	$197.07	$192.04	$190.03

MEDIUM

SQUARE FOOTAGE	1500	1750	2000	2250	2500	2750	3000
COSTS							
FRAME	107.11	104.97	102.83	102.29	99.72	97.17	96.15
INTERIORS	102.45	100.40	98.35	97. 84	95.38	92.95	91.97
OVERHEAD & PROFIT	23.28	22.82	22.35	22.24	21.68	21.12	20.90
TOTAL	$232.85	$228.19	$223.53	$222.37	$216.78	$211.24	$209.03

HIGH

SQUARE FOOTAGE	1500	1750	2000	2250	2500	2750	3000
COSTS							
FRAME	111.98	109.74	107.50	106.94	104.25	101.59	100.52
INTERIORS	107.11	104.97	102.83	102.29	99.72	97.17	96.15
OVERHEAD & PROFIT	24.34	23.86	23.37	23.25	22.66	22.08	21.85
TOTAL	$243.43	$238.56	$233.69	$232.48	$226.64	$220.84	$218.53

ONE STORY DELUXE

LOW

SQUARE FOOTAGE	1500	1750	2000	2250	2500	2750	3000
COSTS							
FRAME	121.72	119.28	116.85	116.24	113.32	110.42	109.26
INTERIORS	116.42	114.10	111.77	111.18	108.39	105.62	104.51
OVERHEAD & PROFIT	26.46	25.93	25.40	25.27	24.63	24.00	23.75
TOTAL	$264.60	$259.31	$254.02	$252.69	$246.34	$240.05	$237.53

MEDIUM

SQUARE FOOTAGE	1500	1750	2000	2250	2500	2750	3000
COSTS							
FRAME	133.89	131.21	128.53	127.86	124.65	121.46	120.19
INTERIORS	128.07	125.51	122.94	122.30	119.23	116.18	114.97
OVERHEAD & PROFIT	29.11	28.52	27.94	27.80	27.10	26.40	26.13
TOTAL	$291.06	$285.24	$279.42	$277.96	$270.98	$264.05	$261.28

HIGH

SQUARE FOOTAGE	1500	1750	2000	2250	2500	2750	3000
COSTS							
FRAME	146.06	143.14	140.22	139.49	135.98	132.50	131.12
INTERIORS	139.71	136.91	134.12	133.42	130.07	126.74	125.42
OVERHEAD & PROFIT	31.75	31.12	30.48	30.32	29.56	28.81	28.50
TOTAL	$317.52	$311.17	$304.82	$303.23	$295.61	$288.05	$285.04

TWO STORY AVERAGE

LOW

SQUARE FOOTAGE	1500	1750	2000	2250	2500	2750	3000
COSTS							
FRAME	83.33	81.67	80.00	79.58	77.58	75.60	74.81
INTERIORS	79.71	78.11	76.52	76.12	74.21	72.31	71.55
OVERHEAD & PROFIT	18.12	17.75	17.39	17.30	16.87	16.43	16.26
TOTAL	**$181.16**	**$177.53**	**$173.91**	**$173.00**	**$168.66**	**$164.35**	**$162.62**

MEDIUM

SQUARE FOOTAGE	1500	1750	2000	2250	2500	2750	3000
COSTS							
FRAME	91.67	89.83	88.00	87.54	85.34	83.16	82.29
INTERIORS	87.68	85.93	84.17	83.73	81.63	79.54	78.71
OVERHEAD & PROFIT	19.93	19.53	19.13	19.03	18.55	18.08	17.89
TOTAL	**$199.27**	**$195.29**	**$191.30**	**$190.30**	**$185.52**	**$180.78**	**$178.89**

HIGH

SQUARE FOOTAGE	1500	1750	2000	2250	2500	2750	3000
COSTS							
FRAME	95.83	93.92	92.00	91.52	89.22	86.94	86.03
INTERIORS	91.67	89.83	88.00	87.54	85.34	83.16	82.29
OVERHEAD & PROFIT	20.83	20.42	20.00	19.90	19.40	18.90	18.70
TOTAL	**$208.33**	**$204.16**	**$200.00**	**$198.96**	**$193.96**	**$189.00**	**$187.02**

TWO STORY DELUXE

LOW

SQUARE FOOTAGE	1500	1750	2000	2250	2500	2750	3000
COSTS							
FRAME	104.16	102.08	100.00	99.48	96.98	94.50	93.51
INTERIORS	99.64	97.64	95.65	95.15	92.76	90.39	89.44
OVERHEAD & PROFIT	22.64	22.19	21.74	21.63	21.08	20.54	20.33
TOTAL	$226.45	$221.92	$217.39	$216.26	$210.82	$205.43	$203.28

MEDIUM

SQUARE FOOTAGE	1500	1750	2000	2250	2500	2750	3000
COSTS							
FRAME	114.58	112.29	110.00	109.43	106.68	103.95	102.86
INTERIORS	109.60	107.41	105.22	104.67	102.04	99.43	98.39
OVERHEAD & PROFIT	24.91	24.41	23.91	23.79	23.19	22.60	22.36
TOTAL	$249.09	$244.11	$239.13	$237.88	$231.90	$225.97	$223.61

HIGH

SQUARE FOOTAGE	1500	1750	2000	2250	2500	2750	3000
COSTS							
FRAME	125.00	122.50	120.00	119.37	116.37	113.40	112.21
INTERIORS	119.56	117.17	114.78	114.18	111.31	108.47	107.33
OVERHEAD & PROFIT	27.17	26.63	26.09	25.95	25.03	24.65	24.39
TOTAL	$271.73	$266.30	$260.87	$259.51	$252.99	$246.52	$243.94

INDEX

Other Estimating References from BNi Building News

The latest estimating costbooks for 2020 from BNi Building News, including the *Square Foot Costbook, General Construction, Conceptual Estimator* and more! Each costbook gives you accurate, detailed costs based on actual projects and [Includes a FREE PDF download version you can customize].

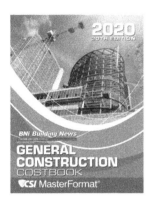

BNi Building News
General Construction
Costbook 2020 with 50-Division CSI MasterFormat

Over 12,000 unit costs provide you with cost coverage for all aspects of construction — from sitework and concrete to doors and painting. The *2020 BNi General Construction Costbook* is broken down into material and labor costs, to allow for maximum flexibility and accuracy in estimating. What's more, you get detailed man-hour tables that let you see the basis for the labor costs based on standard productivity rates.

8½ x 11, $124.95

BNi Building News
PUBLIC WORKS COSTBOOK 2020

Now you can quickly and easily estimate the cost of all types of public works projects involving roads, excavation, drainage systems and much more.

The *BNi Public Works Costbook 2020* is the first place to turn, whether you're preparing a preliminary estimate, evaluating a contractor's bid, or submitting a formal budget proposal. It provides accurate and up-to-date material and labor costs for thousands of cost items, based on the latest national averages and standard labor productivity rates.

Square-foot tables based on the cost-per-square-foot of hundreds of actual projects — invaluable data for quick, ballpark estimates.

8½ x 11, $135.95

BNi Building News
FACILITIES MANAGER'S COSTBOOK 2020

The *BNi Facilities Manager's Costbook 2020* is the first place to turn, whether you're preparing a preliminary estimate, evaluating a contractor's bid, or submitting a formal budget proposal. Labor costs are provided and are based on the prevailing rates for each trade and type of work, PLUS man-hour tables tied to the unit costs, so you can clearly see exactly how the labor costs were calculated and make any necessary adjustments. You also get equipment costs — including rental and operating costs, and square-foot tables based on the cost-per-square-foot of hundreds of actual projects.

8½ x 11, $159.95

BNi Building News
ELECTRICAL COSTBOOK 2020

From meter to duct, conduit to receptacle, The *BNi Electrical Costbook* is the first place to turn, whether you're preparing a preliminary estimate, evaluating a subcontractor's bid, or submitting a formal budget proposal. It puts at your fingertips accurate and up-to-date material and labor costs for thousands of cost items, based on the latest national averages and standard labor productivity rates. What's more, the *2020 BNi Electrical Costbook* includes detailed regional cost modifiers for adjusting your estimate to your local conditions.

8½ x 11, $129.95

BNi Building News
MECHANICAL/ ELECTRICAL COSTBOOK 2020

From pipe to duct to receptacle, this detailed reference book provides extensive coverage of the most technical aspects of building construction. With thousands of current, reliable mechanical and electrical costs at your fingertips, you can estimate quickly and accurately. Geographic Cost Modifiers allow you to tailor your estimates to specific areas of the country.

8½ x 11, $129.95

BNi Building News
HOME BUILDER'S COSTBOOK 2020

Here's the easy way to estimate the cost of all types of residential construction projects! Accurate and up-to-date material and labor costs for thousands of cost items, based on the latest national averages and standard labor productivity rates.

Includes detailed regional cost modifiers for adjusting your estimate to your local conditions. Material costs are included for thousands of items based on current national averages (including allowances for transport, handling and storage).

8½ x 11, $110.95

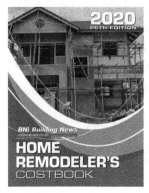

BNi Building News
Home Remodeler's
COSTBOOK 2020

The *2020 BNi Home Remodeler's Costbook* lets you quickly and easily estimate the cost of all types of home remodeling projects, including additions, new kitchens and baths, and much more. The *BNi Home Remodeler's Costbook 2020* is the first place to turn, whether you're preparing a preliminary estimate, evaluating a subcontractor's bid, or submitting a formal budget proposal.

This all-new costbook puts at your fingertips accurate and up-to-date material and labor costs for thousands of cost items, based on the latest national averages and standard labor productivity rates.

8½ x 11, $109.95

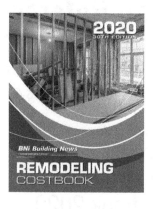

BNi Building News
Remodeling
COSTBOOK 2020

Now you can quickly and easily estimate the cost of all types of home remodeling projects. You'll find yourself turning to the *2020 BNi Remodeling Costbook* again and again, whenever you're preparing a preliminary estimate, evaluating a subcontractor's bid, or submitting a formal budget proposal.

It puts at your fingertips accurate and up-to-date material and labor costs for thousands of cost items. Includes detailed regional cost modifiers for adjusting your estimate to your local conditions.

8½ x 11, $129.95

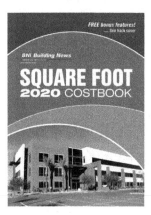

BNi Building News
SQUARE FOOT
COSTBOOK 2020

In this costbook you'll find over 80 detailed square foot cost studies for projects ranging from civic Government Buildings to Hotels to Industrial and Office Buildings to Residential Buildings and so many more. For each building project you get a detailed narrative with background information on the specific project. In addition, you'll receive unit-in-place costs for nearly 15,000 items and materials used in all types of construction. For each item, you can see man-hours, as well as labor/equipment and material costs, all clearly broken out.

8½ x 11, $109.95

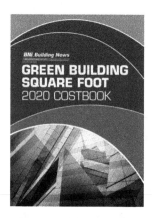

BNi Building News
GREEN BUILDING SQUARE FOOT
COSTBOOK 2020

The new *2020 BNi Green Building Square Foot Costbook* provides you with a comprehensive collection of 57 recent LEED and sustainable building projects along with their actual square foot costs, broken down by CSI MasterFormat section.

For each building, the *2020 BNi Green Building Square Foot Costbook* provides a detailed narrative describing the major features of the actual building, the steps taken to minimize the environmental impact both in its construction and its operation and a square-foot cost breakdown of each building component.

8½ x 11, $109.95

DCR
INTERIORS SQUARE FOOT
COSTBOOK 2020

Unlike other building cost estimating resources, the *2020 DCR Interiors Square Foot Costbook* covers new construction, addition/renovation, adaptive re-use, and tenant build-out.

Each project is broken down by all its interior components presented on a cost-per-square-foot basis. It itemizes the materials used, along with their costs, to assist you in developing a conceptual estimate for interior construction.

8½ x 11, $84.95

DCR
MECHANICAL/ ELECTRICAL SQUARE FOOT
COSTBOOK 2020

Unlike other building cost estimating resources, the *2020 DCR Mechanical/Electrical Square Foot Costbook* breaks down the MEP divisions and itemizes the materials used, along with their costs, in actual projects.

In addition, many of the cost studies in this book feature a variety of "green" technologies, such as hydronic pipe, geothermal heating and cooling, solar water heating, and hybrid ventilation air handlers. It lets you instantly see exactly how MEP costs relate to overall building costs, and how much they can vary from one project to another.

8½ x 11, $84.95

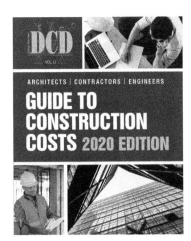

Architects Contractors Engineers
DCD GUIDE TO CONSTRUCTION COSTS
2020 Edition

Find thousands of thoroughly-researched construction material and installation costs you can use when making budgets, checking prices, calculating the impact of change orders or preparing bids. You'll find costs listed for scheduling, testing, temporary facilities, equipment, signage, and more. You'll find costs for every area of construction — from demolition and excavation to material and installation costs for finishes, flooring and much more.

8½ x 11, $72.95

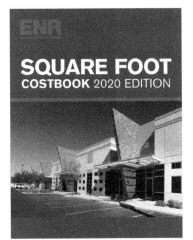

Engineering News-Record
SQUARE FOOT
COSTBOOK 2020

As you know, square-foot costs can vary widely, making them difficult to use for estimating and budgeting. The *2020 ENR Square-Foot Costbook* eliminates this problem by giving you costs that are based on actual projects, not hypothetical models.

This year's projects include:

Restaurants • Offices • Storage buildings • A theatre and studio arts building • A boxing club

A tennis center • A YMCA • A ministries center • And different types of residences

For each building project you get a detailed narrative with background information on the specific project. This lets you put the cost data into context and make appropriate adjustments to your own projections.

Developed in partnership with *Design Cost Data* and *BNi Building News*, this ready-reference costbook also features:

•*Illustrations of each building type • A guide to 5-year cost trends for key building materials*

• *Detailed unit-in-place costs for thousands of items, from asphalt and anchor bolts to vents and wall louvers*

8½ x 11, $84.95

STANDARD ESTIMATING PRACTICE
10th Edition

An invaluable "how to" reference manual on the practice of estimating construction projects.

Standard Estimating Practice presents a standard set of practices and procedures proven to create consistent estimates. From the order of magnitude, to conceptual design, design development, construction documents, to the bid and the various types of contracts you'll run up against. Every step is covered in detail — from specs and plan review to what to expect on bidding day.

Standard Estimating Practice also provides:

- Practical advice for using historical data in determining future production rates
- 14 key elements that will influence production rates on every project
- 10 important considerations when including construction equipment in an estimate
- 7 key costs that need to be included in a direct labor burden

8½ x 11, $99.95

DCR
ARCHITECT'S SQUARE FOOT
COSTBOOK 2020

This manual presents detailed square foot costs for 65 buildings tailored specifically to meet the needs of today's architect. For each project you get a complete cost breakdown of the included systems, so you can easily calculate the impact of modifications and enhancements on your own projects.

This *2020 Architect's Square Foot Costbook* gives you square-foot costs for a wide range of actual projects, from a senior living facility and low-income housing unit to a theater, a restaurant, and a corporate headquarters building. The theme of this year's Architect's Square Foot Costbook is Commercial, Educational, and Residential.

8½ x 11, $84.95

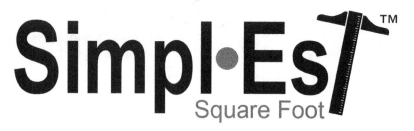

Construction Project Log Book

With *Construction Project Log Book* that task is made as simple as possible.

The 365 Daily Work Log pages let you keep a detailed record for each day of the year. Additional forms such as Accident Reports and numerous checklists help make sure that you're covered.

Document every shipment, machinery rental, delivery, delay, and weather condition — all items that can affect productivity. With the interactive forms you can keep this information on your computer and enter new data daily into the interactive PDF forms. There are forms that actually do the math for you — eliminating typical mistakes.

7 x 9-1/4, $39.95

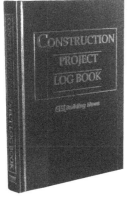

Maintenance Manager's Standard Manual

Since it was first published in 1993, the *Maintenance Manager's Standard Manual* has indeed become the STANDARD reference in the field.

This brand-new Sixth Edition brings it completely up to date, incorporating the latest technology and best practices in all aspects of maintenance management.

Whether you are a facilities manager, engineer, property owner, developer, or anyone else responsible for maintenance operations; not only does it give you all of the essential ingredients for understanding and carrying out successful day-to-day management of maintenance activities, it provides you with an integrated plan for continuous improvement of the maintenance function.

8½ x 11, $99.95

100% Satisfaction Guaranteed!

If not completely satisfied, return the item within 30 days for a complete refund of the purchase price.

This has been the BNi policy for over 70 years.

Is there a title you need and can't find?

Give us a call at 1-888-264-2665; we'll be glad to help.

Order online: www.bnibooks.com

Find hundreds of construction references, forms and contracts to help you with your construction business.

NOTES